Selected Titles in This Series

173 **Yu. A. Davydov, M. A. Lifshits, and N. V. Smorodina,** Local properties of distributions of stochastic functionals, 1998
172 **Ya. G. Berkovich and E. M. Zhmud′,** Characters of finite groups. Part 1, 1998
171 **E. M. Landis,** Second order equations of elliptic and parabolic type, 1998
170 **Viktor Prasolov and Yuri Solovyev,** Elliptic functions and elliptic integrals, 1997
169 **S. K. Godunov,** Ordinary differential equations with constant coefficient, 1997
168 **Junjiro Noguchi,** Introduction to complex analysis, 1998
167 **Masaya Yamaguti, Masayoshi Hata, and Jun Kigami,** Mathematics of fractals, 1997
166 **Kenji Ueno,** An introduction to algebraic geometry, 1997
165 **V. V. Ishkhanov, B. B. Lur′e, and D. K. Faddeev,** The embedding problem in Galois theory, 1997
164 **E. I. Gordon,** Nonstandard methods in commutative harmonic analysis, 1997
163 **A. Ya. Dorogovtsev, D. S. Silvestrov, A. V. Skorokhod, and M. I. Yadrenko,** Probability theory: Collection of problems, 1997
162 **M. V. Boldin, G. I. Simonova, and Yu. N. Tyurin,** Sign-based methods in linear statistical models, 1997
161 **Michael Blank,** Discreteness and continuity in problems of chaotic dynamics, 1997
160 **V. G. Osmolovskiĭ,** Linear and nonlinear perturbations of the operator div, 1997
159 **S. Ya. Khavinson,** Best approximation by linear superpositions (approximate nomography), 1997
158 **Hideki Omori,** Infinite-dimensional Lie groups, 1997
157 **V. B. Kolmanovskiĭ and L. E. Shaĭkhet,** Control of systems with aftereffect, 1996
156 **V. N. Shevchenko,** Qualitative topics in integer linear programming, 1997
155 **Yu. Safarov and D. Vassiliev,** The asymptotic distribution of eigenvalues of partial differential operators, 1997
154 **V. V. Prasolov and A. B. Sossinsky,** Knots, links, braids and 3-manifolds. An introduction to the new invariants in low-dimensional topology, 1997
153 **S. Kh. Aranson, G. R. Belitsky, and E. V. Zhuzhoma,** Introduction to the qualitative theory of dynamical systems on surfaces, 1996
152 **R. S. Ismagilov,** Representations of infinite-dimensional groups, 1996
151 **S. Yu. Slavyanov,** Asymptotic solutions of the one-dimensional Schrödinger equation, 1996
150 **B. Ya. Levin,** Lectures on entire functions, 1996
149 **Takashi Sakai,** Riemannian geometry, 1996
148 **Vladimir I. Piterbarg,** Asymptotic methods in the theory of Gaussian processes and fields, 1996
147 **S. G. Gindikin and L. R. Volevich,** Mixed problem for partial differential equations with quasihomogeneous principal part, 1996
146 **L. Ya. Adrianova,** Introduction to linear systems of differential equations, 1995
145 **A. N. Andrianov and V. G. Zhuravlev,** Modular forms and Hecke operators, 1995
144 **O. V. Troshkin,** Nontraditional methods in mathematical hydrodynamics, 1995
143 **V. A. Malyshev and R. A. Minlos,** Linear infinite-particle operators, 1995
142 **N. V. Krylov,** Introduction to the theory of diffusion processes, 1995
141 **A. A. Davydov,** Qualitative theory of control systems, 1994
140 **Aizik I. Volpert, Vitaly A. Volpert, and Vladimir A. Volpert,** Traveling wave solutions of parabolic systems, 1994
139 **I. V. Skrypnik,** Methods for analysis of nonlinear elliptic boundary value problems, 1994
138 **Yu. P. Razmyslov,** Identities of algebras and their representations, 1994
137 **F. I. Karpelevich and A. Ya. Kreinin,** Heavy traffic limits for multiphase queues, 1994
136 **Masayoshi Miyanishi,** Algebraic geometry, 1994
135 **Masaru Takeuchi,** Modern spherical functions, 1994

(Continued in the back of this publication)

Translations of
MATHEMATICAL
MONOGRAPHS

Volume 173

Local Properties of Distributions of Stochastic Functionals

Yu. A. Davydov
M. A. Lifshits
N. V. Smorodina

American Mathematical Society
Providence, Rhode Island

Ю. А. Давыдов, М. А. Лифшиц, Х. В. Смородина

ЛОКАЛЬНЫЕ СВОЙСТВА РАСПРЕДЕЛЕНИЙ
СТОХАСТИЧЕСКИХ ФУНКЦИОНАЛОВ

Наука · Физматлит
Москва, 1995

Translated from the Russian by
V. E. Nazaĭkinskiĭ and M. A. Shishkova

1991 *Mathematics Subject Classification.* Primary 60–02, 60F17, 60F99, 60G15, 60G30, 60G50, 60H99.

ABSTRACT. This book deals with an important problem of probability theory, namely, the structure of distributions of functionals on trajectories of random processes. By using new methods (the stratification method, the superstructure method, and the method of differential operators), the authors examine whether distributions of functionals of Gaussian processes and Poisson measures have densities with prescribed properties. The absolute continuity of such densities is studied. These methods are used to prove local limit theorems for distributions of a wide class of stochastic functionals.

The book is intended for research mathematicians and graduate students working in probability theory and its applications.

Library of Congress Cataloging-in-Publication Data

Davydov, ÎU. A. (ÎUriĭ Aleksandrovich), 1944–
[Lokal′nye svoĭstva raspredeleniĭ stokhasticheskikh funkt͡sionalov. English]
Local properties of distributions of stochastic functionals / Yu. A. Davydov, M. A. Lifshits, N. V. Smorodina ; [translated from the Russian by V. E. Nazaĭkinskiĭ and M. S. Shishkova].
p. cm. — (Translations of mathematical monographs, ISSN 0065-9282 ; v. 173)
Includes bibliographical references (p. –) and index.
ISBN 0-8218-0584-3 (acid-free paper)
1. Limit theorems (Probability theory) 2. Distribution (Probability theory) 3. Functionals. 4. Stochastic processes. I. Lifshit͡s, M. A. (Mikhail Antol′evich), 1956– . II. Smorodina, N. V. (Natal′i͡a Vasil′evna), 1959– . III. Title. IV. Series.
QA273.67.D39 1998
519.2—dc21 97-44426
CIP

---

10 9 8 7 6 5 4 3 2 1 03 02 01 00 99 98

Contents

Preface

Studying distributions of stochastic functionals (that is, functionals defined on trajectories of random processes) is a problem of great interest and importance in probability theory, in particular in such topics as limit theorems, mathematical statistics, and approximation procedures. The characteristic function method, traditional in probability theory, proved to be ineffective in this problem, and so until the late 1970s there had been only separate results pertaining to a few specific functionals. Noticeable progress in recent years is due to newly developed methods, such as the stratification method, the superstructure method, and the method of differential operators. For distributions of functionals over Gaussian processes and Poisson measures, these methods have provided comprehensive results concerning the absolute continuity and existence of densities with given properties (e.g., boundedness or prescribed differentiability). The new methods are also effective in proving local limit theorems for distributions of a wide class of functionals.

The goal of the present book is to provide an introduction to these problems and methods. The topics covered are mainly related to the authors' original research and have not been presented in monographs earlier. However, we are not striving for maximum generality and completeness, but rather focus on the main ideas and their applications to the most fascinating problems, so as to make the book available to the widest possible readership.

We point out that definitive solutions to many of the problems discussed here have yet to be found, and we hope that this book will inspire further studies in this direction.

Normally, each section is concluded with exercises and some additional information. This information is by no means complete, and is chosen according to the authors' personal tastes and in line with the general scope of the book. As usual, the exercises are intended to aid active assimilation of the main text; hints are given where necessary. The sections are numbered consecutively throughout the book; we adopt the system in which, say, Theorem 5 in §2 is referred to as Theorem 2.5.

A majority of the results presented in the book have been discussed at the seminar on random processes at St. Petersburg State University and the St. Petersburg Branch of the Steklov Mathematical Institute of the Russian Academy of Sciences. We are grateful to all participants of the seminar for their attention to the work, patience, and useful remarks. We are also keenly grateful to I. A. Ibragimov for his constant support of and interest in our work.

The Authors

Introduction

We study distributions of functionals defined on trajectories (sample paths) of random processes. Consider a process $\{\xi(t), t \in T\}$ whose trajectories with probability 1 belong to a function space $\mathbf{X}$ and whose distribution is given by a measure $\mathbf{P}$ on $\mathbf{X}$. A functional f on the trajectories is a measurable mapping of $\mathbf{X}$ into $\mathbf{R}$, and so our problem is to study the measure $\mathbf{P}f^{-1}$, that is, the image of $\mathbf{P}$ under f.

Suppose that ξ is the Brownian motion process on the interval $[0,1]$, $\mathbf{X} = \mathbf{C}[0,1]$, and $\mathbf{P}$ is the Wiener measure (this is one of the most interesting cases). For a long time, performing explicit calculations was the only available method to study the distribution $\mathbf{P}f^{-1}$. For example, it has been successfully applied to the functionals

$$f(x) = \sup x(t), \qquad f(x) = \sup |x(t)|,$$

and

$$f(x) = \int_0^1 |x(t)|^p \, dt$$

for $p = 2$. However, no explicit expression is available for the distribution of the last functional for $p \neq 2$, and even in the early 1970s it was not known whether the measure $\mathbf{P}f^{-1}$ has a density.

Apparently, the first deep results concerning the structure of distributions for a wide class of functionals was obtained by Tsirel′son [**113**]. His method, both geometric and analytical in nature, relied heavily on the Gaussian property and was new in probability theory. Later it has become clear that these results are closely related to the convexity of Gaussian measures, discovered by Borell and Ehrhard [**129, 135**].

Almost at the same time, in the fundamental paper [**146**] P. Malliavin considered hypoelliptic problems. In our terms, he studied the question of whether the distribution $\mathbf{P}f^{-1}$ has a smooth density for the case in which $f(x) = x(1)$ and $\mathbf{P}$ is the distribution of a diffusion process that satisfies a stochastic differential equation of a certain type.

In 1978 Davydov proposed the stratification method, which is based on different ideas and allows one to study the absolute continuity problem for distributions of a large variety of functionals, including nonsmooth functionals. This method was substantially improved by Lifshits [**56**], who applied it to the problem of the existence of a bounded density for $\mathbf{P}f^{-1}$.

Recently, several other approaches to this problem have been proposed, in particular, the method of differential operators due to Smorodina [**104**], which is based on the abstract formalism of differential calculus and is a convenient, powerful tool to attack the problem on the existence of a density with given differentiability for the distribution $\mathbf{P}f^{-1}$.

The stratification method, and especially its modification known as the "superstructure method" (Davydov [**38**]), is very useful in proving local limit theorems for $\mathbf{P}f^{-1}$ when $\mathbf{P}_n$ weakly converges to the limit measure $\mathbf{P}$.

Our book provides a systematic exposition of the stratification method, the superstructure method, and the method of differential operators, as well as of their applications to the above-mentioned problems.

The book consists of five chapters.

The first chapter is introductory. It contains preliminary information from measure theory and the theory of random processes.

In the second chapter the three methods are introduced and their applications are illustrated by simple examples.

The third chapter deals with Gaussian functionals. Some conditions for the existence, boundedness, and smoothness of the distribution density are obtained by means of the stratification method and the method of differential operators. The properties of distributions of convex functionals are derived from the convexity of Gaussian measures. As an important example, the structure of the Gaussian norm distribution is studied in detail.

The fourth chapter deals with Poisson functionals. A special differential calculus is constructed on the Poisson space and the general results obtained are used to study the distributions of smooth functionals and the norm of stable vectors.

The fifth chapter treats strong convergence of distributions. A number of local limit theorems for a wide class of Gaussian functionals and a local invariance principle for sums of independent random variables are proved by the superstructure method. The infinite-dimensional local limit theorem is also considered.

Chapters 3–5 are based on the methods presented in Chapter 2 and do not depend on each other.

Notation

Sets

A^0, $A^{\complement}$, A^-, ∂A, and $\mathcal{L}(A)$ are the interior, the complement, the closure, the boundary, and the linear span of a set A.

A^{ε} is the ε-neighborhood of A.

$A \setminus B$ and $A \times B$ are the difference and the direct product of sets A and B.

$A \triangle B = (A \cup B) \setminus (A \cap B)$ is the symmetric difference of A and B.

$\mathbf{L}^{\perp}$ is the orthogonal complement of a subspace $\mathbf{L}$.

$\mathbf{N} = \{1, 2, \dots\}$ is the set of positive integers.

$\overline{\mathbf{N}} = \mathbf{N} \cup \{\infty\}$ is the extended set of positive integers.

$\mathbf{Z}$ is the set of integers.

$\mathbf{Z}_+ = \{0, 1, \dots\}$ is the set of nonnegative integers.

$\varnothing$ is the empty set.

Spaces

$\mathbf{c}$ is the space of bounded sequences with the norm $\|x\| = \sup_{k \in \mathbf{N}} |x_k|$.

$\mathbf{c}_0$ is the space of sequences tending to zero (with the norm inherited from $\mathbf{c}$).

$\mathbf{C}(T)$ is the space of continuous functions on a metric compactum T with the norm $\|x\| = \sup_{t \in T} |x(t)|$.

$\mathbf{C}[a, b] = \mathbf{C}([a, b])$ is the space of continuous functions on the interval $[a, b]$.

$\mathbf{D}[0, 1]$ is the space of functions that have no second-kind discontinuities on the interval $[0, 1]$; this space is equipped with the Skorokhod topology.

$\mathbf{C}^{\infty}$ is the class of infinitely differentiable functions on $\mathbf{R}^1$.

$\mathbf{C}_0^{\infty}$ is the class of compactly supported functions from $\mathbf{C}^{\infty}$.

$\mathbf{C}_b^0 \subset \mathbf{C}^{\infty}$ is the class of functions bounded together with all derivatives.

$\mathbf{C}_l(D)$ is the class of l times D-differentiable stochastic functionals.

$\mathbf{I}_0$, $\mathbf{I}$, $\tilde{\mathbf{I}}(G, \Pi)$ are the classes of functions integrable with respect to a Poisson random measure.

$\mathbf{H}(X)$ is the tangent space at a point $X \in \mathfrak{X}(G)$, where $\mathfrak{X}(G)$ is the space of configurations on G (§13).

$\mathbf{L}^0(\mathbf{X}, \mathfrak{B})$, or $\mathbf{L}^0(\mathbf{X})$, is the space of measurable functions on a measurable space $(\mathbf{X}, \mathfrak{B})$.

$\mathbf{L}^q(\mathbf{X}, \mathbf{P}, \mathfrak{B}, \mathbf{B})$ is the space of functions with finite norm $\|f\| = \left(\int_{\mathbf{X}} |f|^q \, d\mathbf{P}\right)^{1/q}$ defined on $(\mathbf{X}, \mathfrak{B})$ and with range in a normed space $\mathbf{B}$.

$\mathbf{L}^q(\mathbf{X}, \mathbf{P}, \mathbf{B})$, $\mathbf{L}^q(\mathbf{X}, \mathbf{P})$, and $\mathbf{L}^q(\mathbf{X})$ are abridged notation for $\mathbf{L}^q(\mathbf{X}, \mathbf{P}, \mathfrak{B}, \mathbf{R}^1)$.

$\mathbf{l}^q = \mathbf{L}^q(\mathbf{N})$ is the space of sequences with finite norm $\|x\| = \left(\sum_{k=1}^{\infty} |x_k|^q\right)^{1/q}$.

$\mathcal{L}^j(\mathbf{H}, \mathbf{B})$ is the space of j-linear $\mathbf{B}$-valued forms defined on $\mathbf{H}$.

$\mathcal{L}^j(\mathbf{H}) = \mathcal{L}^j(\mathbf{H}, \mathbf{R}^1)$ is the space of j-linear real-valued forms defined on $\mathbf{H}$.

$\mathcal{L}_2^j(\mathbf{H}, \mathbf{B})$ is the space of j-linear $\mathbf{B}$-valued Hilbert–Schmidt forms defined on $\mathbf{H}$ ($\mathbf{B}$ and $\mathbf{H}$ are Hilbert spaces).

$\mathcal{L}_2^j(\mathbf{H}) = \mathcal{L}_2^j(\mathbf{H}, \mathbf{R}^1)$ is the space of j-linear real-valued Hilbert–Schmidt forms defined on $\mathbf{H}$.

$\mathcal{M}$ and $\mathcal{M}^a$ are the classes of smooth functionals that satisfy the local invariance principle (§18).

$\mathbf{R}^k$ is the k-dimensional Euclidean space.

$\mathbf{R}_+$ and $\mathbf{R}_+^k$ are the positive semiaxis and the positive octant.

$\overline{\mathbf{R}}$ is the extended real axis $\mathbf{R}^1$ (containing the points "plus infinity" and "minus infinity").

$\mathbf{R}^T$ is the space of all functions on a set T.

$\mathbf{R}^\infty = \mathbf{R}^\mathbf{N}$ is the space of all sequences.

$\mathcal{V}(\mathbf{X})$ is the class of lower semicontinuous convex functionals (§8).

$\overline{\mathbf{W}}_1^l$ is the space of smooth distributions on $\mathbf{R}^1$ (§6).

$\overline{\mathbf{W}}^l(D \mid S_0, \dots, S_l)$ or $\overline{\mathbf{W}}^l(S_0, \dots, S_l)$ is the space of l times D-differentiable functionals (§8).

$\mathcal{W}^q(\mathbf{X}, \mathbf{P}, \mathbf{H})$ is the space of smooth $\mathbf{H}$-valued functionals on a space $\mathbf{X}$ that is equipped with a Gaussian measure $\mathbf{P}$ (§8).

$\mathcal{W}^q(\mathbf{X}, \mathbf{P}) = \mathcal{W}^q(\mathbf{X}, \mathbf{P}, \mathbf{R}^1)$.

$\mathbf{Z}(\mathbf{X})$ is the space of finite signed measures on $\mathbf{X}$ with the topology of convergence in variation.

$\mathbf{X}'_\mathbf{P}$, $\mathbf{X}'^A_\mathbf{P}$, and $\mathbf{X}'_{\mathbf{P},0}$ are the spaces of linear measurable, affine measurable, and centered measurable functionals for a Gaussian measure $\mathbf{P}$.

$\mathfrak{X}(G)$ is the space of configurations on G (§13).

Measurability

$\mathfrak{B}_\mathbf{X}$ or $\mathfrak{B}$ is the Borel σ-algebra of a topological space $\mathbf{X}$.

$\mathfrak{B}_k = \mathfrak{B}_{\mathbf{R}^k}$ is the Borel σ-algebra of $\mathbf{R}^k$.

$\mathfrak{C}_0$ is the cylinder algebra (§7).

$\mathfrak{C}$ is the cylinder σ-algebra (§7).

$\sigma(\mathfrak{s})$ is the minimal σ-algebra that contains a class $\mathfrak{s}$ of sets.

Measures

δ_x is the unit measure concentrated at the point x.

λ and λ_k are the Lebesgue measures on $\mathbf{R}^1$ and $\mathbf{R}^k$.

$\mathcal{G}(\mathbf{X})$ is the class of Radon Gaussian measures on a space $\mathbf{X}$.

$\mathcal{G}_0(\mathbf{X}) \subset \mathcal{G}(\mathbf{X})$ is the subclass of centered measures.

μ_A is the restriction of a measure μ to a set A.

$\mu^*(A)$ is the outer measure of the set A.

$\mu_*(A)$ is the inner measure of the set A.

$|\mu|(A)$ is the variation of a signed measure μ on a set A.

$\|\mu\|$ is the total variation of a signed measure μ.

$\mu \ll \nu$ means the absolute continuity of a measure μ with respect to a measure ν.

$\mu \asymp \nu$ means the equivalence (mutual absolute continuity) of measures μ and ν.

$\|\mu - \nu\|$ is the distance in variation between measures μ and ν.

$\mu_n \Rightarrow \mu$ denotes the weak convergence of measures.

$\mu_n \overset{\text{var}}{\to} \mu$ denotes the strong convergence of measures (convergence in variation).

$\mathcal{N}(a, \sigma^2)$ is the Gaussian (normal) distribution in $\mathbf{R}^1$ with mean a and variance σ^2.

$\Phi(r)$ is the distribution function for $\mathcal{N}(0, 1)$.

$\mathbf{P}$ is a probability measure.

$\mathsf{E}_{\mathbf{P}} f$ is the integral (mathematical expectation) of a functional f with respect to a measure $\mathbf{P}$.

$\mathsf{D}_{\mathbf{P}} f = \mathsf{E}_{\mathbf{P}} f^2 - \mathsf{E}^2_{\mathbf{P}} f$ is the variance of a functional f with respect to a measure $\mathbf{P}$.

$\operatorname{supp}(\mathbf{P})$ is the topological support of a measure $\mathbf{P}$.

$N_{\mathbf{P}}$ is the linear support of a measure $\mathbf{P}$.

Miscellany

$\det M$ is the determinant of a matrix M.

$\langle \cdot, \cdot \rangle$ is the bilinear form defined on a pair of spaces in duality.

$(\cdot, \cdot)$ is the inner product on a Hilbert space.

$|\cdot|$ is the norm on a Hilbert space.

$\|\cdot\|_{\mathbf{B}}$ is the norm on a normed space $\mathbf{B}$.

$\mathcal{I}(l)$ is the action functional (§7).

$\sigma_{\mathbf{P}}(l)$ is the admissibility of a vector l with respect to a Gaussian measure $\mathbf{P}$ (§7).

$\operatorname{dom} D$ is the domain of an operator D (§6).

$\dim \mathbf{L}$ is the dimension of $\mathbf{L}$.

$\ker I$ is the kernel of a linear operator I.

$\operatorname{tr} I$ is the trace of I.

I^* is the adjoint of I.

$\mathbf{1}$ is the function identically equal to 1.

$\mathbf{1}_A$ is the indicator function of the set A.

Ell is the ellipsoid of concentration of a Gaussian measure (§10).

ω_x is the modulus of continuity of a function x.

CHAPTER 1

Preliminaries

§1. Random processes and their distributions

A family of random variables $\{\xi_t, t \in T\}$ defined on a probability space $(\boldsymbol{\Omega}, \mathfrak{F}, \mathbf{P})$ will be called a *random process* with parameter set T. The process value at the point t will be denoted by ξ_t or $\xi(t)$; to indicate the dependence on $\omega \in \boldsymbol{\Omega}$, we sometimes write $\xi(t,\omega)$ or $\xi_t(\omega)$. We assume that the trajectories of the process belong to some topological space $\mathbf{X} \subset \mathbf{R}^T$ with probability 1. We also assume that the mapping $\Xi \colon \boldsymbol{\Omega} \to \mathbf{X}$ given by the rule $\omega \to \xi(\,\cdot\,,\omega)$ is $\mathfrak{F}$-measurable and $\mathfrak{B}_\mathbf{X}$-measurable, where $\mathfrak{B}_\mathbf{X}$ is the minimal σ-algebra containing the open subsets of $\mathbf{X}$. The measure $\mathbf{P}\Xi^{-1}$ defined on $\mathfrak{B}_\mathbf{X}$, that is, the image of the measure $\mathbf{P}$ under the mapping Ξ, is called the *distribution* $\mathbf{P}_\xi$ of the process ξ.

Any measurable mapping $f \colon \mathbf{X} \to \mathbf{R}^1$ defines the random variable $\omega \to f(\xi(\omega))$, which is called a *functional* of ξ. Obviously, the distribution of this random variable is just the measure $\mathbf{P}_\xi f^{-1}$ and hence is completely determined by the distribution of the process itself. Therefore, it is reasonable to ignore the specific nature of the measure $\mathbf{P}_\xi$ and study the distribution $\mathbf{P}f^{-1}$ assuming that $\mathbf{P}$ is an arbitrary probability measure on $\mathbf{X}$. In this context, f is called a *stochastic functional.* Usually, $\mathbf{X}$ is one of the following spaces:

- the space $\mathbf{C}(T)$ of continuous functions on a compact set T; this space is equipped with the uniform norm;
- the space $\mathbf{D}(T)$ of right continuous functions that have limits at each point of the interval T; this space is equipped with the Skorokhod topology (for the definition and properties of this space see [**4,** Ch. 3]);
- the space $\mathbf{L}^p(T,\mathfrak{T},\tau)$ of equivalence classes of functions whose pth power ($p \geq 1$) is integrable with respect to the measure τ; this space is equipped with the norm $\|h\|_p = (\int_T |h|^p \, d\tau)^{1/p}$.

The properties of the distribution $\mathbf{P}_\xi$, which characterizes the process as a whole, are the key to studying distributions of stochastic functionals. Since $\mathbf{P}_\xi$ is a measure on the topological space $\mathbf{X}$, we recall some properties of such measures.

Measures on a topological space. Suppose that $\mathbf{X}$ is a Hausdorff space, $\mathfrak{Z} \subset \mathfrak{B}_\mathbf{X}$ is an algebra of Borel sets, and $\mathbf{P} \colon \mathfrak{Z} \to \mathbf{R}_+$ is a monotone set function. Then $\mathbf{P}$ is called a *Radon* function if for all $B \in \mathfrak{Z}$ one has

$$\mathbf{P}\{B\} = \sup\{\mathbf{P}\{Z\},\ Z \subset B,\ Z \in \mathfrak{Z},\ Z \text{ is compact}\}. \tag{1.1}$$

The function $\mathbf{P}$ is said to be *regular* if it satisfies condition (1.1) with compact sets replaced by closed sets. We say that $\mathbf{P}$ is *dense* if (1.1) holds for $B = \mathbf{X}$. Obviously, if $\mathbf{P}$ is a Radon function, then it is regular and dense. If $\mathbf{P}$ is finitely additive, then the converse is also true: the regularity and density of $\mathbf{P}$ imply (1.1).

If $\mathbf{P}$ is countably additive, then it will be called a *measure.* By definition, we consider measures with *finite* values. Occasionally, we use *infinite* measures (with values in $\overline{\mathbf{R}}_+$) and *signed measures* (*charges*) with values in $\mathbf{R}^1$. If $\mathbf{P}$ is countably additive and satisfies one of the above properties, we speak of a *Radon measure*, regular measure, or dense measure, respectively.

For a given $\mathbf{P}$, one can construct the following two functions on the σ-algebra of *all* subsets of $\mathbf{X}$:

$$\mathbf{P}^*\{B\} = \inf\{\mathbf{P}\{Z\},\ Z \in \mathfrak{Z},\ B \subset Z\},$$
$$\mathbf{P}_*\{B\} = \sup\{\mathbf{P}\{Z\},\ Z \in \mathfrak{Z},\ B \supset Z\}.$$

If $\mathbf{P}$ is a measure, then $\mathbf{P}^*$ and $\mathbf{P}_*$ are referred to as the *outer* and the *inner measure*, respectively.

A function h that maps a subset of $\mathbf{X}$ to a measurable space $(\mathbf{Y}, \mathfrak{U})$ is said to be $\mathbf{P}$-*measurable* (where $\mathbf{P}$ is a measure defined on a σ-algebra $\mathfrak{Z}$) if there exist a $\mathfrak{Z}$-measurable function $g\colon \mathbf{X} \to (\mathbf{Y}, \mathfrak{U})$ and a set $Z \in \mathfrak{Z}$ of full measure (that is, $\mathbf{P}\{Z\} = \mathbf{P}\{\mathbf{X}\}$) such that $h = g$ on Z.

A set $B \in \mathbf{X}$ is said to be $\mathbf{P}$-*measurable* if its indicator function $h(x) = \mathbf{1}_B(x)$ is $\mathbf{P}$-measurable.

Let us study conditions under which a function $\mathbf{P}$ defined on an algebra $\mathfrak{Z}$ extends to a Radon measure defined on $\mathfrak{B}_{\mathbf{X}}$. We have the following theorem.

THEOREM 1.1. *If $\mathfrak{Z}$ contains a base of the topology, $\mathbf{P}$ is finitely additive and regular, and $\mathbf{P}^*$ is dense, then there exists a unique Radon measure that is defined on $\mathfrak{B}_{\mathbf{X}}$ and is a continuation of $\mathbf{P}$.*

REMARK. If $\mathbf{P}$ is a Radon measure on $\mathfrak{Z}$, then $\mathbf{P}$ is dense and hence so is $\mathbf{P}^*$; thus, we have the statement of the theorem.

The proof of this and some other properties of measures on topological spaces can be found in [**16,** Ch. 1].

The set of points $x \in \mathbf{X}$ such that any neighborhood of x has a positive measure is called the *topological support* $\operatorname{supp}(\mathbf{P})$ of the measure $\mathbf{P}$. The topological support is closed. If $\mathbf{X}$ has a countable base or $\mathbf{P}$ is a Radon measure, then the topological support is the smallest closed set of full measure: $\mathbf{P}\{\operatorname{supp}(\mathbf{P})\} = \mathbf{P}\{\mathbf{X}\}$.

Measures on a linear space. Let $\mathbf{X}$ and $\mathbf{X}'$ be two linear spaces *in duality.* Thus, there is a bilinear form $\langle x, x'\rangle$, defined on $\mathbf{X} \times \mathbf{X}'$, such that

- for each nonzero $x \in \mathbf{X}$ there exists an $x' \in \mathbf{X}'$ with $\langle x, x'\rangle \neq 0$;
- for each nonzero $x' \in \mathbf{X}'$ there exists an $x \in \mathbf{X}$ with $\langle x, x'\rangle \neq 0$.

We say that the form $\langle\,\cdot\,,\,\cdot\,\rangle$ *brings* $\mathbf{X}$ *and* $\mathbf{X}'$ *into duality.* The following example is typical: $\mathbf{X}$ is a locally convex Hausdorff space and $\mathbf{X}'$ is the space of continuous linear functionals on $\mathbf{X}$.

In the space $\mathbf{X}$ we obtain the *cylinder algebra* $\mathfrak{C}_0$ generated by linear functionals of the form $\langle\,\cdot\,, x'\rangle$ and the *cylinder σ-algebra* $\mathfrak{C} = \sigma(\mathfrak{C}_0)$ that is the minimal σ-algebra containing $\mathfrak{C}_0$.

We can make $\mathbf{X}$ a locally convex Hausdorff space by introducing the set of seminorms

$$q_{x'}(x) = |\langle x, x'\rangle|. \tag{1.2}$$

The corresponding topology is called the *weak* topology since it is the weakest topology in which all functionals of the form $\langle\cdot, x'\rangle$ are continuous. When speaking about the topological properties of $\mathbf{X}$, we usually mean the weak topology. By $\mathfrak{B}_{\mathbf{X}}$ we denote the Borel σ-algebra of the weak topology. We have $\mathfrak{C} \subset \mathfrak{B}_{\mathbf{X}}$ and, in the most interesting cases, $\mathfrak{C} = \mathfrak{B}_{\mathbf{X}}$; however, sometimes the Borel σ-algebra is wider than the cylinder σ-algebra.

In the dual space $\mathbf{X}'$ one can also introduce the weak topology by using the set of seminorms $\{|\langle x, \cdot \rangle|\}$.

Suppose that a measure $\mathbf{P}$ is defined on a measurable space $(\mathbf{X}, \mathfrak{C})$. A vector $a \in \mathbf{X}$ is called the *barycenter* (or the Pettis mean value) of the measure $\mathbf{P}$ if

$$\langle a, x' \rangle = \int_{\mathbf{X}} \langle x, x' \rangle \, \mathbf{P}(dx) \tag{1.3}$$

for all $x' \in \mathbf{X}'$. If $a = 0$, then the measure $\mathbf{P}$ is said to be *centered.*

A linear operator $K\colon \mathbf{X}' \to \mathbf{X}$ is called the *correlation* (covariation) operator of the measure $\mathbf{P}$ if

$$\langle Kx', y' \rangle = \int_{\mathbf{X}} \big(\langle x, x' \rangle - \langle a, x' \rangle\big)\big(\langle x, y' \rangle - \langle a, y' \rangle\big) \, \mathbf{P}(dx) \tag{1.4}$$

for any $x', y' \in \mathbf{X}'$.

Note that in the class of Radon measures there is virtually no difference between measures defined on $\mathfrak{C}$ and on $\mathfrak{B}_{\mathbf{X}}$, since, by Theorem 1.1, each Radon measure defined on $\mathfrak{C}$ can be uniquely extended to $\mathfrak{B}_{\mathbf{X}}$.

The complex-valued function defined on $\mathbf{X}'$ by the relation

$$\psi_{\mathbf{P}}(x') = \int_{\mathbf{X}} \exp\{i\langle x, x' \rangle\} \mathbf{P}\,(dx) \tag{1.5}$$

is called the *characteristic functional* of the measure $\mathbf{P}$.

THEOREM 1.2. *Let $\mathbf{P}$ and $\mathbf{Q}$ be two Radon measures on $\mathfrak{B}_{\mathbf{X}}$ with equal characteristic functionals $\psi_{\mathbf{P}} = \psi_{\mathbf{Q}}$. Then $\mathbf{P} = \mathbf{Q}$.*

PROOF. Since the characteristic functionals coincide, we obviously have the same joint distributions with respect to the measures $\mathbf{P}$ and $\mathbf{Q}$ for any finite set of linear functionals of the form $\langle \cdot, x' \rangle$. This precisely means that $\mathbf{P}$ and $\mathbf{Q}$ coincide on $\mathfrak{C}_0$ and hence on $\mathfrak{C}$. It remains to use the fact that a Radon measure can be uniquely extended from $\mathfrak{C}$ to $\mathfrak{B}_{\mathbf{X}}$. □

The *linear support* $N_{\mathbf{P}}$ of a measure $\mathbf{P}$ is defined as the minimal closed affine subspace of $\mathbf{X}$ for which $\mathbf{P}\{N_{\mathbf{P}}\} = 1$. Obviously, $N_{\mathbf{P}} \supset \operatorname{supp}(\mathbf{P})$. If $\mathbf{P}$ is a Radon measure and the topological support of $\mathbf{P}$ is an affine subspace, then $N_{\mathbf{P}} = \operatorname{supp}(\mathbf{P})$.

Gaussian processes and measures. The distribution $\mathcal{N}(a, \sigma^2)$ with density

$$p(r) = (2\pi)^{-1/2}\sigma^{-1} \exp\{-(r-a)^2/2\sigma^2\} \tag{1.6}$$

is called a *nondegenerate Gaussian* (or *normal*) distribution on $\mathbf{R}^1$.

The distribution δ_a concentrated at a point a is called the *degenerate Gaussian* distribution $\mathcal{N}(a, 0)$. The parameters $a \in \mathbf{R}^1$ and $\sigma^2 \geq 0$ coincide with the mean value and the variance of $\mathcal{N}(a, \sigma^2)$, respectively. The characteristic distribution function of $\mathcal{N}(a, \sigma^2)$ has the form

$$\varphi(u) = \exp\{iau - \sigma^2 u^2/2\}. \tag{1.7}$$

The distribution $\mathcal{N}(0,1)$ is called the *standard* Gaussian distribution; in what follows its distribution function will be denoted by $\Phi(r)$.

A random variable with Gaussian distribution is called *Gaussian.* The notions of a Gaussian random variable and a Gaussian distribution function can naturally be generalized to multidimensional and infinite-dimensional linear spaces. The required generalization can be obtained by using the following construction. Let $\langle \mathbf{X}, \mathbf{X}' \rangle$ be a pair of spaces in duality. A measure $\mathbf{P}$ defined on the measurable space $(\mathbf{X}, \mathfrak{C})$ is said to be *Gaussian* if the distribution with respect to $\mathbf{P}$ of each linear functional $\langle \cdot\,, x' \rangle$ is Gaussian.

A random process $\{\xi_t, t \in T\}$ is said to be *Gaussian* if the joint distribution of any finite set $\{\xi_{t_1}, \dots, \xi_{t_n}\}$ is a Gaussian measure on $\mathbf{R}^n$ (in particular, the distribution ξ_t belongs to the family $\mathcal{N}(a, \sigma^2)$). The notions of a Gaussian process and a Gaussian measure are related as follows: under the natural choice of duality, *the distribution $\mathbf{P}_\xi$ of a Gaussian process ξ is a Gaussian measure in the space $\mathbf{X}$ of trajectories.* In Chapter 3 we consider numerous examples of such measures and their most important properties.

§2. Convergence of probability measures

Weak convergence. The theory of weak convergence is considered in detail in several textbooks (e.g., see [**4, 24**]). Here we present some basic definitions and results that are not commonly known but will be useful for our further goals.

A sequence of (probability) measures $\{\mathbf{P}_n\}$ defined on the σ-algebra $\mathfrak{B}_\mathbf{X}$ of Borel subsets of a metric space $\mathbf{X}$ is said to be *weakly convergent* to a measure $\mathbf{P}$ if for any continuous bounded function h on $\mathbf{X}$ we have

$$\int_\mathbf{X} h\, d\mathbf{P}_n \to \int_\mathbf{X} h\, d\mathbf{P}.$$

We denote weak convergence by the symbol $\Rightarrow$. The following theorem contains equivalent definitions of weak convergence [**4,** §§2, 5].

THEOREM 2.1. *The weak convergence $\mathbf{P}_n \Rightarrow \mathbf{P}$ is equivalent to each of the following properties.*

1) $\overline{\lim}\, \mathbf{P}_n\{F\} \le \mathbf{P}\{F\}$ *for any closet set F;*

2) $\underline{\lim}\, \mathbf{P}_n\{G\} \le \mathbf{P}\{G\}$ *for any open set G;*

3) $\lim \mathbf{P}_n\{A\} = \mathbf{P}\{A\}$ *for any set A with $\mathbf{P}\{\partial A\} = 0$;*

4) *for any $\mathbf{P}$-almost everywhere continuous functional $f \colon \mathbf{X} \to \mathbf{R}^1$ the distributions $\mathbf{P}_n f^{-1}$ weakly converge to $\mathbf{P} f^{-1}$.*

We have an even more general statement.

Let h_n and h be measurable mappings of $\mathbf{X}$ to a separable metric space $\mathbf{Y}$. Let $\mathbf{X}_0$ be the set of $x \in \mathbf{X}$ for which the relation $h_n(x_n) \to h(x)$ is violated for some sequence (x_n) that converges to x. Then (see [**4,** Theorem 5.5]) the relations $\mathbf{P}_n \Rightarrow \mathbf{P}$ and $\mathbf{P}\{\mathbf{X}_0\} = 0$ imply $\mathbf{P}_n h_n^{-1} \Rightarrow \mathbf{P} h^{-1}$.

This statement readily yields the following proposition.

PROPOSITION 2.1. *Let $\mathbf{Y} = \mathbf{R}^1$. Assume that h_n and h are bounded by the same constant. If $\mathbf{P}_n \Rightarrow \mathbf{P}$ and $\mathbf{P}\{\mathbf{X}_0\} = 0$, then*

$$\int_\mathbf{X} h_n\, dP_n \to \int_\mathbf{X} h\, dP. \tag{2.1}$$

PROOF. Suppose that $c > 0$ is a constant that majorizes h_n and h. Let $\mathcal{P}_n = \mathbf{P}_n h_n^{-1}$ and $\mathcal{P} = \mathbf{P}h^{-1}$. By our assumptions, $\mathcal{P}_n \Rightarrow \mathcal{P}$. The function $\varphi(t) = -c\mathbf{1}_{(-\infty,c]}(t) + t\mathbf{1}_{(-c,c)}(t) + c\mathbf{1}_{[c,\infty)}(t)$ is continuous and bounded on $\mathbf{R}^1$. Therefore,

$$\int_{\mathbf{R}^1} \varphi\, d\mathcal{P}_n \to \int_{\mathbf{R}^1} \varphi\, d\mathcal{P}.$$

Since $\int_{\mathbf{R}^1} \varphi\, d\mathcal{P}_n = \int_{\mathbf{R}^1} t\mathcal{P}_n\,(dt) = \int_{\mathbf{X}} h_n\, d\mathbf{P}_n$ and $\int_{\mathbf{R}^1} \varphi\, d\mathcal{P} = \int_{\mathbf{X}} h\, d\mathbf{P}$, we arrive at (2.1). □

The following theorem is very useful.

THEOREM 2.2. *Suppose that* $\mathbf{X}$ *is separable. Let* $\{f_\vartheta\}$ *be a family of real functions that are uniformly bounded on* $\mathbf{X}$ *and equicontinuous at each point of a set* Y *of* $\mathbf{P}$*-full measure. If* $\mathbf{P}_n \Rightarrow \mathbf{P}$, *then*

$$\int_{\mathbf{X}} f_\vartheta\, d\mathbf{P}_n \to \int_{\mathbf{X}} f_\vartheta\, d\mathbf{P}$$

uniformly with respect to ϑ.

PROOF. Let $c > 0$ be a constant that majorizes the functions f_ϑ. For any $\varepsilon > 0$ and $x \in Y$, we can find an open ball V_x centered at x such that

$$\sup_\vartheta \omega_{f_\vartheta}(V_x) < \varepsilon. \tag{2.2}$$

By reducing the radius of this ball, we can ensure that $\mathbf{P}\{\partial V_x\} = 0$. Obviously, the family $(V_x)_{x\in Y}$ is a cover of Y. Since $\mathbf{X}$ is separable, we can take a countable subcover (V_k), $V_k = V_{x_k}$. Let us choose N so that $\mathbf{P}(V) > 1 - \varepsilon/c$, where $V = \bigcup_1^N V_k$, and consider the functions

$$\widetilde{f}_\vartheta(x) = \sum_{k=1}^N f_\vartheta(x_k)\mathbf{1}_{U_k}(x),$$

where $U_k = V_k \setminus \bigcup_{j=1}^{k-1} V_j$.

Then we have

$$\left|\int_{\mathbf{X}} f_\vartheta\, d\mathbf{P}_n - \int_{\mathbf{X}} \widetilde{f}_\vartheta\, d\mathbf{P}_n\right| \le \sum_{k=1}^N \omega_{f_\vartheta}(V_k)\mathbf{P}_n\{U_k\} + \mathbf{P}_n\{V^\complement\}\cdot c \le \varepsilon + c\mathbf{P}_n\{V^\complement\}.$$

We also have a similar inequality with $\mathbf{P}_n$ replaced by $\mathbf{P}$.

On the other hand, since $\mathbf{P}\{\partial U_k\} = 0$, we have

$$\lim_{n\to\infty} \int_{\mathbf{X}} \widetilde{f}_\vartheta\, d\mathbf{P}_n = \sum_1^N f_\vartheta(x_k) \lim_{n\to\infty} \mathbf{P}_n\{U_k\} = \int_{\mathbf{X}} \widetilde{f}_\vartheta\, d\mathbf{P}.$$

The set $V^\complement$ is closed, and therefore,

$$\overline{\lim_n}\, \mathbf{P}_n(V^\complement) \le \mathbf{P}(V^\complement) \le \varepsilon/c.$$

Thus,

$$\overline{\lim_n}\, \sup_\vartheta \left|\int_{\mathbf{X}} f_\vartheta\, d\mathbf{P}_n - \int_{\mathbf{X}} f_\vartheta\, d\mathbf{P}\right| \le 2\big(\varepsilon + c\,\overline{\lim_n}\, \mathbf{P}_n\{V^\complement\}\big) \le 4\varepsilon.$$

Since ε is arbitrarily small, we obtain the desired statement. □

This theorem can be used to study the convergence of integrals of functions assuming values in Banach spaces.

THEOREM 2.3. *Let h be a measurable mapping of a separable space $\mathbf{X}$ to a Banach space $(\mathbf{B}, \|\cdot\|)$. We assume that h is bounded with respect to the norm and continuous $\mathbf{P}$-almost everywhere. If $\mathbf{P}_n \Rightarrow \mathbf{P}$, then*

$$\left\| \int_{\mathbf{X}} h\, d\mathbf{P}_n - \int_{\mathbf{X}} h\, d\mathbf{P} \right\| \to 0.$$

PROOF. Let us consider the family $\{h_{y^*}\}$, $y^* \in \mathrm{Ball}_1^*$, of functions $h_{y^*}(x) = \langle h(x), y^* \rangle$, where Ball_1^* is the closed unit ball in $\mathbf{B}^*$. It follows from the assumptions of the theorem that this family is uniformly bounded and $\mathbf{P}$-almost everywhere equicontinuous. By Theorem 2.1, we have

$$\begin{aligned} &\sup_{y \in \mathrm{Ball}_1^*} \left| \int_{\mathbf{X}} \langle h(x), y^* \rangle \mathbf{P}_n\,(dx) - \int_{\mathbf{X}} \langle h(x), y^* \rangle\, \mathbf{P}\,(dx) \right| \\ &\quad = \sup_{y^* \in \mathrm{Ball}_1^*} \left| \left\langle \int_{\mathbf{X}} h\, d\mathbf{P}_n - \int_{\mathbf{X}} h\, d\mathbf{P}, y^* \right\rangle \right| = \left\| \int_{\mathbf{X}} h\, d\mathbf{P}_n - \int_{\mathbf{X}} h\, d\mathbf{P} \right\| \to 0. \quad \square \end{aligned}$$

Just as in the finite-dimensional case, the weak convergence of measures in a Banach space implies the convergence of their characteristic functionals.

THEOREM 2.4. *Let $\mathbf{P}_n$ and $\mathbf{P}$ be measures in a separable Banach space $(\mathbf{X}, \|\cdot\|)$, and let ψ_n and ψ be their characteristic functionals. If $\mathbf{P}_n \Rightarrow \mathbf{P}$, then $\psi_n(x^*) \to \psi(x^*)$ uniformly with respect to $x^* \in \mathrm{Ball}^*$ for any ball $\mathrm{Ball}^* \subset X^*$.*

PROOF. We see that the family of functions $\{\exp(i\langle\, \cdot\, , x^* \rangle), x^* \in \mathrm{Ball}^*\}$ is uniformly bounded and equicontinuous. Therefore, the desired statement follows from Theorem 2.2. $\square$

Note that, in contrast with the case of $\mathbf{R}^n$, the converse statement does not hold in the infinite-dimensional case.

Convergence in variation. Let μ be a finite charge (signed measure) on $\mathfrak{B}_{\mathbf{X}}$. The measure $|\mu|$ defined by the relation

$$|\mu|(A) = \sup_{(A_n)} \sum_n |\mu(A_n)|,$$

where the least upper bound is taken over all countable partitions of $\mathbf{X}$, is called the *variation* of μ.

The value $\|\mu\| = |\mu|(\mathbf{X})$ is called the *total variation* of μ.

The space of all finite charges μ on $\mathfrak{B}_{\mathbf{X}}$ equipped with the norm $\|\mu\|$ is a Banach space and is denoted by $\mathbf{Z}(\mathbf{X})$. The convergence with respect to the norm of this space is called *convergence in variation* or *strong convergence* and is denoted by the symbol $\xrightarrow{\text{var}}$.

If the charge μ is absolutely continuous with respect to the measure ν and if $h = d\mu/d\nu$, then $\|\mu\| = \int_{\mathbf{X}} |h|\, d\nu = \|h\|_{\mathbf{L}^1(\mathbf{X},\nu)}$. Therefore, in the absolutely continuous case (that is, for $\mu_n \ll \nu$ and $\mu \ll \nu$) the convergence $\mu_n \xrightarrow{\text{var}} \mu$ is equivalent to the convergence of the densities $h_n = d\mu_n/d\nu$ to $h = d\mu/d\nu$ in the norm of $\mathbf{L}^1(\mathbf{X}, \nu)$. (This property is generalized in Problem 2.4.) The total variation cannot be made larger by applying a mapping.

THEOREM 2.5. *Let f be a measurable mapping of $(\mathbf{X}, \mathfrak{A})$ to $(\mathbf{Y}, \mathfrak{Y})$, and let $\nu = \mu f^{-1}$. Then $\|\nu\| \leq \|\mu\|$.*

PROOF. If sets A_n form a partition of $\mathbf{Y}$, then the sets $f^{-1}(A_n)$ form a partition of $\mathbf{X}$. Therefore,

$$\sum_n |\nu(A_n)| = \sum_n |\mu f^{-1}(A_n)| = \sum_n |\mu(f^{-1}(A_n))| \leq \|\mu\|,$$

and the desired inequality follows. □

THEOREM 2.6. *Let $(\mathbf{X}, \mathfrak{A}, \mu) = (\mathbf{X}_1, \mathfrak{A}_1, \mu_1) \times (\mathbf{X}_2, \mathfrak{A}_2, \mu_2)$. Then $\|\mu\| = \|\mu_1\| \cdot \|\mu_2\|$.*

PROOF. We set $\nu_1 = |\mu_1|$ and $\nu_2 = |\mu_2|$. Obviously, $\mu_1 \ll \nu_1$ and $\mu_2 \ll \nu_2$. Suppose that $h_1 = d\mu_1/d\nu_1$ and $h_2 = d\mu_2/d\nu_2$. One can readily verify that $\mu_1 \times \mu_2 \ll \nu_1 \times \nu_2$ and

$$\frac{d(\mu_1 \times \mu_2)}{d(\nu_1 \times \nu_2)}(x, y) = h_1(x)h_2(x), \qquad x \in \mathbf{X}_1, \quad y \in \mathbf{X}_2.$$

Therefore,

$$\begin{aligned} \|\mu_1 \times \mu_2\| &= \int_{\mathbf{X}_1 \times \mathbf{X}_2} |h_1(x)h_2(y)| \nu_1\,(dx) \nu_2\,(dy) \\ &= \int_{\mathbf{X}_1} |h_1|\,d\nu_1 \cdot \int_{\mathbf{X}_2} |h_2|\,d\nu_2 = \|\mu_1\| \cdot \|\mu_2\|. \quad \square \end{aligned}$$

Let $\mathbf{X}$ be a linear topological space, and let μ_1 and μ_2 be two measures on $\mathfrak{B}_{\mathbf{X}}$. It follows from the preceding two theorems that $\|\mu_1 \times \mu_2\| \leq \|\mu_1\| \cdot \|\mu_2\|$. Indeed, the convolution $\mu_1 * \mu_2$ is equal to μf^{-1}, where $\mu = \mu_1 \times \mu_2$ and $f(x_1, x_2) = x_1 + x_2$. Therefore,

$$\|\mu_1 * \mu_2\| = \|\mu f^{-1}\| \leq \|\mu\| = \|\mu_1\| \cdot \|\mu_2\|.$$

Let us recall the definition of the *mixture* of measures. Let $(\mathbf{Y}, \mathfrak{Y}, \nu)$ be a measure space, and let $\{\mu_y\}_{y \in \mathbf{Y}}$ be a family of charges on $(\mathbf{X}, \mathfrak{A})$ such that for each $A \in \mathfrak{A}$ the function $y \to \mu_y(A)$ is measurable. The charge

$$\mu(A) = \int_{\mathbf{Y}} \mu_y(A) \nu\,(dy)$$

is called the *mixture of charges* $\{\mu_y\}$ with respect to the measure ν. From this definition and from the definition of total variation we readily obtain

$$\|\mu\| \leq \int_{\mathbf{Y}} \|\mu_y\| \nu\,(dy). \tag{2.3}$$

The strong convergence of measures implies the weak convergence. Although the converse is obviously false, we have the following statement.

THEOREM 2.7. *Let $\mathbf{P}_n \Rightarrow \mathbf{P}$ and $\mathbf{Q}_n \Rightarrow \mathbf{Q}$. Then*

$$\|\mathbf{P} - \mathbf{Q}\| \leq \varliminf_n \|\mathbf{P}_n - \mathbf{Q}_n\|. \tag{2.4}$$

PROOF. Suppose F is a closed set and $\delta > 0$. We can choose a small $\varepsilon > 0$ so that $\mathbf{P}\{\partial(F^\varepsilon)\} = \mathbf{Q}\{\partial(F^\varepsilon)\} = 0$ and $\mathbf{Q}\{F^\varepsilon\} \le \mathbf{Q}\{F\} + \delta$. Then

$$\begin{aligned}\mathbf{P}\{F\} - \mathbf{Q}\{F\} &\le \mathbf{P}\{F^\varepsilon\} - \mathbf{Q}\{F^\varepsilon\} + \delta \\ &= \lim_n \big(\mathbf{P}_n\{F^\varepsilon\} - \mathbf{Q}_n\{F^\varepsilon\}\big) + \delta \le \overline{\lim_n} \|\mathbf{P}_n - \mathbf{Q}_n\| + \delta.\end{aligned}$$

Since δ is arbitrarily small, we have

$$\mathbf{P}\{F\} - \mathbf{Q}\{F\} \le \overline{\lim_n} \|\mathbf{P}_n - \mathbf{Q}_n\|.$$

To pass from closed to arbitrary sets, it suffices to use the regularity formula

$$\mathbf{P}\{A\} = \sup \big\{\mathbf{P}\{F\},\, F \subset A,\, F \text{ is closed}\big\}.$$

This formula holds for all $A \in \mathfrak{B}_\mathbf{X}$ provided that $\mathbf{X}$ is a metric space. □

PROBLEM 2.1. Let $\mathbf{P}_n$, $n \in \overline{\mathbf{N}}$, be a sequence of Gaussian measures in a separable Banach space $\mathbf{X}$, and let a_n be the barycenter of the measure $\mathbf{P}_n$. If $\mathbf{P}_n \Rightarrow \mathbf{P}_\infty$, then $a_n \to a_\infty$ in the weak topology generated by the duality $\langle \mathbf{X}, \mathbf{X}^* \rangle$.

PROBLEM 2.2. Suppose $\mathbf{P}_1$ and $\mathbf{P}_2$ are finite measures, $\mathbf{Q} = (\mathbf{P}_1 + \mathbf{P}_2)/2$, $z_1 = d\mathbf{P}_1/d\mathbf{Q}$, and $z_2 = d\mathbf{P}_2/d\mathbf{Q}$. Then

$$\|\mathbf{P}_1 - \mathbf{P}_2\| = \mathsf{E}_\mathbf{Q}|z_1 - z_2| = 2\mathsf{E}_\mathbf{Q}|1 - z_2|,$$

where $\mathsf{E}_\mathbf{Q}$ denotes the integral with respect to the measure $\mathbf{Q}$.

PROBLEM 2.3. Suppose that $\mathbf{P}_1$ and $\mathbf{P}_2$ are probability measures, $\mathbf{Q}$ is a probability measure with respect to which $\mathbf{P}_1$ and $\mathbf{P}_2$ are absolutely continuous (for example, $\mathbf{Q} = (\mathbf{P}_1 + \mathbf{P}_2)/2$), $z_1 = d\mathbf{P}_1/d\mathbf{Q}$, and $z_2 = d\mathbf{P}_2/d\mathbf{Q}$. The *Hellinger distance* $\rho(\mathbf{P}_1, \mathbf{P}_2)$ between $\mathbf{P}_1$ and $\mathbf{P}_2$ is defined by

$$\rho^2(\mathbf{P}_1, \mathbf{P}_2) = \frac{1}{2} \int (\sqrt{z_1} - \sqrt{z_2})^2 \, d\mathbf{Q}.$$

Verify that this distance is well defined, i.e., that $\rho(\mathbf{P}_1, \mathbf{P}_2)$ is independent of the choice of $\mathbf{Q}$, and prove the inequalities

$$2\rho^2(\mathbf{P}_1, \mathbf{P}_2) \le \|\mathbf{P}_1 - \mathbf{P}_2\| \le 2\sqrt{2}\rho(\mathbf{P}_1, \mathbf{P}_2).$$

PROBLEM 2.4. Let $(\mu_n, n \in \overline{\mathbf{N}})$ be a sequence of finite measures, $\mu_n \overset{\text{var}}{\to} \mu_\infty$. Prove that for an arbitrary measure ν the density of the continuous component of μ_n with respect to ν converges (in the metric of $\mathbf{L}^1(\mathbf{X}, \nu)$) to the density of the absolutely continuous component of μ_∞ with respect to ν.

PROBLEM 2.5. Let (ξ_n) be a sequence of random elements of a separable space $\mathbf{X}$. Suppose that the sequence $(\mathbf{P}_n)$ of their distributions weakly converges, $\mathbf{P}_n \Rightarrow \mathbf{P}$. Let $\{f_\vartheta\}$ be a family of functions defined on $\mathbf{X}$, equicontinuous $\mathbf{P}$-almost everywhere, and satisfying

$$\sup_x |f_\vartheta(x)| \le H(x)$$

for some function H and all ϑ. If the family of random variables $(H(\xi_n))$ is uniformly integrable, then

$$\int f_\vartheta \, d\mathbf{P}_n \to \int f_\vartheta \, d\mathbf{P} \qquad \text{uniformly with respect to } \vartheta.$$

This statement is a generalization of Theorem 2.2.

§3. Measurable partitions and systems of conditional measures

Suppose that $(\mathbf{X}, \mathfrak{B})$ is a measurable space, $\mathbf{P}$ a probability measure on $\mathfrak{B}$, and Γ a partition of $\mathbf{X}$ into disjoint subsets γ.

We intend to define measures $\mathbf{P}_\gamma$ on the elements of Γ so that $\mathbf{P}$ can be represented as a mixture of the $\mathbf{P}_\gamma$. This representation enables us to study the properties of $\mathbf{P}$ by analyzing measures of relatively simple structure.

By $\mathbf{X}/\Gamma$ we denote the *quotient space* generated by Γ, that is, the set whose elements are the elements of Γ. Let $\pi\colon \mathbf{X} \to \mathbf{X}/\Gamma$ be the canonical projection; it takes each $x \in \mathbf{X}$ to the element $\gamma \in \Gamma$ that contains x.

We equip the set $\mathbf{X}/\Gamma$ with the σ-algebra $\mathfrak{B}_{\mathbf{X},\Gamma}$ of sets $A \subset \mathbf{X}/\Gamma$ such that $\pi^{-1}(A) \in \mathfrak{B}$. Then on the measurable space $(\mathbf{X}/\Gamma, \mathfrak{B}_{\mathbf{X},\Gamma})$ we can consider the measure $\mathbf{P}_\Gamma\{A\} = \mathbf{P}\{\pi^{-1}(A)\}$, which will be called the *quotient measure.*

Finally, let us consider the σ-subalgebra $\mathfrak{B}^\Gamma \subset \mathfrak{B}$ that consists of the sets $\pi^{-1}(A)$, $A \in \mathfrak{B}_{\mathbf{X},\Gamma}$. The elements of $\mathfrak{B}^\Gamma$ are $\mathfrak{B}$-measurable subsets of $\mathbf{X}$ that respect the partition Γ, that is, entirely contain each element of the partition with which they have a nonempty intersection.

A system of probability measures $\{\mathbf{P}_\gamma\}$, $\gamma \in \Gamma$, defined on $\mathfrak{B}$ is called a *system of conditional measures* for $\mathbf{P}$ with respect to Γ if for any $B \in \mathfrak{B}$ the function $\gamma \to \mathbf{P}_\gamma\{B\}$ is $\mathfrak{B}_{\mathbf{X},\Gamma}$-measurable and

$$\mathbf{P}\{B \cap \pi^{-1}(A)\} = \int_A \mathbf{P}_\gamma\{B\}\mathbf{P}_\Gamma\,(d\gamma) \tag{3.1}$$

for any $A \in \mathfrak{B}_{\mathbf{X},\Gamma}$.

For $A = \mathbf{X}/\Gamma$ formula (3.1) becomes the well-known total probability formula

$$\mathbf{P}\{B\} = \int_{\mathbf{X}/\Gamma} \mathbf{P}_\gamma\{B\}\mathbf{P}_\Gamma\,(d\gamma). \tag{3.2}$$

It is just this formula that allows us to say that $\mathbf{P}$ can be represented as a mixture of the measures $\mathbf{P}_\gamma$; moreover, the quotient measure is used as the weight.

Let us show how the term "a family of conditional measures" is related to the notion of conditional probability, which is more frequently used in probability theory. The conditional probability of an event B with respect to a σ-algebra $\mathcal{D}$ is defined as the $\mathcal{D}$-measurable function such that

$$\mathbf{P}\{BD\} = \int_D \mathbf{P}\{B \mid \mathcal{D}\}(x)\mathbf{P}\,(dx), \qquad D \in \mathcal{D}.$$

By comparing this definition with (3.1), we see that the function $\mathbf{P}_{\pi(\cdot)}(B)$ is a version of the conditional probability $\mathbf{P}\{B \mid \mathfrak{B}^\Gamma\}$.

The natural question of whether for all $\mathbf{X}$, $\mathbf{P}$, and Γ there exists a family of conditional measures is not simple. The following example shows that sometimes the answer is negative.

Suppose that $\mathbf{X} = \mathbf{R}^{\mathbf{Z}}$, $\mathbf{P}$ is a distribution on $\mathbf{X}$ corresponding to a sequence of independent identically distributed random variables, and Γ is the partition of $\mathbf{X}$ into classes of translation-equivalent sequences, that is, $x \overset{\Gamma}{\sim} y$ if and only if for some integer n and for all integer k we have $x_k = y_{n+k}$. If the distribution of each random variable is not concentrated at a single point, then in the above situation there does not exist a family of conditional measures. Indeed, a typical equivalence class $\gamma \in \mathbf{X}/\Gamma$ is an infinite sequence of distinct elements of $\mathbf{X}$, and the conditional measure $\mathbf{P}_\gamma$ on this class must be finite and translation-equivalent. Obviously, these two properties contradict each other.

Nevertheless, under some restrictions on $\mathbf{X}$ and Γ it is possible to prove the existence of a conditional measure.

Suppose that $\mathbf{X}$ is a complete separable metric space and $\mathfrak{B} = \mathfrak{B}_{\mathbf{X}}$ is the Borel σ-algebra. Let Γ be a partition of $\mathbf{X}$ into the preimages of points for some measurable mapping of $(\mathbf{X}, \mathfrak{B})$ to a complete separable metric space. This partition will be called *measurable.*

THEOREM 3.1. *If Γ is a measurable partition of the space $\mathbf{X}$ and $\mathbf{P}$ is a probability measure on $(\mathbf{X}, \mathfrak{B})$, then there exists a family of conditional measures $\{\mathbf{P}_\gamma\}_{\gamma\in\mathbf{X}/\Gamma}$; for $\mathbf{P}_\Gamma$-almost all γ the conditional measure is concentrated on the corresponding element of the partition, $\mathbf{P}_\gamma\{\pi^{-1}(\gamma)\} = 1$.*

The proof of Theorem 3.1 can be found in [**78**, Ch. VI, §46].

Now let us consider the properties of a family of conditional measures. The most important property is uniqueness.

PROPOSITION 3.1. *If $\{\mathbf{P}^1_\gamma\}$ and $\{\mathbf{P}^2_\gamma\}$ are two families of conditional measures for $\mathbf{P}$ with respect to a measurable partition Γ, then the measures $\mathbf{P}^1_\gamma$ and $\mathbf{P}^2_\gamma$ coincide for $\mathbf{P}_\Gamma$-almost all γ.*

PROOF. For $B \in \mathfrak{B}$ and $A \in \mathfrak{B}^\Gamma$, formula (3.1) implies

$$\int_A \mathbf{P}^1_\gamma\{B\}\mathbf{P}_\Gamma(d\gamma) = \int_A \mathbf{P}^2_\gamma\{B\}\mathbf{P}_\Gamma(d\gamma).$$

Since A is arbitrary, for a given B we have $\mathbf{P}^1_\gamma\{B\} = \mathbf{P}^2_\gamma\{B\}$ for $\mathbf{P}_\Gamma$-almost all γ. Hence, for each countable system of sets, the measures $\mathbf{P}^1_\gamma$ and $\mathbf{P}^2_\gamma$ coincide for $\mathbf{P}_\Gamma$-almost all γ. Recall that we consider measures on the Borel σ-algebra of a separable metric space.

In this σ-algebra there is a countable class of sets such that the values of a measure on the set of this class uniquely determine the measure. Therefore, for γ from a set of $\mathbf{P}_\Gamma$-full measure we have $\mathbf{P}^1_\gamma = \mathbf{P}^2_\gamma$. □

PROPOSITION 3.2. *If g is an integrable function on the space $(\mathbf{X}, \mathfrak{B}, \mathbf{P})$, then*

$$\int_{\mathbf{X}} g(x)\mathbf{P}\,(dx) = \int_{\mathbf{X}/\Gamma}\int_{\mathbf{X}} g(x)\mathbf{P}_\gamma\,(dx)\,\mathbf{P}_\Gamma\,(dx). \tag{3.3}$$

PROOF. For $g = \mathbf{1}_B$, (3.3) can be reduced to (3.2). One can readily pass to step functions, nonnegative functions, and arbitrary integrable functions. □

Now let us show how one can study the images of measures under mappings by using conditional measures.

Let $f\colon (\mathbf{X}, \mathfrak{B}) \to (\mathbf{Y}, \mathfrak{U})$ be a measurable mapping.

PROPOSITION 3.3. *For $A \in \mathfrak{U}$ we have*

$$\mathbf{P}f^{-1}(A) = \int_{\mathbf{X}/\Gamma} \mathbf{P}_\gamma f^{-1}(A)\mathbf{P}_\Gamma\,(d\gamma). \tag{3.4}$$

PROOF. Formula (3.4) is the special case of (3.2) for $B = f^{-1}(A)$. □

The measures $\mathbf{P}_\gamma f^{-1}(\cdot)$ defined on $(\mathbf{Y}, \mathfrak{U})$ are called the *conditional distributions* of the functional f with respect to the measure $\mathbf{P}$ and the partition Γ.

We also need an analog of formula (3.4) for the density of the distribution of a functional. Let μ be a measure on $(\mathbf{Y}, \mathfrak{U})$. We assume that there exists a jointly measurable function $p\colon (\gamma, y) \to p_\gamma(y)$ such that for $\mathbf{P}_\Gamma$-almost all $\gamma \in \mathbf{X}/\Gamma$ the function $p_\gamma(\cdot)$ is a version of the density of the measure $\mathbf{P}_\gamma f^{-1}$ with respect to the measure μ on the σ-algebra $\mathfrak{U}$:

$$p_\gamma(y) = \frac{d\mathbf{P}_\gamma f^{-1}}{d\mu}(y). \tag{3.5}$$

PROPOSITION 3.4. *The function*

$$p(y) = \int_{\mathbf{X}/\Gamma} p_\gamma(y)\mathbf{P}_\Gamma\,(d\gamma) \tag{3.6}$$

is a version of the density of the measure $\mathbf{P}f^{-1}$ with respect to the measure μ on the σ-algebra $\mathfrak{U}$.

PROOF. For $A \in \mathfrak{U}$ we use (3.4), substitute (3.5), then change the order of integration and use the definition of p:

$$\begin{aligned}\mathbf{P}f^{-1}(A) &= \int_{\mathbf{X}/\Gamma} \mathbf{P}_\gamma f^{-1}(A)\mathbf{P}_\Gamma\,(d\gamma)\\ &= \int_{\mathbf{X}/\Gamma}\int_A p_\gamma(y)\mu\,(dy)\mathbf{P}_\Gamma\,(d\gamma) = \int_A p(y)\mu\,(dy).\end{aligned}$$

This shows that p is the density of $\mathbf{P}f^{-1}$ with respect to μ. □

Relation (3.6) allows us to obtain some information about the density of the distribution of functionals f from the information about the density of conditional distributions p_γ. In the subsequent chapters, formula (3.6) serves as one of the main tools.

Now let us use the terminology of conditional distributions to show that the functional f is independent of the σ-algebra $\mathfrak{B}^\Gamma$.

THEOREM 3.2. *Let $f\colon (\mathbf{X}, \mathfrak{B}) \to (\mathbf{Y}, \mathfrak{U})$ be a measurable functional with values in a separable metric space $\mathbf{Y}$. If Γ is a measurable partition of $\mathbf{X}$, then for f to be independent of $\mathfrak{B}^\Gamma$ it is necessary and sufficient that $\mathbf{P}_\gamma f^{-1} = \mathbf{P}f^{-1}$ for $\mathbf{P}_\Gamma$-almost all $\gamma \in \mathbf{X}/\Gamma$.*

PROOF. First, we verify the sufficiency. Let

$$\begin{aligned} &A_1 \in \mathfrak{U}, && B_1 = f^{-1}(A_1) \in \mathfrak{B},\\ &A_2 \in \mathfrak{B}_{\mathbf{X},\Gamma}, && B_2 = \pi^{-1}(A_2) \in \mathfrak{B}^\Gamma.\end{aligned}$$

We apply formula (3.1) to B_1 and A_2. Then we use the assumptions of the theorem and, finally, factor out the constant coefficient from the integral:

$$\begin{aligned}\mathbf{P}\{B_1B_2\} &= \mathbf{P}\{B_1 \cap \pi^{-1}(A_2)\} = \int_{A_2} \mathbf{P}_\gamma\{f^{-1}(A_1)\}\mathbf{P}_\Gamma\,(d\gamma)\\ &= \int_{A_2} \mathbf{P}\{f^{-1}(A_1)\}\mathbf{P}_\Gamma\,(d\gamma) = \mathbf{P}\{f^{-1}(A_1)\}\mathbf{P}_\Gamma\{A_2\} = \mathbf{P}\{B_1\}\mathbf{P}\{B_2\}.\end{aligned}$$

The relation $\mathbf{P}\{B_1B_2\} = \mathbf{P}\{B_1\}\mathbf{P}\{B_2\}$ shows that the σ-algebra $\mathfrak{B}^\Gamma$ and the σ-algebra generated by the functional f are independent.

The necessity of the assumptions of the theorem follows from the same chain of relations:

$$\int_{A_2} \mathbf{P}_\gamma\{f^{-1}(A_1)\}\mathbf{P}_\Gamma(d\gamma) = \int_{A_2} \mathbf{P}\{f^{-1}(A_1)\}\mathbf{P}_\Gamma(d\gamma).$$

By literally repeating the proof of Proposition 3.1, we prove that the measures $\mathbf{P}_\gamma f^{-1}$ and $\mathbf{P}f^{-1}$ are equal for $\mathbf{P}_\Gamma$-almost all γ. □

Now we assume that two measures $\mathbf{P}$ and $\mathbf{Q}$, $\mathbf{Q} \ll \mathbf{P}$, are given on $\mathfrak{B}$. Obviously, then $\mathbf{Q}_\Gamma \ll \mathbf{P}_\Gamma$. Let us find the relation between $\mathbf{P}_\gamma$ and $\mathbf{Q}_\gamma$.

THEOREM 3.3. *For $\mathbf{P}_\Gamma$-almost all γ we have $\mathbf{Q}_\gamma \ll \mathbf{P}_\gamma$ and*

$$\frac{d\mathbf{Q}_\gamma}{d\mathbf{P}_\gamma}(x) = \frac{d\mathbf{Q}}{d\mathbf{P}}(x)\left[\frac{d\mathbf{Q}_\Gamma}{d\mathbf{P}_\Gamma}(\pi(x))\right]^{-1} \tag{3.7}$$

for $\mathbf{P}_\gamma$-almost all x.

PROOF. It suffices to verify that the measures

$$\mathbf{Q}_\gamma(dx) = \frac{d\mathbf{Q}}{d\mathbf{P}}(x)\left[\frac{d\mathbf{Q}_\Gamma}{d\mathbf{P}_\Gamma}\pi(x)\right]^{-1}\mathbf{P}_\gamma(dx)$$

satisfy the following relation of the form (3.1):

$$\mathbf{Q}\{B\cap\pi^{-1}(A)\} = \int_A \mathbf{Q}_\gamma\{B\}\mathbf{Q}_\Gamma(d\gamma)$$

for $B \in \mathfrak{B}$ and $A \in \mathfrak{B}_{\mathbf{X},\Gamma}$.

First, we express $\mathbf{Q}$ in terms of $\mathbf{P}$:

$$\mathbf{Q}\{B\cap\pi^{-1}(A)\} = \int_{\mathbf{X}} g(x)\mathbf{P}(dx),$$

where $g(x) = \mathbf{1}_{\{B\cap\pi^{-1}(A)\}}(x)\frac{d\mathbf{Q}}{d\mathbf{P}}(x)$. Applying (3.3), we obtain

$$\begin{aligned}\int_{\mathbf{X}} g(x)\mathbf{P}(dx) &= \int_{\mathbf{X}/\Gamma}\int_{\mathbf{X}} g(x)\mathbf{P}_\gamma(dx)\mathbf{P}_\Gamma(d\gamma)\\ &= \int_A\int_{\mathbf{X}} \mathbf{1}_B(x)\frac{d\mathbf{Q}}{d\mathbf{P}}(x)\mathbf{P}_\gamma(dx)\mathbf{P}_\Gamma(d\gamma)\\ &= \int_A \mathbf{Q}_\gamma\{B\}\frac{d\mathbf{Q}_\Gamma}{d\mathbf{P}_\Gamma}(\gamma)\mathbf{P}_\Gamma(d\gamma) = \int_A \mathbf{Q}_\gamma\{B\}\mathbf{Q}_\Gamma(d\gamma). \quad \square\end{aligned}$$

CHAPTER 2

Methods for Studying Distributions of Functionals

§4. Stratifications of measures

General outline of the method. A stratification of a probability measure is a construction intended for studying distributions of functionals of a random process with the use of the total probability formula. The main idea is to consider measurable partitions of the probability space such that the elements of these partitions have a simple structure. In what follows, these elements are always finite-dimensional sets.

In the traditional approach, the total probability formula is commonly used in the form

$$\mathbf{P}\{A\} = \int_{\mathbf{X}} \mathbf{P}\{A \mid \xi = r\} \mathbf{P}_\xi\,(dr),$$

where ξ is a random variable with distribution $\mathbf{P}_\xi$ defined on a probability space $\mathbf{X}$. However, the conditional measure $\mathbf{P}\{\,\cdot \mid \xi = r\}$ is often no easier to study than the original measure $\mathbf{P}$. No wonder that this formula proves useless in many cases, and one has to use partitions of quite different type, for which the elements, rather than (as in the traditional approach) the quotient space, are finite-dimensional.

Let us outline the subsequent considerations. The distribution of a random process is treated as a Borel measure $\mathbf{P}$ on a topological space $\mathbf{X}$ whose points are the process trajectories. Let $f\colon (\mathbf{X}, \mathfrak{B}_\mathbf{X}) \to (\mathbf{R}^m, \mathfrak{B}^m)$ be a measurable functional ranging in the Euclidean space $\mathbf{R}^m$. We shall study its distribution, that is, the measure $\mathbf{P}f^{-1}$ on $\mathfrak{B}^m$, in three stages.

First, we choose a partition Γ of $\mathbf{X}$ for which it is easy to study the conditional measures $\mathbf{P}_\gamma$. To this end, the partition elements are chosen to be sets of simple geometric structure, like lines, rays, segments, smooth curves, or multidimensional planes. These elements will be called *strata.* (In most cases, the elements are one-dimensional, so we speak of "fibers" rather than "strata.") The strata are chosen so that f will be a smooth function with nondegenerate derivative along each stratum.

Second, we apply the mapping f to the conditional measures $\mathbf{P}_\gamma$. This is a comprehensively studied finite-dimensional problem for which the results are well known; we present them in the sequel. At the second stage of our investigation, we obtain some information about the conditional distributions $\mathbf{P}_\gamma f^{-1}$, the images of $\mathbf{P}_\gamma$ under f.

At the third stage, we estimate the quotient measure $\mathbf{P}_\Gamma$ and use formula (3.4) to obtain some information about $\mathbf{P}f^{-1}$ from the information obtained at the second stage. By using (3.5) instead of (3.4), we can study the properties of the *density* of the distribution $\mathbf{P}f^{-1}$.

Our immediate goal is to construct a wide class of partitions suitable for our scheme and study the related conditional measures.

Admissible semigroups. Let $(\mathbf{X}, \mathfrak{B}_{\mathbf{X}}, \mathbf{P})$ be a topological space equipped with a Borel measure. A measurable mapping $G\colon \mathbf{X} \to \mathbf{X}$ is said to be *admissible* if $\mathbf{P}G^{-1} \ll \mathbf{P}$.

A family of mappings $\{G_c\}$, $c \in \mathbf{R}^m_+$, will be called an *admissible semigroup* if

1) G_0 is the identity mapping;

2) $G_{c_1} \circ G_{c_2} = G_{c_1+c_2}$;

3) for any $c \in \mathbf{R}^m_+$, the mapping G_c is admissible and one-to-one;

4) for each $x \in \mathbf{X}$, the mapping $c \mapsto G_c(x)$ is one-to-one.

We say that two elements $x_1, x_2 \in \mathbf{X}$ are equivalent if

$$G_{c_1}(x_1) = G_{c_2}(x_2)$$

for some $c_1, c_2 \in \mathbf{R}^m_+$.

PROBLEM 4.1. Prove that this is an equivalence relation on $\mathbf{X}$.

Let Γ be the partition given by this equivalence relation, and let $\pi\colon \mathbf{X} \to \mathbf{X}/\Gamma$ be the canonical projection. The equivalence classes $\pi^{-1}(\gamma)$, $\gamma \in \mathbf{X}/\Gamma$, will be called the *orbits* of the semigroup G. This is natural, since the orbits are invariant under each mapping $G_c(\,\cdot\,)$.

The following two statements allow us to introduce a useful real parametrization of the orbits.

PROPOSITION 4.1. *Let $x_1 \sim x_2$. Then there exists a unique $c(x_1, x_2) \in \mathbf{R}^m$ such that $G_{c_1}(x_1) = G_{c_2}(x_2)$ implies $c_1 = c_2 + c(x_1, x_2)$.*

PROOF. Let $G_{c_1}(x_1) = G_{c_2}(x_2)$ and $G_{c_3}(x_1) = G_{c_4}(x_2)$. Then $c_1 - c_2 = c_3 - c_4$. Indeed, by the definition of an admissible semigroup, we have

$$\begin{aligned} G_{c_2+c_3}(x_2) &= G_{c_3} \circ G_{c_2}(x_2) = G_{c_3} \circ G_{c_1}(x_1) = G_{c_3+c_1}(x_1) \\ &= G_{c_1} \circ G_{c_3}(x_1) = G_{c_1} \circ G_{c_4}(x_2) = G_{c_4+c_1}(x_2). \end{aligned}$$

Moreover, property 4) of an admissible semigroup implies

$$c_2 + c_3 = c_4 + c_1; \qquad c_2 - c_1 = c_4 - c_3.$$

Thus, $c(x_1, x_2) = c_1 - c_2$ depends only on x_1 and x_2. □

PROPOSITION 4.2. *Let $\pi^{-1}(\gamma)$ be an orbit of $\{G_c\}$, and let $x \in \pi^{-1}(\gamma)$. Then $J_x\colon y \mapsto c(x, y)$ is a one-to-one mapping of $\pi^{-1}(\gamma)$ onto a λ-measurable set $C_x \subset \mathbf{R}^m$. This mapping intertwines the action of $\{G_c\}$ on $\pi^{-1}(\gamma)$ with the action of the semigroup of translations on C_x:*

$$J_x G_c(y) = J_x(y) + c. \tag{4.1}$$

PROOF. Let us verify that the mapping is one-to-one. Suppose that

$$\begin{aligned} &c(x, y') = c(x, y''); \\ &G_{c_1}(x) = G_{c'}(y'); \qquad c_1 = c' + c(x, y'); \\ &G_{c_2}(x) = G_{c''}(y''); \qquad c_2 = c'' + c(x, y'). \end{aligned}$$

Then

$$\begin{aligned} G_{c'+c''}(y') &= G_{c''} \circ G_{c'}(y') = G_{c''} \circ G_{c_1'}(x) = G_{c''+c'+c(x,y')}(x) \\ &= G_{c'+c_2}(x) = G_{c'} \circ G_{c_2}(x) = G_{c'+c''}(y''). \end{aligned}$$

Since the mapping $G_{c'+c''}(\cdot)$ is one-to-one, it follows that $y' = y''$, and thus $J_x(\cdot)$ is one-to-one.

Now let us prove (4.1). Suppose that $G_{c_1}(x) = G_{c_2}(y)$ and $J_x(y) = c(x,y) = c_1 - c_2$. Then

$$G_{c+c_1}(x) = G_c \circ G_{c_1}(x) = G_c \circ G_{c_2}(y) = G_{c_2}(G_c(y)),$$

and hence, $J_x(G_c(y)) = c - c_1 - c_2 = J_x(y) + c$. It follows from (4.1) that C_x is measurable, which completes the proof. □

PROBLEM 4.2. Prove that choosing a different initial point x in the same orbit results only in a linear change of J_x and in a translation of the set C_x:

$$J_{x_1}(\cdot) = J_{x_2}(\cdot) + c(x_1, x_2), \tag{4.2}$$

$$C_{x_1} = C_{x_2} + c(x_1, x_2). \tag{4.3}$$

In addition to the algebraic and set-theoretic properties 1)–4), we impose the following topological condition of continuity:

5) J_x is a homeomorphism of $\pi^{-1}(\gamma)$ and C_x if $\pi^{-1}(\gamma)$ is equipped with the topology inherited from $\mathbf{X}$.

Let λ^m be the Lebesgue measure on $\mathbf{R}^m$. Then we can define a measure λ_γ on $\mathfrak{B}_{\mathbf{X}}$ by setting

$$\lambda_\gamma(B) = \lambda^m\big(J_x(B \cap \pi^{-1}(\gamma))\big). \tag{4.4}$$

By condition 5), this is well defined. Obviously, λ_γ is concentrated on $\pi^{-1}(\gamma)$.

Note that λ_γ is independent of the reference point $x \in \pi^{-1}(\gamma)$ used in (4.4) by virtue of (4.2).

PROPOSITION 4.3. *The measure λ_γ is G_c-invariant. If $B \in \mathfrak{B}_{\mathbf{X}}$ and $c \in \mathbf{R}^m_+$, then $G_c^{-1}(B) \in \mathfrak{B}_{\mathbf{X}}$ and*

$$\lambda_\gamma(B) = \lambda_\gamma\big(G_c^{-1}(B)\big). \tag{4.5}$$

PROOF. Since the orbit $\pi^{-1}(\gamma)$ is G_c-invariant, we have

$$G_c^{-1}(B) \cap \pi^{-1}(\gamma) = G_c^{-1}(B \cap \pi^{-1}(\gamma)). \tag{4.6}$$

Relations (4.1) and (4.6) imply

$$J_x\big(G_c^{-1}(B) \cap \pi^{-1}(\gamma)\big) = J_x\big(G_c^{-1}(B \cap \pi^{-1}(\gamma))\big) = J_x(B \cap \pi^{-1}(\gamma)) - c.$$

Therefore, since the Lebesgue measure is invariant, we obtain

$$\begin{aligned}\lambda_\gamma(G_c^{-1}(B)) &= \lambda^m\big(J_x\big(G_c^{-1}(B) \cap \pi^{-1}(\gamma)\big)\big) = \lambda^m\big(J_x(B \cap \pi^{-1}(\gamma)) - c\big)\\ &= \lambda^m\big(J_x(B \cap \pi^{-1}(\gamma))\big) = \lambda_\gamma(B). \quad \square\end{aligned}$$

Calculation of conditional measures. Now let us additionally assume that $\mathbf{X}$ is *metrizable, complete*, and *separable* and that the partition into G_c-orbits is *measurable.* Then the system of conditional measures $\{\mathbf{P}_\gamma\}$ is well defined (see Theorem 3.1). We set $\mathbf{P}^c = \mathbf{P}G_c^{-1}$. Since the semigroup G is admissible, we have $\mathbf{P}^c \ll \mathbf{P}$.

Let $p^c = d\mathbf{P}^c/d\mathbf{P}$ be the corresponding density.

THEOREM 4.1. *Let the partition Γ of a complete separable metric space $\mathbf{X}$ into orbits of an admissible semigroup $\{G_c\}$, $c \in \mathbf{R}_+^m$, be measurable. Then for $\mathbf{P}_\Gamma$-almost all γ, the conditional measure $\mathbf{P}_\gamma$ is absolutely continuous with respect to the invariant measure λ_γ, and the density is given by*

$$\frac{d\mathbf{P}_\gamma}{d\lambda_\gamma}(G_u(x)) = K_\gamma[p^u(G_u(x))]^{-1} \tag{4.7}$$

for $\mathbf{P}_\gamma$-almost all x. The normalizing factor K_γ is constant on γ.

REMARK. In this theorem $\mathbf{X}$ can be replaced by any open set $V \subset \mathbf{X}$, since by changing the metric, we can make this set a *complete* separable space. In this case, we must assume that the mapping G_c takes V to V. Prior to proving the theorem, let us present a criterion for absolute continuity of a measure on $\mathbf{R}^m$ in terms of properties of its translates.

Let μ be a finite measure on $(\mathbf{R}^m, \mathfrak{B}^m)$. By μ^u, $u \in \mathbf{R}^m$, we denote the translate of μ given by $\mu^u(A) = \mu(A - u)$.

LEMMA 4.1. *Let $U \subset \mathbf{R}^m$ be a set of positive Lebesgue measure, $\lambda^m(U) > 0$, and let $\mu^u \ll \mu$ for almost all $u \in U$. Let us choose a representative of the function $\rho(u,v) = d\mu^u/d\mu(v)$ jointly measurable in (u,v).*

Then $\mu \ll \lambda^m$, and for μ-almost all $v \in \mathbf{R}^m$ the density $M(\cdot) = d\mu/d\lambda^m(\cdot)$ satisfies the relation

$$M(v+u) = M(v)\left[\frac{d\mu^u}{d\mu}(v+u)\right]^{-1} \tag{4.8}$$

simultaneously for λ^m-almost all $u \in U$.

PROOF OF LEMMA 4.1. Suppose that ν is a nonzero absolutely continuous measure concentrated on $-U$. Then the convolution $\mu * \nu$ satisfies

$$\mu * \nu(A) = \int_{-U} \mu(A-u)\nu(du) = \int_{-U} \mu^u(A)\nu(du) = \int_U \mu^{-u}(A)\nu(-du). \tag{4.9}$$

We can rewrite the relation $\mu^u \ll \mu$ as $\mu \ll \mu^{-u}$. Therefore, it follows from (4.9) that $\mu \ll \mu * \nu$. On the other hand, $\mu * \nu \ll \lambda^m$, since $\nu \ll \lambda^m$.

Thus, $\mu \ll \mu*\nu \ll \lambda^m$. Let $M(\cdot)$ be the density of μ with respect to λ^m. Then the function $(u,v) \mapsto M(v)/M(v+u)$ is a measurable representation of the density $d\mu(v)/d\mu^{-u}$.

Indeed, after the linear change $w = v+u$ we obtain

$$\begin{aligned}\int_A \frac{M(v)}{M(v+u)}\mu^{-u}(dv) &= \int_{A+u} \frac{M(w-u)}{M(w)}\mu(dw) = \int_{A+u} \frac{M(w-u)}{M(w)}M(w)\lambda^m(dw)\\ &= \int_{A+u} M(w-u)\lambda^m(dw) = \int_A M(v)\lambda^m(dv) = \mu(A).\end{aligned}$$

Another measurable representation of the density $d\mu(v)/d\mu^{-u}$ can be constructed from the density $\rho(u,v)$ mentioned in the conditions of Lemma 4.1. Namely, by a linear change of variable we prove that the function $\rho(u, v+u)$ may serve as such a density:

$$\int_A \rho(u, v+u)\mu^{-u}(dv) = \int_{A+u} \rho(u,w)\mu(dw) = \mu^u(A+u) = \mu(A).$$

It is easy to verify that any two measurable representations of $d\mu(v)/d\mu^{-u}$ coincide on $U \times \mathbf{R}^m$ almost everywhere with respect to the measure $\lambda^m \times \mu$.

By equating the two representatives, we obtain

$$\rho(u, v+u) = \frac{M(v)}{M(v+u)}. \tag{4.10}$$

Hence, for μ-almost all v, (4.10) holds for λ^m-almost all $u \in U$. □

PROOF OF THEOREM 4.1. First of all note that the quotient measures $(\mathbf{P}^c)_\Gamma$ and the families of conditional measures $\{(\mathbf{P}^c)_\gamma\}$ corresponding to the measure $\mathbf{P}^c$ and the partition Γ satisfy the relations

$$(\mathbf{P}^c)_\Gamma = \mathbf{P}_\Gamma; \qquad (\mathbf{P}^c)_\gamma = \mathbf{P}_\gamma G_c^{-1}. \tag{4.11}$$

The first relation readily follows from the fact that the orbits are G_c-invariant: if $A \in \mathfrak{B}_{\mathbf{X},\Gamma}$, then

$$(\mathbf{P}^c)_\Gamma(A) = \mathbf{P}^c(\pi^{-1}(A)) = \mathbf{P}(G_c^{-1}\pi^{-1}(A)) = \mathbf{P}(\pi^{-1}(A)) = \mathbf{P}_\Gamma(A).$$

The second relation in (4.11) can be obtained by straightforward verification of (3.1): for any $B \in \mathfrak{B}$ and $A \in \mathfrak{B}_{\mathbf{X},\Gamma}$, we have

$$\begin{aligned}\int_A \mathbf{P}_\gamma G_c^{-1}(B)(\mathbf{P}^c)_\Gamma\,(d\gamma) &= \int_A \mathbf{P}_\gamma(G_c^{-1}(B))\mathbf{P}_\Gamma\,(d\gamma) = \mathbf{P}(G_c^{-1}(B) \cap \pi^{-1}(A)) \\ &= \mathbf{P}(G_c^{-1}(B \cap \pi^{-1}(A))) = \mathbf{P}^c(B \cap \pi^{-1}(A)).\end{aligned}$$

It follows that the measures $\mathbf{P}_\gamma G_c^{-1}$ satisfy the definition of conditional measures for the measure $\mathbf{P}^c$ with respect to the partition Γ.

Furthermore, by applying Theorem 3.3 to the measures $\mathbf{P}^c$ and $\mathbf{P}$ and by taking into account the fact that the quotient measures coincide, from (3.7) we obtain the following expression for the density of the conditional measure:

$$\frac{d\mathbf{P}_\gamma G_c^{-1}}{d\mathbf{P}_\gamma}(x) = \frac{d(\mathbf{P}^c)_\gamma}{d\mathbf{P}_\gamma}(x) = \frac{d\mathbf{P}^c}{d\mathbf{P}}(x) = p^c(x). \tag{4.12}$$

Let us use the parametrization from Proposition 4.2. By this representation, the space $(\mathbf{X}, \mathfrak{B}, \lambda_\gamma)$ is equivalent to some subset of $(\mathbf{R}^m, \mathfrak{B}^m, \lambda^m)$, and moreover, the isomorphism J_x takes the semigroup G_c to a semigroup of translations. One can apply Lemma 4.1 to the measure $\mu = \mathbf{P}_\gamma J_x^{-1}$, since the condition $\mu^c \ll \mu$ is equivalent to $\mathbf{P}_\gamma G_c^{-1} \ll \mathbf{P}_\gamma$ and, by (4.11), the latter relation is equivalent to $(\mathbf{P}^c)_\gamma \ll \mathbf{P}_\gamma$, which was established in Theorem 3.3.

By Lemma 4.1, we have $\mathbf{P}_\gamma J_x^{-1} \ll \lambda^m$, which is equivalent to $\mathbf{P}_\gamma \ll \lambda^m J_x = \lambda_\gamma$.

Thus, we have proved the absolute continuity of the conditional measure. Let us calculate the corresponding density. Let us apply formula (4.8) to the measure $\mu = \mathbf{P}_\gamma J_x^{-1}$. We choose some v that satisfies (4.8). Without loss of generality, we can assume that $v = 0$; otherwise, we could redefine the reference point x by taking $x' = G_v(x)$ as a new reference point on the orbit.

Taking into account the relation between the semigroup of translations and the semigroup G_c, we rewrite (4.8) in the form

$$\frac{d\mathbf{P}_\gamma J_x^{-1}}{d\lambda^m}(u) = \frac{d\mathbf{P}_\gamma J_x^{-1}}{d\lambda^m}(0)\left[\frac{d\mathbf{P}_\gamma G_u^{-1} J_x^{-1}}{d\mathbf{P}_\gamma J_x^{-1}}(u)\right]^{-1}.$$

By setting $K_\gamma = d\mathbf{P}_\gamma J_x^{-1}(0)/d\lambda^m$, we obtain

$$\frac{d\mathbf{P}_\gamma J_x^{-1}}{d\lambda^m}(u) = K_\gamma \left[\frac{d\mathbf{P}_\gamma G_u^{-1} J_x^{-1}}{d\mathbf{P}_\gamma J_x^{-1}}(u)\right]^{-1}.$$

Now, returning from $\mathbf{R}^m$ to the space $(\mathbf{X}, \mathfrak{B}_\mathbf{X}, \lambda_\gamma)$, we can substitute $u = J_x G_u(x)$ into the result and rewrite it as follows:

$$\frac{d\mathbf{P}_\gamma}{d\lambda_\gamma}(G_u(x)) = K_\gamma \left[\frac{d\mathbf{P}_\gamma G_u^{-1}}{d\mathbf{P}_\gamma}(G_u(x))\right]^{-1}.$$

Finally, by substituting (4.12) into the right-hand side, we obtain the desired formula (4.7). □

REMARK. The statement and the above proof of Theorem 4.1 remain valid if we abandon the topological properties (metrizability, completeness, and separability of $\mathbf{X}$) and only assume that there exists a unique system of conditional measures for the partition of $\mathbf{X}$ into orbits.

Calculation of conditional distributions. Following the general scheme of the stratification method, we first calculate conditional measures and then study their images, that is, conditional distributions. Since the conditional measures obtained in Theorem 4.1 are concentrated on finite-dimensional sets, we need special tools for studying transformations of measures in $\mathbf{R}^m$. In what follows we present related analytical results. Let λ^m be the Lebesgue measure in $\mathbf{R}^m$, $\lambda = \lambda^1$.

THEOREM 4.2. *Let Δ be a bounded open set in $\mathbf{R}^1$, $f\colon \Delta \to \mathbf{R}^1$ a λ-measurable function, and B a λ-measurable subset of Δ. Suppose that f is differentiable λ-almost everywhere on B.*

Under these conditions, $\lambda_B f^{-1} \ll \lambda$ if and only if

$$\lambda\{c \in B \mid f'(c) = 0\} = 0. \tag{4.13}$$

If (4.13) *holds, then the density $p = d\lambda_B f^{-1}/d\lambda$ has the form*

$$p(u) = \sum_{c \in f^{-1}\{u\} \cap B} |f'(c)|^{-1}. \tag{4.14}$$

(If $f'(c)$ does not exist, we set $|f'(c)| = \infty$.)

We shall prove the m-dimensional generalization of this theorem.

THEOREM 4.3. *Let Δ be an open set in $\mathbf{R}^m$, $f\colon \Delta \to \mathbf{R}^m$ a λ^m-measurable function, and B a λ^m-measurable subset of Δ. Suppose that f is Fréchet differentiable λ-almost everywhere on B.*

Under these conditions, $\lambda_B^m f^{-1} \ll \lambda^m$ if and only if

$$\lambda^m\{c \in B \mid \det Df(c) = 0\} = 0, \tag{4.15}$$

where Df is the matrix of first-order partial derivatives of f. If (4.15) *holds, then the density $p = d\lambda_B^m f^{-1}/d\lambda$ has the form*

$$p(u) = \sum_{c \in f^{-1}\{u\} \cap B} |\det Df(c)|^{-1}. \tag{4.16}$$

PROOF. Let $\overline{B} \subset B$ be the subset of points of density of B where the derivative Df is defined. Then $\lambda^m\{B \setminus \overline{B}\} = 0$. Furthermore [**111,** Ch. III, §3.1], there exist λ^m-measurable disjoint sets B_n such that $\overline{B} = \bigcup_n B_n$ and the restriction of f to each B_n satisfies the Lipschitz condition. We extend f from B_n to a Lipschitz function defined on $\mathbf{R}^m$. Then [**111,** Ch. III, §3.2] for any λ^m-measurable nonnegative function g we have

$$\int_{\mathbf{R}^m} g(c)|\det Df(c)|\lambda^m(dc) = \int_{\mathbf{R}^m} \Big(\sum_{c\in f^{-1}\{u\}} g(c)\Big)\lambda^m(du).$$

By taking $g = \mathbf{1}_{\theta_n}$, where $\theta_n = \{c \in B_n\colon \det Df(c) = 0\}$, we obtain $\lambda^m(f(\theta_n)) = 0$. Hence, condition (4.15) is necessary for the measure $\lambda^m_B f^{-1}$ to be absolute continuous. Next, we can use the above relation for the function

$$g(c) = \mathbf{1}_{\{f^{-1}(A)\cap B_n\}}|\det Df(c)|^{-1},$$

where A is an arbitrary λ^m-measurable set. We obtain

$$\lambda^m\{f^{-1}(A)\cap B_n\} = \int_A \Big(\sum_{c\in f^{-1}\{u\}} |\det Df(c)|^{-1}\Big)\lambda^m(du).$$

By summing over n, we arrive at (4.16):

$$\lambda^m_B f^{-1}(A) = \lambda^m\{f^{-1}(A)\cap \overline{B}\} = \sum_n \lambda^m\{f^{-1}(A)\cap B_n\} = \int_A p(u)\lambda^m(du). \quad \square$$

The above theorems can be generalized to the case in which f is applied to some measure other than λ^m.

PROPOSITION 4.4. *Let Δ, B, and f satisfy the assumptions of Theorem 4.3, and let μ be a finite measure such that $\mu \ll \lambda^m$. If condition (4.15) holds, then $\mu_B f^{-1} \ll \lambda^m$ and the density $p = d\mu_B f^{-1}/d\lambda^m$ has the form*

$$p(u) = \sum_{c\in f^{-1}\{u\}\cap B} |\det Df(c)|^{-1}\frac{d\mu}{d\lambda^m}(c). \tag{4.17}$$

In particular, for $m = 1$, formula (4.13) implies $\mu_B f^{-1} \ll \lambda$ and

$$p(u) = \sum_{c\in f^{-1}\{u\}\cap B} |f'(c)|^{-1}\frac{d\mu}{d\lambda}(c). \tag{4.18}$$

The following theorems give sufficient conditions under which the image of a measure continuously depends on the mapping.

THEOREM 4.4. *Let f_n, $n \in \overline{\mathbf{N}}$, be convex functions defined on a bounded interval $\Delta \subset \mathbf{R}^1$, and let μ be a Borel measure such that $\mu \ll \lambda_\Delta$. Suppose that*

$$f'_\infty(c) \neq 0 \qquad \text{and} \qquad f_\infty(c) = \lim_n f_n(c)$$

for λ-almost all $c \in \Delta$. Then

$$\mu f_n^{-1} \overset{\text{var}}{\to} \mu f^{-1}. \tag{4.19}$$

The space $\mathbf{C}^{(1)}(\Delta)$ of continuously differentiable functions on a bounded interval $\Delta \subset \mathbf{R}^1$ will be equipped with the norm $\|f\|_1 = \sup_\Delta(|f| + |f'|)$.

THEOREM 4.5. *Let $\mathcal{P}$ be a family of measures on the σ-algebra $\mathfrak{B}_\Delta$ such that the densities of all $\mu \in \mathcal{P}$ are equicontinuous. Suppose that $f \in \mathbf{C}^1(\Delta)$ and $f' \neq 0$ almost everywhere. Then*

$$\lim_{\delta\to 0} \sup_{\mu\in\mathcal{P}} \sup_{g} \left\{ \|\mu f^{-1} - \mu g^{-1}\| \mid \|f-g\|_1 \leq \delta \right\} = 0.$$

PROBLEM 4.3. Construct an example showing that closeness of f and g in the C^1 metric cannot be replaced by closeness in C^0.

Absolute continuity of the distribution of a functional. By way of illustration, we show how the stratification method allows one to prove the absolute continuity of the distribution of a functional.

PROPOSITION 4.5. *Let $\{G_c\}$, $c \in \mathbf{R}^1$, be an admissible semigroup defined on the space $(\mathbf{X}, \mathfrak{B}_\mathbf{X}, \mathbf{P})$, and let $f\colon (\mathbf{X}, \mathfrak{B}_\mathbf{X}) \to \mathbf{R}^1$ be a measurable functional. Suppose that for $\mathbf{P}$-almost all $x \in \mathbf{X}$ the derivative of f along the orbit is defined and*

$$D_G f(x) = \lim_{c\to 0} c^{-1}[f(G_c(x)) - f(x)] \neq 0. \tag{4.20}$$

Then $\mathbf{P}f^{-1} \ll \lambda^1$.

PROOF. Let Γ be the partition of $\mathbf{X}$ into orbits of G_c. Let us consider an orbit $\gamma \in \mathbf{X}/\Gamma$. By Theorem 4.1, the corresponding conditional measure $\mathbf{P}_\gamma$ is absolutely continuous with respect to the invariant measure λ_γ. By (4.20), we can apply Proposition 4.4 to the measure $\mu = \mathbf{P}_\gamma$ and to the restriction of f to γ. Thus, $\mathbf{P}_\gamma f^{-1} \ll \lambda$ and, by Proposition 3.3, $\mathbf{P}f^{-1} \ll \lambda$, as desired. □

For a given functional f, it is not always easy to satisfy (4.20) by choosing only one semigroup G_c. More often, the desired semigroup can be constructed locally, that is, in a neighborhood of x for $\mathbf{P}$-almost all $x \in \mathbf{X}$. Furthermore, it is convenient to replace condition (4.20), which simultaneously deals with f and G, by somewhat stronger conditions imposed on f and G separately. Let us present the corresponding definitions and results.

Let $\mathbf{X}$ be a linear space. Let $B \subset \mathbf{X}$ be a Borel subset with respect to the original topology $\tau_\mathbf{X}$ and let $H \subset X$ be a subset equipped with a topology τ_H. We say that a functional $f\colon \mathbf{X} \to \mathbf{R}^1$ is (B, H)-*smooth* if for each $x \in B$ there exists a direction $l \in H$ such that

$$D_l f(x) = \lim_{c\to 0} c^{-1}[f(x+cl) - f(x)] \neq 0 \tag{4.21}$$

and the function $(y, h) \mapsto D_h f(y)$ is defined and $(\tau_\mathbf{X} \times \tau_H)$-continuous in some $(\tau_\mathbf{X} \times \tau_H)$-neighborhood of (x, l). By $\mathcal{M}(B, H)$ we denote the class of (B, H)-smooth functionals.

Let a Borel measure $\mathbf{P}$ and a seminorm $\|\cdot\|$ be given on $\mathbf{X}$. We say that the measure $\mathbf{P}$ has a (B, H)-*sufficient class of admissible mappings* if for any $x \in B$ and $l \in H$ there exist an ordered system of $\tau_\mathbf{X}$-neighborhoods V^α of x, a system of $\mathbf{P}$-admissible semigroups G^α defined on V^α, and a system of vector fields $l^\alpha\colon V^\alpha \to H$ such that:

1) for $\mathbf{P}$-almost all $y \in V^\alpha$, one has

$$\lim_{c\to 0} c^{-1}\|G_c^\alpha(y) - y - cl^\alpha(y)\| = 0; \tag{4.22}$$

2) the mapping $y \mapsto l^\alpha(y)$ is continuous on $V^\alpha \cap B$ with respect to $\tau_{\mathbf{X}}$ and τ_H;
3) τ_H-$\lim_\alpha l^\alpha(x) = l$.

THEOREM 4.6. *Let $(\mathbf{X}, \tau_X)$ be a topological space, $\mathbf{P}$ a Borel measure on $\mathbf{X}$, and $\|\cdot\|$ a seminorm on $\mathbf{X}$. Let B and H be subsets of $\mathbf{X}$ equipped with topologies $\tau_{\mathbf{X}}$ and τ_H, respectively, and let $f\colon (\mathbf{X}, \mathfrak{B}_{\mathbf{X}}) \to \mathbf{R}^1$ be a measurable functional. Suppose that*:

1) *the measure $\mathbf{P}$ has a (B,H)-sufficient class of admissible mappings*;
2) *the functional f satisfies a local Lipschitz condition on B with respect to $\|\cdot\|$*;
3) $f \in \mathcal{M}(B,H)$;
4) *either the topology $\tau_{\mathbf{X}}$ has a countable base, or $\mathbf{P}$ is a Radon measure.*

Then $\mathbf{P}_B f^{-1} \ll \lambda$.

PROOF. Under the assumptions of the theorem, one can readily construct a cover of B by $\tau_{\mathbf{X}}$-open sets W^α such that Proposition 4.5 can be applied to the measure $\mathbf{P}^\alpha = \mathbf{P}_{W^\alpha \cap B}$. By taking a countable subcover $\{W^n\} \subset \{W^\alpha\}$, we obtain

$$\mathbf{P}_B f^{-1} \ll \sum_{n=1}^{\infty} 2^{-n} \mathbf{P}^n f^{-1} \ll \lambda. \quad \square$$

REMARK. If we use linear semigroups $G_c^\alpha(x) = x + cl^\alpha$, then condition 2) can be omitted.

The first assumption in the statement of the theorem deals only with the measure $\mathbf{P}$, and the other two only with the functional f. This simplifies their verification in specific cases.

Bounds for the density of a sum of independent variables. We shall illustrate the proposed methods by a well-known problem from the theory of summation of random variables.

Let ξ_i, $i \geq 1$, be independent identically distributed random variables. Set $S_n = \sum_{i=1}^n \xi_i$. If the distribution of S_n has a density, we denote this density by $p_n(\cdot)$. Gnedenko proved the following uniform local limit theorem [**26; 51,** Ch. IV].

THEOREM 4.7. *Suppose that $\mathsf{E}\xi_i = 0$ and $\mathsf{D}\xi_i = \sigma^2 > 0$. Then*

$$\sup_r \left| p_n(r\sigma\sqrt{n})\sigma\sqrt{n} - \frac{1}{\sqrt{2\pi}} e^{-r^2/2} \right| \underset{n}{\to} 0 \tag{4.23}$$

if and only if p_N is bounded for some N.

In contrast with a majority of limit theorems, Gnedenko's condition is imposed on the distribution of the sum of several terms rather than on the distribution of a single term. Therefore, it is natural to ask whether it is possible to find a condition on the distribution of an individual summand guaranteeing (4.23). A partial answer is given by the following theorem.

THEOREM 4.8. *If*

$$\mathsf{E}p_1^\varepsilon(\xi) = \int p_1^{1+\varepsilon}(r)\, dr < \infty \tag{4.24}$$

for some $\varepsilon > 0$, then p_n is bounded for $n > (1+\varepsilon)\varepsilon^{-1}$.

PROOF. We use the stratification technique. Let $\mathbf{P}$ be the measure generated by the random vector $(\xi_1, \ldots, \xi_n)$ on the space $\mathbf{X} = R^n$. By x we denote elements of $\mathbf{X}$, and by $x_1, \ldots, x_n$ the components of x. Then

$$\frac{d\mathbf{P}}{d\lambda^n}(x) = \prod_{i=1}^{n} p_1(x_i).$$

We construct a partition of $\mathbf{X}$ as follows. Let $a_0 = 0$ and $a_k = 2^{k-1}$, $k \in \mathbf{N}$. We set $X_k = \{x \in \mathbf{X} \colon \min_i p_1(x_i) \in (a_k, a_{k+1}]\}$.

For $x \in \mathbf{X}_k$ we define

$$I(x) = \max\{i \colon p_1(x_i) \in (a_k, a_{k+1}]\}.$$

In $\mathbf{X}_k$ we introduce the partition generated by the following equivalence relation:

$$x \sim y := \begin{cases} x, y \in \mathbf{X}_k, \\ I(x) = I(y), \\ x_j = y_j \quad \text{for } j \neq I(x). \end{cases}$$

Since $\mathbf{X} = \bigcup_{k=0}^{\infty} \mathbf{X}_k$, we obtain a partition of the entire space $\mathbf{X}$ into one-dimensional sets, each being parallel to one of the coordinate axes. We denote this partition by U, its elements by u, and the quotient and conditional measures by $\mathbf{P}_U$ and $\{\mathbf{P}_u\}$, respectively.

One can readily calculate the density of the conditional measure. Suppose that $x \in u \in \mathbf{X}_k/U$ and $I(x) = i$. Then

$$p_u(x) = \frac{d\mathbf{P}_u}{d\lambda}(x) = c_k p_1(x_i), \tag{4.25}$$

where the normalizing factor is

$$c_k = \left[\int_{\{y | a_k \leq p_1(y) \leq a_{k+1}\}} p_1(y)\lambda(dy)\right]^{-1}. \tag{4.26}$$

We must study the density p_n of the image of $\mathbf{P}$ under the mapping $S_n \colon \mathbf{X} \ni x \mapsto \sum_{i=1}^{n} x_i \in \mathbf{R}^1$. The total probability formula (3.6) implies

$$p_n(r) = \int \mathbf{P}_U(du) \frac{d\mathbf{P}_u S_n^{-1}}{d\lambda}(r).$$

Since each class u lies on a straight line parallel to some coordinate axis and the mapping S_m along each of these lines acts as a translation, it follows that the measure $\mathbf{P}_u S_m^{-1}$ is a translate of $\mathbf{P}_u$. Therefore, by (4.15), we obtain

$$\begin{aligned} \sup_{r \in \mathbf{R}^1} p_n(r) &\leq \int \sup_r \frac{d\mathbf{P}_u S_m^{-1}}{d\lambda}(r) \mathbf{P}_U(du) \\ &= \int \sup_{x \in u} p_u(x) \mathbf{P}_U(du) \leq \sum_{k=0}^{\infty} \mathbf{P}\{\mathbf{X}_k\} c_k a_{k+1}. \end{aligned} \tag{4.27}$$

We also note that

$$\begin{aligned}\mathbf{P}\{\mathbf{X}_k\} &= \mathbf{P}\{x \mid a_k < \min_i p_1(x_i) \le a_{k+1}\} \\ &= \mathbf{P}\{x \mid a_k < \min_i p_1(x_i)\} - \mathbf{P}\{x \mid a_{k+1} < \min_i p_1(x_i)\} \\ &= \Big(\int_{\{y \mid p_1(y) > a_k\}} p_1(y)\lambda(dy)\Big)^n - \Big(\int_{\{y \mid p_1(y) > a_{k+1}\}} p_1(y)\lambda(dy)\Big)^n \\ &= \mathbf{P}^n\{p_1(\xi) > a_k\} - \mathbf{P}^n\{p_1(\xi) > a_{k+1}\}.\end{aligned}$$

Taking into account the elementary estimate

$$a^n - b^n \le n|a-b|a^{n-1},$$

from (4.17) we obtain

$$\sup_{r\in\mathbf{R}^1} p_n(r) \le \sum_{k=0}^{\infty} n c_k^{-1}\mathbf{P}^{n-1}\{p_1(\xi) > a_k\} c_k a_{k+1}.$$

By the Chebyshev inequality, we have

$$\int_{\{y \mid p_1(y) > a_k\}} p_1(y)\lambda(dy) \le a_k^{-\varepsilon}\int p_1^{1+\varepsilon}(y)\,dy.$$

Since $a_k = 2^k$, we finally obtain

$$\begin{aligned}\sup_{r\in\mathbf{R}^1} p_n(r) &\le n\int_{\mathbf{R}^1} p_1^{1+\varepsilon}(y)\,dy\Big(1 + \sum_{k=1}^{\infty} a_{k+1}a_k^{-(n-1)\varepsilon}\Big) \\ &= 2n\int_{\mathbf{R}^1} p_1^{1+\varepsilon}(y)\,dy\Big(1 + 2\sum_{k=0}^{\infty} 2^{k(1-(n-1)\varepsilon)}\Big) \\ &= n\big(1 + 2(1 - 2^{(1-(n-1)\varepsilon)})^{-1}\big)\int_{\mathbf{R}^1} p_1^{1+\varepsilon}(y)\,dy < \infty. \quad \square\end{aligned}$$

Note that we divided the space $\mathbf{X}$ into countably many parts and constructed different partitions for different parts. This scheme is typical for virtually all applications of the stratification method to density estimates.

§5. Superstructure

If one deals with proving absolute continuity or verifying strong convergence, the stratification method often requires much more information about conditional distributions than is actually needed to solve the problem. The *superstructure* method, presented in this section, is easier and more effective in these problems. The use of this method is related to an extension of the space of sample functions.

The idea of the method. Let $(\mathbf{X}, \mathfrak{A}, \mathbf{P})$ be a probability space, and let f be a functional. We study the absolute continuity of the distribution $\mathbf{P}f^{-1}$. The idea of the superstructure method is to choose an auxiliary family of measures $\mathbf{Q}^\varepsilon$ and functionals F_ε with the following two properties.[1] First, the measures $\mathbf{Q}^\varepsilon F_\varepsilon^{-1}$ converge in variation to $\mathbf{P}f^{-1}$ as $\varepsilon \to 0$. Second, the measures $\mathbf{Q}^\varepsilon F_\varepsilon^{-1}$ are

[1]In general, for different ε, these measures and functionals may be defined on different spaces.

absolutely continuous, and the absolute continuity can be verified by the simplest version of the stratification method.

First, we consider a somewhat simplified situation. Assume that we are given a family $\{G_c, c \in [0,a]\}$ of transformations of the space $\mathbf{X}$ such that the action of these transformations on $\mathbf{P}$ is small for small parameter values:

$$\mathbf{P}G_c^{-1} \overset{\mathrm{var}}{\to} \mathbf{P} \quad \text{as} \quad c \to 0. \tag{5.1}$$

We use these transformations to construct the families $\{Q^\varepsilon\}$ and $\{F_\varepsilon\}$. To this end, on the direct product $(Y_\varepsilon, \mathfrak{Y}_\varepsilon) = ([0,\varepsilon], \mathfrak{B}_{[0,\varepsilon]}) \times (\mathbf{X}, \mathfrak{A})$, $\varepsilon \in [0,a]$, we introduce the measure $\mathbf{Q}^\varepsilon = \frac{1}{\varepsilon}\lambda \times \mathbf{P}$ and the functional $F_\varepsilon \colon \mathbf{Y}_\varepsilon \to \mathbf{R}$, $F_\varepsilon(c,x) = f(G_c x)$.

Note that since $\{x \mid (c,x) \in F_\varepsilon^{-1}(A)\} = G_c^{-1}(f^{-1}(A))$, we have

$$\mathbf{Q}^\varepsilon F_\varepsilon^{-1}(A) = \Big(\frac{1}{\varepsilon}\lambda \times \mathbf{P}\Big)\{F_\varepsilon^{-1}(A)\} = \frac{1}{\varepsilon}\int_0^\varepsilon \mathbf{P}\{G_c^{-1}(f^{-1}(A))\}\,dc$$

for $A \in \mathfrak{B}^1$. Hence, by (5.1),

$$\begin{aligned}\|\mathbf{P}f^{-1} - \mathbf{Q}^\varepsilon F_\varepsilon^{-1}\| &= \left\|\frac{1}{\varepsilon}\int_0^\varepsilon \mathbf{P}f^{-1}\,dc - \frac{1}{\varepsilon}\int_0^\varepsilon \mathbf{P}G_c^{-1}f^{-1}\,dc\right\| \\ &\le \frac{1}{\varepsilon}\int_0^\varepsilon \|\mathbf{P} - \mathbf{P}G_c^{-1}\|\,dc \to 0 \quad \text{as} \quad \varepsilon \to 0.\end{aligned}$$

Thus, for small ε the measures $\mathbf{Q}^\varepsilon F_\varepsilon^{-1}$ are close in variation to the measure $\mathbf{P}f^{-1}$. At the same time, by considering the partition of the space into fibers "parallel" to the segment $[0,\varepsilon]$ and using the Fubini theorem, we see that

$$\mathbf{Q}^\varepsilon F_\varepsilon^{-1} = \frac{1}{\varepsilon}\int_{\mathbf{X}} \lambda\varphi_x^{-1}\mathbf{P}\,(dx),$$

where $\varphi_x(c) = f(G_c x)$, $c \in [0,\varepsilon]$. This relation means that the measure $\mathbf{Q}^\varepsilon F_\varepsilon^{-1}$ is a mixture of measures $\lambda\varphi_x^{-1}$, that is, of images of the Lebesgue measure under the restrictions of f to the "orbits" $c \to G_c x$, $c \in [0,\varepsilon]$. We can study the absolute continuity of the measures $\lambda\varphi_x^{-1}$ by the methods developed in §4. Thus, if, say, for $\mathbf{P}$-almost all x the function $c \mapsto \varphi_x(c)$ has a nonzero derivative almost everywhere on $[0,\varepsilon]$, then by Theorem 4.2 the measures $\lambda\varphi_x^{-1}$ are absolutely continuous for $\mathbf{P}$-almost all x. It follows that the measures $\mathbf{Q}^\varepsilon F_\varepsilon^{-1}$ are absolutely continuous, and eventually, $\mathbf{P}f^{-1}$ is absolutely continuous.

Theorems on absolute continuity. By slightly changing the above argument, we obtain the following result.

THEOREM 5.1. *Suppose that a family of mappings* $\{G_c,\ c \in [0,a]\}$ *of the space* $\mathbf{X}$ *satisfies property* (5.1), *and moreover, for* $\mathbf{P}$*-almost all* x *there exists a number* $\varepsilon_x > 0$ *such that the derivative* $\varphi_x'(c)$ *exists and is nonzero for almost all* $c \in [0,\varepsilon_x]$; *here, as above,* $\varphi_x(c) = f(G_c x)$.

Then $\mathbf{P}f^{-1} \ll \lambda$.

PROOF. We set $X_n = \{x \mid \varepsilon_x \ge 1/n\}$. The representation $X_n = \{x \mid g(x) = 1/n\}$, where

$$g(x) = \int_0^{1/n} \mathbf{1}_{\mathbf{R}\setminus\{0\}}(\varphi_x'(c))\,dc,$$

implies that the sets X_n are measurable. Furthermore, the sets X_n form an increasing sequence and $\mathbf{P}\{X_n^c\} \to 0$ as $n \to \infty$. By $\mathbf{P}_n$ we denote the restriction of $\mathbf{P}$ to the σ-algebra $\mathfrak{A} \cap X_n$. Consider the measure $\mathbf{Q}_n = n\lambda \times \mathbf{P}_n$ on the direct product of $[0, 1/n]$ by X_n. As before, let $F(c, x) = f(G_c x)$. Then, in a similar way, we obtain

$$\|\mathbf{Q}_n f^{-1} - \mathbf{P} f^{-1}\| = \left\| n \int_0^{1/n} \mathbf{P}_n G_c^{-1} f^{-1}\, dc - n \int_0^{1/n} \mathbf{P} f^{-1}\, dc \right\|$$

$$\leq n \int_0^{1/n} \|\mathbf{P}_n G_c^{-1} - \mathbf{P}\|\, dc \leq n \int_0^{1/n} \|\mathbf{P} G_c^{-1} - \mathbf{P}\|\, dc + \mathbf{P}\{X_n^c\}.$$

Thus, $\mathbf{Q}_n F^{-1} \overset{\text{var}}{\to} \mathbf{P} f^{-1}$ as $n \to \infty$. Note that

$$\mathbf{Q}_n F^{-1} = n \int_{X_n} \lambda \varphi_x^{-1} \mathbf{P}\,(dx).$$

By the construction of X_n and by the assumptions of this theorem and Theorem 4.2, the measures $\lambda \varphi_x^{-1}$ are absolutely continuous for $x \in X_n$. Hence, $\mathbf{Q}_n F^{-1} \ll \lambda$. It follows that $\mathbf{P} f^{-1} \ll \lambda$. □

Theoretically, the structure of f may vary strongly on different parts of the space $\mathbf{X}$. Therefore, it may be very difficult to find a suitable unified family of mappings (G_c). To make the above argument effective, we must "localize" it. The price we have to pay for this is imposing some additional conditions on the space $\mathbf{X}$ and the mappings G_c.

THEOREM 5.2. *Suppose that* $(\mathbf{X}, d)$ *is a separable metric space,* $\mathfrak{A} = \mathfrak{B}_{\mathbf{X}}$*, and* $\mathbf{P}$ *is a probability measure on* $\mathfrak{A}$*. Suppose that for* $\mathbf{P}$*-almost all* $x \in \mathbf{X}$*, there exists an open neighborhood* V *and a family* $\{G_c, c \in [0, a_x]\}$ *of mappings of the space* $\mathbf{X}$ *such that*

1) $\mathbf{P} G_c^{-1} \overset{\text{var}}{\to} \mathbf{P}$ *as* $c \to 0$;

2) $G_c x \to x$ *with respect to* $\mathbf{P}$ *as* $c \to 0$;

3) *for* $\mathbf{P}$*-almost all* $y \in V$*, there exists an* $\varepsilon_y > 0$ *such that* $\varphi'_y(c)$ *exists and is nonzero for almost all* $c \in [0, \varepsilon_y]$*; here, as before,* $\varphi_y(c) = f(G_c y)$*.*

Then $\mathbf{P} f^{-1} \ll \lambda$*.*

REMARK. If there is a family of mappings (G_c) such that condition 1) is satisfied and condition 3) holds for $\mathbf{P}$-almost all $y \in \mathbf{X}$, then Theorem 5.2 remains valid even without the separability assumption (see also Problem 5.2).

PROOF. By $\mathbf{X}_0$ we denote the set of x described in the statement of the theorem. By assumption, $\mathbf{P}\{\mathbf{X}_0\} = 1$. The corresponding open neighborhoods form a cover of $\mathbf{X}_0$. The separability of the space allows us to choose a countable subcover (V_k). Since

$$\mathbf{P} f^{-1} \leq \sum_k \mathbf{P}_{V_k} f^{-1},$$

it suffices to prove the absolute continuity of each measure $\mathbf{P}_{V_k} f^{-1}$. To this end, we need the following statement.

LEMMA 5.1. *Let* $\mathbf{P}$ *be a probability measure on a Borel* σ*-algebra of a metric space* $\mathbf{X}$*. Suppose that a family of measurable mappings* $\{G_c \colon \mathbf{X} \to \mathbf{X}\}$ *possesses properties* 1) *and* 2) *from Theorem* 5.2*. Then for any measurable set* A *one has*

$$\mathbf{P}_A G_c^{-1} \overset{\text{var}}{\to} \mathbf{P}_A \qquad \text{as} \quad c \to 0. \tag{5.2}$$

PROOF. First, we assume that $\mathbf{P}\{\partial A\} = 0$. The inequality

$$\begin{aligned}\sup_{B\in\mathfrak{A}} |\mathbf{P}\{A\cap G_c^{-1}(B)\} - \mathbf{P}\{A\cap B\}| &\le \sup_{B\in\mathfrak{A}} |\mathbf{P}\{A\cap B\} - \mathbf{P}\{G_c^{-1}(A\cap B)\}| \\ &+ \sup_{B\in\mathfrak{A}} |\mathbf{P}\{A\cap G_c^{-1}(B)\} - \mathbf{P}\{G_c^{-1}(A)\cap G_c^{-1}(B)\}|\end{aligned}$$

implies

$$\|\mathbf{P}_A G_c^{-1} - \mathbf{P}_A\| \le \|\mathbf{P}G_c^{-1} - \mathbf{P}\| + \mathbf{P}\{A\triangle G_c^{-1}(A)\}. \tag{5.3}$$

For each $\varepsilon \to 0$, we have

$$\begin{aligned}\mathbf{P}\{A\triangle G_c^{-1}(A)\} &= \int_{\mathbf{X}} |\mathbf{1}_A(x) - \mathbf{1}_A(G_c x)|\,\mathbf{P}(dx) \\ &\le \mathbf{P}\{x \mid d(x, G_c x) \ge \varepsilon\} + \mathbf{P}\{A^\varepsilon \cap (A^c)^\varepsilon\},\end{aligned}$$

since the condition $d(x, G_c x) < \varepsilon$ implies $\mathbf{1}_A(x) = \mathbf{1}_A(G_c x)$ for $x \notin A^\varepsilon \cap (A^c)^\varepsilon$.

Thus, we obtain

$$\overline{\lim_{\varepsilon\to 0}}\, \mathbf{P}\{A\triangle G_c^{-1}(A)\} \le \mathbf{P}\{A^\varepsilon \cap (A^c)^\varepsilon\},$$

and $\mathbf{P}\{A^\varepsilon \cap (A^c)^\varepsilon\} \to \mathbf{P}\{A^- \cap (A^c)^-\} = \mathbf{P}\{\partial A\} = 0$ as $\varepsilon \to 0$. Therefore, $\lim_{\varepsilon\to 0} \mathbf{P}\{A\triangle G_c^{-1}(A)\} = 0$. Inequality (5.3) and condition 1) of the theorem readily prove (5.2).

The general case follows from the fact that for an arbitrary Borel set A and for all $\varepsilon > 0$, there exists a set $B \in \mathfrak{A}$ such that $\mathbf{P}\{\partial B\} = 0$ and $\mathbf{P}\{A\triangle B\} < \varepsilon$ (see Problem 5.1).

Indeed,

$$\begin{aligned}\|\mathbf{P}_A G_c^{-1} - \mathbf{P}_A\| &\le \|\mathbf{P}_A G_c^{-1} - \mathbf{P}_B G_c^{-1}\| + \|\mathbf{P}_B G_c^{-1} - \mathbf{P}_B\| + \|\mathbf{P}_B - \mathbf{P}_A\| \\ &\le 2\varepsilon + \|\mathbf{P}_B G_c^{-1} - \mathbf{P}_B\|.\end{aligned}$$

Now, using the special case, we obtain (5.2) from the last inequality. □

Now let us prove the theorem. By Lemma 5.1, the measure $\mathbf{P}_{V_k}$ satisfies the assumptions of Theorem 5.1. Theorem 5.2 is thereby proved. □

To use the stratification method, we need some families of mappings (G_c) that possess the group structure and whose action on the measure $\mathbf{P}$ is quasi-invariant, that is, $\mathbf{P}G_c^{-1} \sim \mathbf{P}$. In the superstructure method, the assumption about the group structure is not necessary, and the quasi-invariance property is replaced by the assumption that the measures $\mathbf{P}G_c^{-1}$ and $\mathbf{P}$ are close in variation for small c. Moreover, in the stratification method one must take care that the partitions into orbits be measurable. Here this is not necessary.

Application to semistable processes. The following application illustrates fairly well the possibilities of the superstructure method.

Recall that a process $\xi = (\xi_t, t \in \mathbf{R}_+)$ is said to be *semistable* of order $\alpha > 0$ if for all $a \ge 0$ the processes (ξ_{at}) and $(a^\alpha \xi_t)$ have the same finite-dimensional distributions.

We assume that the trajectories of the process ξ lie in the space $\mathbf{C}[0,\infty)$ with probability 1. (In this connection, see Problem 5.3.)

By $\mathbf{P}$ we denote the distribution of the process ξ in $\mathbf{C}[0,\infty)$ and consider the functional

$$f(x) = \sup_{[0,1]} x(t), \qquad x \in \mathbf{C}[0,\infty).$$

Since the trajectories are semistable, we readily see that $\xi_0 = 0$ with probability 1. Therefore, the distribution $\mathbf{P}f^{-1}$ is concentrated on the half-line $[0,\infty)$.

THEOREM 5.3. *ppose that the distribution of the random variable ξ_1 has a density. Then the distribution $\mathbf{P}f^{-1}$ has the following structure: at the zero point it may have an atom, and on $(0,\infty)$ it is absolutely continuous.*

Note that the mapping $S\colon \mathbf{C}(\mathbf{R}_+) \to \mathbf{C}(\mathbf{R}^1)$ defined by the relation

$$Sx(t) = e^{-t\alpha} x(e^t), \qquad t \in \mathbf{R}^1, \tag{5.4}$$

takes a semistable process ξ of order α to the strictly stationary process $\eta = S\xi$ and that this correspondence is one-to-one (see Problem 5.5). Let $p = \mathbf{P}f^{-1}\{0\}$. Since the process ξ is semistable, we readily obtain

$$\mathbf{P}\{\sup_{[0,T]} \xi(t) = 0\} = \mathbf{P}\{\sup_{[0,1]} \xi(Tt) = 0\} = \mathbf{P}\{\sup_{[0,1]} \xi(t) = 0\}.$$

for all $T > 0$. Therefore,

$$p = \mathbf{P}\{\sup_{[0,1]} \xi(t) = 0\} = \mathbf{P}\{\sup_{\mathbf{R}_+} \xi(t) \le 0\} = \mathbf{P}\{\sup_{\mathbf{R}^1} \eta(s) \le 0\}.$$

So, the fact that the distribution $\mathbf{P}f^{-1}$ has an atom of value p is equivalent to the fact that with probability p the trajectories of the stationary process η lie under the x-axis. For $0 < p < 1$, this is possible only if the process η is not ergodic. Thus, we have the following assertion.

COROLLARY 1. *If the semistable process ξ is generated by an ergodic stationary process for which the distribution of $\eta(0)$ is absolutely continuous and not concentrated on $(-\infty, 0]$, then the distribution of $\mathbf{P}f^{-1}$ is absolutely continuous.*

Let us point out another useful result.

COROLLARY 2. *Suppose that the distribution of the variable $\xi(1)$ is absolutely continuous. Then the distribution of the random variable*

$$g(\xi) = \sup_{[0,1]} |\xi(t)|$$

is also absolutely continuous.

The proof readily follows from the fact that $|\xi|$, as well as ξ, is semistable. Here we do not need the ergodicity of the corresponding stationary process, since the inequality $\mathbf{P}\{\sup_{[0,1]} |\xi(t)| = 0\} > 0$ would imply $\mathbf{P}\{\xi(1) = 0\} > 0$, but this contradicts the absolute continuity of the distribution of $\xi(1)$.

PROOF OF THEOREM 5.3. Let $\pi_1(x) = x(1)$ and $x \in \mathbf{C}[0,\infty)$. We set

$$X_1 = \{x \mid f(x) = \pi_1(x)\}, \qquad X_2 = X_1^{\complement} \cap \{x \mid f(x) > 0\}, \qquad X_3 = \{x \mid f(x) = 0\}.$$

Obviously, the atom at the origin of the distribution $\mathbf{P}f^{-1}$ equals $\mathbf{P}\{X_3\}$, and the restriction of $\mathbf{P}f^{-1}$ to the half-line $(0,\infty)$ coincides with the measure

$\mathbf{P}_{X_1 \cup X_2} f^{-1}$, which is equal to $\mathbf{P}_{X_1} f^{-1} + \mathbf{P}_{X_2} f^{-1}$. By assumption, $\mathbf{P}\pi_1^{-1} \ll \lambda$. Therefore,

$$\mathbf{P}_{X_1} f^{-1} = \mathbf{P}_{X_1} \pi_1^{-1} \ll \mathbf{P}\pi_1^{-1} \ll \lambda.$$

Thus, it remains to show that $\mathbf{P}_{X_2} f^{-1} \ll \lambda$. To this end, we use Theorem 5.2. In the space $\mathbf{X} = \mathbf{C}[0, \infty)$, we consider the mappings

$$G_c x(t) = \frac{x(t(1+c))}{(1+c)^\alpha}, \qquad t \in \mathbf{R}_+, \quad c \in [0,1]. \tag{5.5}$$

Since ξ is semistable, we readily find that the measure $\mathbf{P}$ is invariant under these mappings,

$$\mathbf{P}G_c^{-1} = \mathbf{P}.$$

Hence, assumption 1) in Theorem 5.2 is trivially satisfied.

It follows from the continuity of the trajectories that $G_c x \to x$ as $c \to 0$ for each $x \in \mathbf{X}$; thus, assumption 2) is also satisfied. Obviously, the set X_2 is measurable (and even open, since f and π_1 are continuous). Therefore, we can use Lemma 5.1, whence it follows that $\mathbf{P}_{X_2} G_c^{-1} \overset{\text{var}}{\to} \mathbf{P}_{X_2}$ as $c \to 0$.

Now we note that for any $x \in X_2$ and for all sufficiently small c we have

$$\varphi_x(c) = f(G_c x) = \sup_{t \in [0,1]} \frac{x(t(1+c))}{(1+c)^\alpha} = (1+c)^{-\alpha} f(x),$$

and hence, $\varphi_x'(c) = -\alpha f(x)(1+c)^{-\alpha-1}$. Since $f(x) > 0$ for $x \in X_2$, it follows that $\varphi_x'(c) \neq 0$ for all sufficiently small c. Thus, by applying Theorem 5.1 to the measure $\mathbf{P}_{X_2}$, we can prove that $\mathbf{P}_{X_2} f^{-1}$ is absolutely continuous. □

We have used the orbits $\{G_c x, c \in [0,1]\}$ of the mappings (5.5), which are "pieces" of the orbits of the mappings $\widetilde{G}_s$:

$$\widetilde{G}_s x(t) = e^{-s\alpha} x(te^s), \qquad s \in \mathbf{R}^1, \quad t \in \mathbf{R}_+. \tag{5.6}$$

Although the family $\{\widetilde{G}_s\}$ is a group, the partition Γ into orbits of this group is not measurable (see Problem 5.5). Thus, one can hardly use Γ for studying the distributions $\mathbf{P}f^{-1}$ within the framework of the stratification method.

The advantages of the superstructure method are obvious as long as we study strong convergence. These problems are treated in Chapter 5.

PROBLEM 5.1. Let $\mathbf{X}$ be a metric space, $\mathbf{P}$ a finite measure on $\mathfrak{B}_{\mathbf{X}}$, and $\mathfrak{L}$ the algebra of $\mathbf{P}$-continuous sets. Prove that $\sigma(\mathfrak{L}) = \mathfrak{B}_{\mathbf{X}}$.

PROBLEM 5.2. Show that the space $\mathbf{X}$ in Theorem 5.2 may not be separable if the measure $\mathbf{P}$ is assumed to be dense.

PROBLEM 5.3. Verify that the statement of Theorem 5.3 remains valid if we assume that almost all trajectories of the process ξ lie in $\mathbf{D}[0, \infty)$, that is, with probability 1 they are right continuous and have left limits at each point.

PROBLEM 5.4. Let (ξ_t), $t \in \mathbf{R}^1$, be a measurable process stationary in the narrow sense. Suppose that the random variable

$$\tau = \int_a^b h(\xi_t)\, dt,$$

where h is a measurable function and $-\infty < a < b < \infty$, is well defined. We also assume that $h(\xi_a) \neq h(\xi_b)$ with probability 1. Prove that the distribution of τ is absolutely continuous.

(*Hint*: in the space $\mathbf{X} = \mathbf{L}^0(\mathbf{R}^1)$ consider the family of translation mappings

$$G_c x(t) = x(t+c), \qquad c \in [0,1], \quad t \in \mathbf{R}^1.$$

Use Theorem 5.1 and the fact that $\mathbf{P}G_c^{-1} = \mathbf{P}$, where $\mathbf{P}$ is the distribution of ξ in $\mathbf{X}$.)

PROBLEM 5.5. Prove the following properties of the mapping $S\colon \mathbf{C}(R_+) \to \mathbf{C}(R^1)$ defined in (5.4).

a) If ξ is a continuous semistable process, then $\eta = S(\xi)$ is a continuous strictly stationary process.

b) The mapping S takes the group of mappings $(\widetilde{G}_c, c \in \mathbf{R}^1)$ defined by (5.6) into the group of translations

$$V_c x(t) = S\widetilde{G}_c S^{-1} x(t) = x(t+c), \qquad c \in \mathbf{R}^1.$$

c) The partition of the space $\mathbf{C}(R^+)$ into orbits of the group (V_c), as well as the partition of the space $\mathbf{C}(R_+)$ into orbits of the group $(\widetilde{G}_c)$, is measurable. (*Hint*: assuming the contrary, prove that if we identify the orbits with $\mathbf{R}^1$, then the conditional measures on the orbits must be invariant under translations; this is impossible for probability measures.)

§6. Differential operators

Let $f\colon \Omega \to \mathbf{R}^1$ be a random variable on a probability space $(\Omega, \mathfrak{F}, \mathbf{P})$. In this section, we present a rather general approach to obtaining sufficient conditions for the density of the distribution $\mathbf{P}f^{-1}$ to be smooth. In the simplest case $\Omega = \mathbf{R}^1$ and $d\mathbf{P}(x) = p(x)\,dx$, the sufficient conditions consist in smoothness of p and f and the nonvanishing of f'.

In what follows, we show that the same holds for a quite general situation. Namely, for the existence of a smooth density it suffices to require that the random variable f and the measure $\mathbf{P}$ be differentiable in some generalized sense and that the "generalized" derivative of f does not vanish too often.

Main definitions. Let $(\Omega, \mathfrak{F})$ be a measurable space, and let $\mathbf{R}^\Omega$ be the space of all possible mappings $\Omega \to \mathbf{R}^1$.

A partial mapping $D\colon \mathbf{R}^\Omega \to R^\Omega$ with domain $\operatorname{dom} D$ will be called a *differential operator* (of the first order) if

1) the set $\operatorname{dom} D$ is closed with respect to smooth superpositions, that is, for any $n \geq 1$, any C^∞-function $\varphi\colon \mathbf{R}^n \to \mathbf{R}$, and any $g_1, \dots, g_n \in \operatorname{dom} D$, we have $\varphi(g_1, \dots, g_n) \in \operatorname{dom} D$;

2) the conventional differentiation rule for the superposition is valid:

$$D\varphi(g_1, \dots, g_n) = \sum_{i=1}^n \partial_i \varphi(g_1, \dots, g_n) D g_i.$$

The usual differentiation operator and the operator of differentiation along a vector field are differential operators in the sense of this definition.

We shall consider examples in which $\operatorname{dom} D \subset \mathbf{L}^0(\Omega, \mathfrak{F})$ and a probability measure $\mathbf{P}$ is fixed on a measurable space $(\Omega, \mathfrak{F})$. In this situation, we additionally assume that the mapping D does not distinguish functions coinciding $\mathbf{P}$-almost everywhere.

We define spaces of *l-times D-differentiable* functionals $\mathbf{C}_l(D)$ by the formulas $\mathbf{C}_0(D) = \mathbf{R}^\Omega$, $\mathbf{C}_1(D) = \operatorname{dom} D$, and $\mathbf{C}_l(D) = D^{-1}(\mathbf{C}_{l-1}(D))$. Obviously, $\mathbf{C}_l(D)$ is the domain of the operator $D^l = D \circ D \circ \cdots \circ D$.

Let $\mathbf{L}$ be a linear functional with domain $\operatorname{dom} L \subset \mathbf{R}^\Omega$. A linear functional DL with domain $\operatorname{dom}(DL) = \{g \in \operatorname{dom} D \mid Dg \in \operatorname{dom} L\}$ defined by $(DL)(g) = -L(Dg)$ will be called the *D-derivative* of $\mathbf{L}$.

Obviously, $\operatorname{dom}\,(D^l L) = \{g \in \operatorname{dom}\,(D^l) \mid D^l g \in \operatorname{dom}\, L\}$ and $(D^l L)(g) = (-1)^l L(D^l g)$, $l = 1, 2, \ldots$.

We mostly consider the class of linear functionals defined by signed measures,

$$L_\mu(g) = \int_\Omega g\, d\mu, \qquad \operatorname{dom} L_\mu = L_1(|\mu|).$$

In what follows, we do not distinguish between a measure and the linear functional corresponding to this measure, and write μ instead of L_μ.

Let $\widetilde{\Omega}$ be a nonempty set. Each surjective mapping $\xi\colon \widetilde{\Omega} \to \Omega$ defines a σ-algebra $\widetilde{\mathfrak{F}} = \xi^{-1}(\mathfrak{F})$ on the set $\widetilde{\Omega}$.

By using ξ, we transfer various objects (differential operators, linear functionals (in particular, measures), etc.) to the measurable space $(\widetilde{\Omega}, \widetilde{\mathfrak{F}})$. If D is a linear differential operator with $\operatorname{dom} D \subset \mathbf{R}^\Omega$, then the differential operator $(D\circ\xi)(g\circ\xi) = Dg\circ\xi$ is defined on the set $\operatorname{dom} D\circ\xi = \{g\circ\xi \mid g \in \operatorname{dom} D\}$; if L is a linear functional with $\operatorname{dom} L \subset \mathbf{R}^\Omega$, then the linear functional $(L \circ \xi)(g \circ \xi) = L(g)$ is defined on the set $\operatorname{dom} L \circ \xi = \{g \circ \xi \mid g \in \operatorname{dom} L\}$, and so forth. The surjectivity of ξ guarantees that these notions are well defined.

Now let us introduce spaces of smooth measures. In accordance with the theory of distributions, by $\overline{\mathbf{W}}_1^l$ ($l \geq 1$) we denote the class of distributions on $\mathbf{R}^1$ whose lth derivative is a finite charge. If a probability measure belongs to the class $\overline{\mathbf{W}}_1^l$, then it is absolutely continuous with respect to Lebesgue measure and its distribution function has $l-1$ absolutely continuous bounded derivatives; the lth-order derivative is defined almost everywhere and is a function of bounded variation. Accordingly, for $l \geq 2$ the density of such a measure has $l-2$ absolutely continuous derivatives, and its $(l-1)$st-order derivative is a function of bounded variation (for details, see [**28**]).

For example, if a distribution belongs to the class $\overline{\mathbf{W}}_1^l$, then it is absolutely continuous and its density is a function of bounded variation (hence, the density is bounded).

Now let us define some analogs of the classes $\overline{\mathbf{W}}_1^l$ in the most general situation. In this case, the role of distributions is played by D-differentiable linear functionals.

Let $S \subset \mathbf{R}^\Omega$ be a linear set. We say that a linear functional L belongs to the class $\overline{\mathbf{W}}^0(S)$ if $\operatorname{dom} L \supset S$ and L is defined on S by the formula $L(g) = \int_\Omega g\, d\mu$, where μ is a measure. The measure μ will be called the *representing measure* of L. Note that this measure is not uniquely determined in general. More generally, let $S_0, \ldots, S_l$ be linear subsets of $\mathbf{C}_0(D), \ldots, \mathbf{C}_l(D)$, respectively, such that $DS_j \subset S_{j-1}$ ($j = 1, \ldots, l$). We say that a functional L belongs to the class $\overline{\mathbf{W}}^l(S_0, \ldots, S_l) = \overline{\mathbf{W}}^l(D \mid S_0, \ldots, S_l)$ if $D^j L \in \overline{\mathbf{W}}^0(S_j)$, $j = 0, \ldots, l$. Then the relations

$$\int Dg\, d\mu_{j-1} = -\int g\, d\mu_j, \qquad g \in S_j, \tag{6.1}$$

where μ_j is the representing measure of $\mathbf{D}^j L$, are naturally called *formulas of integration by parts*. We point out that in this case the formulas of integration by parts readily follow from the definition of the classes $\overline{\mathbf{W}}^l$.

The fact that the measure $\mathbf{P}$ (*treated as a linear functional*) *belongs to one of the classes* $\overline{\mathbf{W}}^l$ *shows how differentiable this measure is.*

Let us prove a transformation property of $\overline{\mathbf{W}}^l$ under a mapping of spaces.

PROPOSITION 6.1. *If* $L \in \overline{\mathbf{W}}^l(D \mid S_0, \dots, S_l)$, *then* $L \circ \xi \in \overline{\mathbf{W}}^l(D \circ \xi \mid S_0 \circ \xi, \dots, S_l \circ \xi)$.

PROOF. Let $g \in S_j$. Then

$$\begin{aligned}[(D\circ\xi)^j(L\circ\xi)](g\circ\xi) &= (-1)^j(L\circ\xi)((D\circ\xi)^j(g\circ\xi)) = (-1)^j(L\circ\xi)(D^j g\circ\xi)\\ &= (-1)^j L(D^j g) = D^j L(g) = \int_\Omega g\, d\mu_j = \int_\Omega (g\circ\xi)\, d(\mu_j\circ\xi).\end{aligned}$$

Thus, $\mu_j\xi$ is the representing measure of $(D\circ\xi)^j(L\circ\xi)$. □

Sufficient condition for smoothness. Let $f\colon \Omega \to \mathbf{R}^1$ be a random variable on a probability space $(\Omega, \mathfrak{F}, \mathbf{P})$. We consider the measure $\mathcal{P} = \mathbf{P}f^{-1}$ (the distribution function f) as a distribution on $\mathbf{R}^1$. We present sufficient conditions under which it belongs to the classes $\overline{\mathbf{W}}^l_1$.

We need some auxiliary notation. For $l = 1, 2, \dots$ and $i = 1, \dots, l$, we set

$$I(l,i) = \Big\{\mathbf{k} = (k_1, \dots, k_l) \in \mathbf{Z}^l_+ \;\Big|\; \sum_{j=1}^{l} jk_j = l - i\Big\}.$$

We also set $|\mathbf{k}| = k_1 + \cdots + k_l$. If D is a differential operator and $g \in \mathbf{C}_{l+1}(D)$, then by $S_{l,i}(D,g)$, $i = 0, \dots, l$, we denote the set of functions of the form

$$\varphi_1(g)\varphi_2(Dg)(Dg)^{-l-|\mathbf{k}|}(D^2 g)^{k_1}\dots(D^{l+1}g)^{k_l},$$

where $\mathbf{k} \in I(l,i)$, $\varphi_1 \in \mathbf{C}_0^\infty$, and $\varphi_2 \in \mathbf{C}_b^\infty$ vanishes in a neighborhood of zero.

THEOREM 6.1. *Let* $f\colon \Omega \to \mathbf{R}^1$ *be a random variable. Assume that for some differential operator* D *and some* $l \geq 1$, *the following conditions hold*:

1) $f \in \mathbf{C}_{l+1}(D)$ *and* $\mathbf{P}\{Df \neq 0\} = 1$;
2) $\mathbf{P} \in \overline{\mathbf{W}}^l(D \mid S_0, \dots, S_l)$ *for some* $S_i \supset S_{l,i}(D,f)$, $i = 0, \dots, l$;
3) *for each* $i = 0, \dots, l$ *and each* $\mathbf{k} \in I(l,i)$, *the integral*

$$\int |Df|^{-l-|\mathbf{k}|}\,|D^2 f|^{k_1}\dots|D^{l+1}f|^{k_l}\, d|\mu_i| \tag{6.2}$$

is finite (*here* μ_i *is the representing measure of the functional* $D^i\mathbf{P}$ *on* S_i).

Then $\mathbf{P}f^{-1} \in \overline{\mathbf{W}}^l_1$.

The proof is based on the following result.

LEMMA 6.1. *Suppose that there exists a constant $c > 0$ such that*

$$|\mathsf{E}\varphi^{(l)}(f)| \le c \sup|\varphi|. \tag{6.3}$$

for any $\varphi \in \mathbf{C}_0^\infty$.

Then $\mathcal{P} = \mathbf{P}f^{-1} \in \overline{\mathbf{W}}_1^l$.

PROOF OF LEMMA 6.1. Let us transform the expression under the modulus sign on the left-hand side in (6.3):

$$\mathsf{E}\varphi^{(l)}(f) = \int \varphi^{(l)}\, d\mathcal{P} = (-1)^l(\varphi, \mathcal{P}^{(l)}),$$

where $(\varphi, \mathcal{P}^{(l)})$ is the value of the lth derivative of $\mathcal{P}$ (in the sense of distributions) on the element φ. Now, from (6.3) and from the Riesz theorem about the representation of a linear functional by an integral, it follows that the functional $\mathcal{P}^{(l)}$ is a finite charge. □

PROOF OF THEOREM 6.1. Let us verify the assumptions of Lemma 6.1. We choose an auxiliary sequence of $\mathbf{C}^\infty$-functions $\psi_n\colon \mathbf{R}^1 \to [0,1]$ such that

a) $\psi_n(r) = 0$ for $r \in [-\frac{1}{2n}, \frac{1}{2n}]$;

b) $\psi_n(r) = 1$ for $r \notin [-\frac{1}{n}, \frac{1}{n}]$;

c) $\sup_{n,r} |r^j \psi_n^{(j)}(r)| < \infty$ for each $j = 1, 2, \ldots$.

PROBLEM 6.1. Prove that such sequences exist.

Let us consider the integral $\int \varphi^{(l)}(f)\psi_n(Df)\, d\mathbf{P}$. By using the representation $\varphi^{(j)}(f) = D(\varphi^{(j-1)}f)(Df)^{-1}$, $j = 1, \ldots, l$, the differentiation rules, and the definition of the derivative of a linear functional, we successively transfer the differentiation from φ to ψ_n and $\mathbf{P}$. As a result, we obtain

$$\int \varphi^{(l)}(f)\psi_n(Df)\, d\mathbf{P} = (-1)^l \sum_{i=0}^{l} D^i\mathbf{P}\Big\{\varphi(f) \sum_{\mathbf{k}\in I(l,i)} A_{\mathbf{k}}(D^2 f)^{k_1} \ldots (D^{l+1}f)^{k_l} \sum_{j=0}^{|\mathbf{k}|} A_{j,\mathbf{k}}(Df)^{j-l-|\mathbf{k}|}\psi_n^{(j)}(Df)\Big\}.$$

Here $A_{\mathbf{k}}$ and $A_{j,\mathbf{k}}$ are absolute constants whose explicit form is unessential.

By using the representing measures of the functionals $D^i\mathbf{P}$, we obtain

$$\int \varphi^{(l)}(f)\psi_n(Df)\, d\mathbf{P} = (-1)^l \sum_{i=0}^{l} \sum_{\mathbf{k}\in I(l,i)} A_{\mathbf{k}} \sum_{j=0}^{|\mathbf{k}|} A_{j,\mathbf{k}} \times \int \varphi(f)\psi_n^{(j)}(Df)(Df)^{j-l-|\mathbf{k}|}(D^2f)^{k_1}\ldots(D^{l+1}f)^{k_l}\, \mu_i. \tag{6.4}$$

Now let us pass to the limit as $n \to \infty$ on the right-hand side in (6.4). First, let us show that the terms with $j \ne 0$ tend to zero. Indeed, the functions $\psi_n^{(j)}(Df)(Df)^j$ are bounded uniformly with respect to n, and the measure of the set of points at which $\psi_n^{(j)} \ne 0$ tends to zero as $n \to \infty$. This passage to the limit is valid, since the integrals (6.2) are finite. Similarly, we can pass to the limit in the terms with $j = 0$, thus obtaining

$$
\begin{aligned}
\int \varphi^{(l)}(f)\, d\mathbf{P} = (-1)^l \sum_{i=0}^{l} \sum_{\mathbf{k}\in I(l,i)} A_{\mathbf{k}} A_{0,\mathbf{k}} \\
\times \int \varphi(f)(Df)^{-l-|\mathbf{k}|}(D^2 f)^{k_1} \dots (D^{l+1} f)^{k_l}\, d\mu_i.
\end{aligned}
\tag{6.5}
$$

Obviously, the desired bound (6.3) follows form the identity (6.5). □

REMARK. Let us apply Theorem 6.1 in the simplest case $l = 1$. We need f to be differentiable twice (condition 1)) and $\mathbf{P}$ to be differentiable once (condition 2)). Furthermore, we need the finiteness of the integrals $\int D^2 f(Df)^{-2}\, d\mathbf{P}$ and $\int (Df)^{-1}\, d\mu_1$, where μ_1 is the representing measure of the functional $D\mathbf{P}$. Under these conditions, $\mathcal{P}$ has a density of bounded variation. Our further goal is to find differential operators that can be used in Theorem 6.1 for some (interesting for applications) classes of stochastic functionals. Chapters 3 and 4 partially deal with this problem. In Chapter 3, we construct such operators for Gaussian functionals, and in Chapter 4, for functionals of sample functions of a Poisson random measure.

CHAPTER 3

Gaussian Functionals

§7. Gaussian measures on linear spaces

Basic characteristics. The definitions of a Gaussian measure and its characteristics, such as the barycenter, the correlation operator, and the characteristic functional, were given in §1.

Let $\langle \mathbf{X}, \mathbf{X}' \rangle$ be a pair of linear spaces in duality. We denote the class of all Radon measures defined on $(\mathbf{X}, \mathfrak{C})$ by $\mathcal{G}(\mathbf{X})$ and the subclass of centered Radon Gaussian measures by $\mathcal{G}_0(\mathbf{X})$.

PROPOSITION 7.1. *Each measure $\mathbf{P} \in \mathcal{G}(\mathbf{X})$ has a barycenter and a self-adjoint correlation operator continuous in the weak topology.*

PROOF. Let $\overline{\mathbf{X}}$ be the completion of $\mathbf{X}$ in the weak topology, $i\colon \mathbf{X} \to \overline{\mathbf{X}}$ the natural embedding, and $\overline{\mathbf{P}} = \mathbf{P}i^{-1}$ the extension of $\mathbf{P}$ to $\overline{\mathbf{X}}$. We choose a compact set $\Delta \subset \mathbf{X}$ such that $\mathbf{P}\{\mathbf{X} \setminus \Delta\} \leq 1/3$ and consider the convex closed hull $\overline{\Delta}$ of the compact set $i(\Delta)$ in $\overline{\mathbf{X}}$. Since $\overline{\mathbf{X}}$ is complete, it follows that $\overline{\Delta}$ is compact in $\overline{\mathbf{X}}$ [**87,** Ch. III]. Now for any functional $x' \in \mathbf{X}'$ satisfying the condition

$$\sup_{\bar{x} \in \overline{\Delta}} |\langle \bar{x}, x' \rangle| \leq \varepsilon, \tag{7.1}$$

we have

$$\overline{\mathbf{P}}\{\bar{x} \in \overline{\mathbf{X}}\colon |\langle \bar{x}, x' \rangle| \geq \varepsilon\} \leq \overline{\mathbf{P}}\{\overline{\mathbf{X}} \setminus \overline{\Delta}\} \leq \overline{\mathbf{P}}\{\overline{\mathbf{X}} \setminus i(\Delta)\} \leq 1/3. \tag{7.2}$$

Let $\mathcal{N}(\alpha, \sigma^2)$ be the distribution of the functional $\langle \cdot, x' \rangle$ with respect to the measure $\overline{\mathbf{P}}$. Then, (7.2) implies the estimate

$$\Phi\left(\frac{\varepsilon - \alpha}{\sigma}\right) - \Phi\left(\frac{-\varepsilon - \alpha}{\sigma}\right) \geq \frac{2}{3}. \tag{7.3}$$

In particular, $\Phi\big((\varepsilon - \alpha)/\sigma\big) \geq 1/2$, and hence, $\alpha \leq \varepsilon$; similarly, we obtain $\alpha \geq -\varepsilon$. Thus, (7.1) implies $|\alpha| \leq \varepsilon$. This argument shows that the linear functional

$$\mathcal{A}\colon x' \to \alpha = \mathsf{E}_{\overline{\mathbf{P}}}\langle \cdot, x' \rangle \tag{7.4}$$

defined on $\mathbf{X}'$ is continuous in the Mackey topology corresponding to the pair $\langle \overline{\mathbf{X}}, \mathbf{X}' \rangle$ (that is, in the topology of uniform convergence of functionals on convex compact subsets of $\overline{\mathbf{X}}$). The Mackey topology is the strongest topology on $\mathbf{X}'$ for which the space of continuous linear functionals coincides with $\overline{\mathbf{X}}$ [**87,** Ch.III]. In particular, the functional $\mathcal{A}$ is generated by some element $\bar{a} \in \overline{\mathbf{X}}$:

$$\langle \bar{a}, x' \rangle = \mathcal{A}(x') = \mathsf{E}_{\overline{\mathbf{P}}}\langle \cdot, x' \rangle = \mathsf{E}_{\overline{\mathbf{P}}}\langle \cdot, x' \rangle. \tag{7.5}$$

Thus, $\bar{a}$ is the barycenter of $\overline{\mathbf{P}}$. Let us show that $\bar{a} = i(a)$ for some $a \in \mathbf{X}$, which implies that a is the barycenter of $\mathbf{P}$.

The linear space $i(\mathbf{X})$ is $\overline{\mathbf{P}}$-measurable with $\overline{\mathbf{P}}\{i(\mathbf{X})\} = 1$. Since a Gaussian measure is symmetric with respect to its barycenter, we also have $\overline{\mathbf{P}}\{2\bar{a}-i(\mathbf{X})\} = 1$. Therefore, the intersection of the spaces $i(\mathbf{X})$ and $2\bar{a} - i(\mathbf{X})$ is not empty: for some $x_1, x_2 \in \mathbf{X}$, we have $i(x_1) = 2\bar{a} - i(x_2)$. Thus, $a = (x_1 + x_2)/2$, which proves the existence of a barycenter for $\mathbf{P}$.

Now let us choose a $y' \in \mathbf{X}'$ and consider the functional $\mathcal{K}\colon \mathbf{X}' \to \mathbf{R}^1$ given by

$$\mathcal{K}(x') = \mathsf{E}_{\mathbf{P}}\langle \cdot - a, x'\rangle\langle \cdot - a, y'\rangle = \mathsf{E}_{\overline{\mathbf{P}}}\langle \cdot - \bar{a}, x'\rangle\langle \cdot - \bar{a}, y'\rangle. \tag{7.6}$$

If x' satisfies (7.1), then it follows from (7.3) and the expression for the Gaussian density that $\sigma = \left(\mathsf{E}_{\overline{\mathbf{P}}}\langle \cdot - \bar{a}, x'\rangle^2\right)^{1/2} \le 2\varepsilon$. By using the Hölder inequality, we obtain

$$|\mathcal{K}(x')|^2 \le \sigma^2 \mathsf{E}_{\overline{\mathbf{P}}}\langle \cdot - \bar{a}, y'\rangle^2 \le 4\varepsilon^2 \mathsf{E}_{\overline{\mathbf{P}}}\langle \cdot - \bar{a}, y'\rangle^2. \tag{7.7}$$

Hence, $\mathcal{K}$, as well as $\mathcal{A}$, is continuous on $\mathbf{X}'$ in the Mackey topology, and for some element $\overline{K}y'$ of the space $\overline{\mathbf{X}}$ we have

$$\langle \overline{K}y', x'\rangle = \mathcal{K}(x') = \mathsf{E}_{\overline{\mathbf{P}}}\langle \cdot - \bar{a}, x'\rangle\langle \cdot - \bar{a}, y'\rangle.$$

The operator $\overline{K}\colon \mathbf{X}' \to \overline{\mathbf{X}}$ thus obtained is the correlation operator of the measure $\overline{\mathbf{P}}$. Obviously, $\overline{K}$ is linear and self-adjoint. To construct the correlation operator of the measure $\mathbf{P}$, it suffices to verify the inclusion $\overline{K}(\mathbf{X}') \subset i(\mathbf{X})$ and then set $K = i^{-1}\overline{K}$. Thus, let us take some $x' \in \mathbf{X}'$ and verify that $l = \overline{K}x' \in i(\mathbf{X})$. By comparing the characteristic functionals, one can easily prove that the translate of the measure $\overline{\mathbf{P}}$ by a vector l coincides with the measure

$$\overline{\mathbf{P}}_l(d\bar{x}) = \exp\left\{\langle \bar{x} - \bar{a}, x'\rangle - \frac{1}{2}\langle l, x'\rangle\right\}\overline{\mathbf{P}}(dx).$$

Thus, $\overline{\mathbf{P}}_l \ll \overline{\mathbf{P}}$, and hence, $\overline{\mathbf{P}}_l\{i(\mathbf{X})+l\} = \overline{\mathbf{P}}\{i(\mathbf{X})\} = 1$. Therefore, the intersection of the spaces $i(\mathbf{X})+l$ and $i(\mathbf{X})$ is not empty. It follows that $l \in i(\mathbf{X})$ and the operator K is well defined by $Kx' = i^{-1}\overline{K}x' = i^{-1}l$.

Obviously, K is self-adjoint and continuous. □

PROPOSITION 7.2. *The characteristic functional of the Gaussian measure* $\mathbf{P}$ *with barycenter* a *and correlation operator* K *has the form*

$$\varphi_{\mathbf{P}}(x') = \mathsf{E}_{\mathbf{P}} \exp\{i\langle \cdot, x'\rangle\} = \exp\left\{i\langle a, x'\rangle - \frac{1}{2}\langle Kx', x'\rangle\right\}. \tag{7.8}$$

PROOF. By the definitions of a barycenter and a correlation operator, the linear functional $\langle \cdot, x'\rangle$ has the distribution $\mathcal{N}(\langle a, x'\rangle, \langle Kx', x'\rangle)$. Therefore, (7.8) follows from the one-dimensional formula (1.7). □

Spaces of measurable functionals. The definition of a Gaussian measure implies that all functionals of the form $\langle \cdot, x'\rangle$ are square integrable. Therefore, $\mathbf{X}' \subset \mathbf{L}^2(\mathbf{X}, \mathfrak{C}, \mathbf{P})$. The closure of $\mathbf{X}'$ in $\mathbf{L}^2(\mathbf{X}, \mathfrak{C}, \mathbf{P})$ is called the *space* $\mathbf{X}'_{\mathbf{P}}$ *of measurable linear functionals.* The elements of $\mathbf{X}'_{\mathbf{P}}$ are called *measurable linear functionals.* A functional $z \in \mathbf{X}'_{\mathbf{P}}$ treated as a random variable on $(\mathbf{X}, \mathfrak{C}, \mathbf{P})$ will be denoted by $\langle \cdot, z\rangle$.

By adding constants to measurable linear functionals, we obtain *the space* $\mathbf{X}'^A_{\mathbf{P}} = \mathcal{L}\{\mathbf{X}'_{\mathbf{P}}, \mathbf{1}(\cdot)\}$ *of measurable affine functionals.* The spaces $\mathbf{X}'_{\mathbf{P}}$ and $\mathbf{X}'^A_{\mathbf{P}}$

inherit the Hilbert structure from $\mathbf{L}^2(\mathbf{X}, \mathfrak{C}, \mathbf{P})$. By $(\cdot,\cdot)$ we denote the corresponding inner product and by $|\cdot|$ the L^2-norm (it may well happen that $|x'| = 0$ for some nonzero $x' \in \mathbf{X}'$).

Let us define a linear operator $I^*\colon \mathbf{X}' \to \mathbf{X}_{\mathbf{P}}'^{A}$ by setting

$$I^*x'(\cdot) = \langle \cdot, x'\rangle - \mathsf{E}_{\mathbf{P}}\langle \cdot, x'\rangle. \tag{7.9}$$

The closure of the linear manifold $I^*(\mathbf{X}')$ in $\mathbf{X}_{\mathbf{P}}'^{A}$ is called the *space of centered linear functionals* and is denoted by $\mathbf{X}'_{\mathbf{P},0}$.

PROPOSITION 7.3. *Let $\mathbf{P} \in \mathcal{G}(\mathbf{X})$. Then the following assertions hold.*

1) *The operator I^* is continuous on $\mathbf{X}'$ in the Mackey topology of the pair $\langle \overline{\mathbf{X}}, \mathbf{X}'\rangle$, where $\overline{\mathbf{X}}$ is the completion of $\mathbf{X}$.*

2) *The adjoint operator $I\colon \mathbf{X}_{\mathbf{P}}'^{A} \to \mathbf{X}$ exists and is defined by the formula*

$$\langle Iz, y'\rangle = (z, I^*y') \tag{7.10}$$

for all $y' \in \mathbf{X}'$ and $z \in \mathbf{X}_{\mathbf{P}}'^{A}$.

The operator I is continuous in the weak topologies of $\mathbf{X}$ and $\mathbf{X}_{\mathbf{P}}'^{A}$. The kernel of I coincides with the one-dimensional space of constants in $\mathbf{X}_{\mathbf{P}}'^{A}$.

3) *The correlation operator K of the measure P can be represented in the form $K = II^*$. In particular, for $x', y' \in \mathbf{X}'$ one has*

$$\langle Kx', y'\rangle = \langle II^*x', y'\rangle = (I^*x', I^*y'). \tag{7.11}$$

4) *The space $\mathbf{X}'_{\mathbf{P}}$ is separable.*

PROBLEM 7.1. Prove statements 1)–3). (*Hint*: to prove 1), use the estimates from the proof of Proposition 7.1.)

PROOF OF STATEMENT 4). Suppose that $\Delta \subset \mathbf{X}$ is a compact set and $\mathbf{P}\{\Delta\} > 0$. For $M > 0$, we set

$$\mathbf{X}'(M) = \{x' \in \mathbf{X}'\colon \sup_{x\in\Delta} |\langle x, x'\rangle| \le M\}.$$

Since each continuous functional x' is bounded on Δ, we have $\bigcup_{M>0} \mathbf{X}'(M) = \mathbf{X}'$. To prove that $I^*(\mathbf{X}'(M))$ is separable, we construct a sequence of finite-dimensional spaces $L_n \subset \mathbf{X}'_{\mathbf{P}}$ that approximate $I^*(\mathbf{X}'(M))$.

We set $L_0 = \{0\}$ and proceed by induction. At the nth step, we choose an $x'_n \in \mathbf{X}'(M)$ so that

$$2 \inf_{z\in L_{n-1}} |I^*x'_n - z| \ge \sigma_n = \sup_{x'\in\mathbf{X}'(M)} \inf_{z\in L_{n-1}} |I^*x' - z|$$

and set $L_n = \mathcal{L}\{I^*x'_n, L_{n-1}\}$.

Let us show that $\sigma = \lim_{n\to\infty} \sigma_n = 0$. Suppose that $\sigma > 0$. Then

$$\inf_{z\in L_{n-1}} |I^*x'_n - z| \ge \sigma_n/2 \ge \sigma/2$$

for all $n \geqslant 1$. Hence,

$$\mathbf{P}\{\Delta\} \le \lim_{m\to\infty} \mathbf{P}\{\max_{1\le n\le m} |\langle \cdot, x'_n\rangle| \le M\} \le \lim_{m\to\infty} \big(\mathcal{N}(0, \sigma^2/4)\{[-M, M]\}\big)^m = 0,$$

which contradicts the choice of Δ. Hence, $\sigma_n \to 0$, which means that the spaces L_n approximate the set $I^*(\mathbf{X}'(M))$. It follows that $I^*(\mathbf{X}'(M))$ is separable in $\mathbf{X}'_{\mathbf{P}}$. Therefore, the spaces $I^*(\mathbf{X}')$, $\mathbf{X}'_{\mathbf{P},0}$, and $\mathbf{X}'_{\mathbf{P}}$ are also separable. $\square$

PROBLEM 7.2. Let $\mathbf{P}, \mathbf{Q} \in \mathcal{G}(\mathbf{X})$ and $\mathbf{Q} \ll \mathbf{P}$. Prove that $\mathbf{X}'_\mathbf{Q} \subset \mathbf{X}'_\mathbf{P}$, $\mathbf{X}'^A_\mathbf{Q} \subset \mathbf{X}'^A_\mathbf{P}$, and

$$\sup_{z \in \mathbf{X}'^A_\mathbf{Q}} |z|_\mathbf{P} / |z|_\mathbf{Q} < \infty. \tag{7.12}$$

Admissible translations and directions. Each vector $l \in \mathbf{X}$ generates a one-parameter transformation group on $\mathbf{X}$, which is called a *translation group*: $G^l = \{G^l_c, c \in \mathbf{R}^1\}$,

$$G^l_c(x) = x + cl. \tag{7.13}$$

A vector $l \in \mathbf{X}$ is called an *admissible translation* for a measure $\mathbf{P}$ if the measure $\mathbf{P}^l = \mathbf{P}(G^l_1)^{-1}$ is absolutely continuous with respect to $\mathbf{P}$. A vector l defines an *admissible direction* for $\mathbf{P}$ if the entire group G^l is admissible, that is, for each $c \in \mathbf{R}^1$ the translation cl is admissible.

By $\mathbf{H_P}$ we denote the set of all vectors that define admissible directions for $\mathbf{P}$. This set is a vector space (prove this fact!) and is invariant under measure translations: $\mathbf{H_P} = \mathbf{H}_{\mathbf{P}^l}$ for all $l \in \mathbf{X}$. The space $\mathbf{H_P}$ is often called the *core* of the measure $\mathbf{P}$.

For a Gaussian measure, the space $\mathbf{H_P}$ is quite large. Therefore, in specific applications, we always have a large supply of admissible directions.

The following theorem relates admissible translations, directions, and centered linear functionals.

THEOREM 7.1. *Let $\mathbf{P} \in \mathcal{G}(\mathbf{X})$. Then the operator I defined in (7.10) is a linear isomorphism between the spaces $\mathbf{X}'_{\mathbf{P},0}$ and $\mathbf{H_P}$.*

If $z \in \mathbf{X}'_{\mathbf{P},0}$ and $l = Iz \in \mathbf{H_P}$, then

$$\frac{d\mathbf{P}^l}{d\mathbf{P}}(x) = \exp\left\{\langle x, z\rangle - \frac{1}{2}|z|^2\right\}. \tag{7.14}$$

The set of admissible translations of the measure $\mathbf{P}$ coincides with $\mathbf{H_P}$, that is, each admissible translation of a Gaussian measure defines an admissible direction.

PROOF. First, note that the operator I is one-to-one on the subspace $\mathbf{X}'_{\mathbf{P},0} \subset \mathbf{X}'^A_\mathbf{P}$, since the kernel of I is the space of constants, which is orthogonal to $\mathbf{X}'_{\mathbf{P},0}$.

Suppose that l is an admissible translation. Let us show that there exists a $z \in \mathbf{X}'_{\mathbf{P},0}$ such that $Iz = l$. For each $x' \in \mathbf{X}'$, by $\mu_{x'}$ and $\nu_{x'}$ we denote the distributions of the functional $\langle \cdot, x'\rangle$ with respect to $\mathbf{P}$ and $\mathbf{P}_l$, respectively. These distributions are Gaussian; namely, $\mu_{x'} = \mathcal{N}(\langle a, x'\rangle, \langle Kx', x'\rangle)$ and $\nu_{x'} = \mathcal{N}(\langle a+l, x'\rangle, \langle Kx', x'\rangle)$.

Since $\mathbf{P}^l \ll \mathbf{P}$, it follows that for any $\varepsilon > 0$ there exists a $\delta > 0$ such that $\mathbf{P}\{A\} < \delta$ implies $\mathbf{P}^l\{A\} < \varepsilon$ for any A. In particular, if $B \in \mathbf{R}^1$ is an arbitrary measurable set, then $\mu_{x'}\{B\} < \delta$ implies $\nu_{x'}\{B\} < \varepsilon$. The latter inequality means that the parameters of $\mu_{x'}$ and $\nu_{x'}$ satisfy

$$\sup_{x' \in \mathbf{X}'} |\langle l, x'\rangle| \langle Kx', x'\rangle^{-1/2} < \infty.$$

By (7.11), it follows that

$$\sup_{x' \in \mathbf{X}'} |\langle l, x'\rangle| \, |I^* x'|^{-1} < \infty.$$

Hence, the linear functional $\langle l, \cdot\rangle$ admits a continuous extension from $I^*(\mathbf{X}')$ to $\mathbf{X}'_{\mathbf{P},0}$. Therefore, for some $z \in \mathbf{X}'_{\mathbf{P},0}$ and all $x' \in \mathbf{X}'$, we have

$$\langle l, x'\rangle = (z, I^*x') = \langle Iz, x'\rangle,$$

whence it follows that $l = Iz$.

Now let $l = Iz$, $z \in \mathbf{X}'_{\mathbf{P},0}$. To show that l is an admissible translation, it suffices to verify that $\mathbf{P}^l$ coincides with the measure

$$\mathbf{Q}(dx) = \exp\Big\{\langle x, z\rangle - \frac{1}{2}|z|^2\Big\}\mathbf{P}(dx).$$

Let us prove that the characteristic functionals of these measures coincide. By Proposition 7.2, we have

$$\begin{aligned} \mathsf{E}_{\mathbf{P}^l} \exp\{i\langle\cdot, x'\rangle\} &= \mathsf{E}_{\mathbf{P}} \exp\{i\langle\cdot + l, x'\rangle\} \\ &= \exp\Big\{-\frac{1}{2}\langle Kx', x'\rangle + i\langle a + l, x'\rangle\Big\}. \end{aligned} \tag{7.15}$$

Suppose that $\mathcal{R}$ is the correlation matrix of the joint distribution of the functionals $\langle\cdot, x'\rangle$ and $\langle\cdot, z\rangle$ with respect to $\mathbf{P}$. Then

$$\begin{aligned} \mathsf{E}_{\mathbf{Q}} \exp\{i\langle\cdot, x'\rangle\} &= \mathsf{E}_{\mathbf{P}} \exp\Big\{i\langle\cdot, x'\rangle + \langle\cdot, z'\rangle - \frac{1}{2}|z|^2\Big\} \\ &= \exp\Big\{i\langle a, x'\rangle - \frac{1}{2}|z|^2\Big\}(\det \mathcal{R})^{-1/2} \\ &\quad \times \iint_{\mathbf{R}^2} \exp\Big\{ir_1 + r_2 - \frac{1}{2}(\mathcal{R}^{-1}r, r)\Big\}\, dr_1 dr_2 \\ &= \exp\Big\{i\langle a, x'\rangle - \frac{1}{2}|z|^2 - \frac{1}{2}[\mathcal{R}_{11} + 2\mathcal{R}_{12}(i) + \mathcal{R}_{22}(-i)^2]\Big\}. \end{aligned}$$

Into this equation, we substitute the correlations

$$\mathcal{R}_{11} = \langle Kx', x'\rangle, \qquad \mathcal{R}_{22} = |z|^2,$$
$$\mathcal{R}_{12}a = \mathsf{E}_{\mathbf{P}}\langle\cdot, z\rangle\langle\cdot - a, x'\rangle = (z, I^*x') = \langle Iz, x'\rangle = \langle l, x'\rangle.$$

After some calculations we obtain

$$\mathsf{E}_{\mathbf{Q}} \exp\{i\langle\cdot, x'\rangle\} = \exp\Big\{i\langle a, x'\rangle - \frac{1}{2}\langle Kx', x'\rangle + i\langle l, x'\rangle\Big\},$$

which coincides with (7.15). By Theorem 1.2, the coincidence of measures follows from the coincidence of the corresponding characteristic functionals. Therefore, $\mathbf{Q} = \mathbf{P}^l$.

Thus, we have proved that the set of admissible translations coincides with $I^*(\mathbf{X}'_{\mathbf{P},0})$. □

The case of a centered measure $\mathbf{P}$ must be considered separately. In this case, $\mathbf{X}'_{\mathbf{P}} = \mathbf{X}'_{\mathbf{P},0}$, and by Theorem 7.1, there is an isomorphism between the space $\mathbf{H}_{\mathbf{P}}$ of admissible translations and the space $\mathbf{X}'_{\mathbf{P}}$ of measurable linear functionals.

Let us introduce two convenient notions. Let $z \in \mathbf{X}'_{\mathbf{P},0}$ and $l = Iz \in \mathbf{H}_{\mathbf{P}}$. Then the number

$$\sigma_{\mathbf{P}}(l) = |z|^{-1} \tag{7.16}$$

is called the *admissibility* of the translation l, and the number

$$\mathcal{I}(l) = |z|^2 \tag{7.17}$$

is referred to as the *action functional.* If $l \in \mathbf{X} \setminus \mathbf{H_P}$, then we adopt the natural convention that $\sigma_{\mathbf{P}}(l) = 0$ and $\mathcal{I}(l) = \infty$.

PROBLEM 7.3. Let $\mathbf{P}, \mathbf{Q} \in \mathcal{G}(\mathbf{X})$, $\mathbf{P} \ll \mathbf{Q}$, and $\mathbf{Q} \ll \mathbf{P}$. Prove that $\mathbf{H_P} = \mathbf{H_Q}$ and

$$0 < \inf_{l \in \mathbf{H_P}} \{\sigma_{\mathbf{P}}(l)/\sigma_{\mathbf{Q}}(l)\} \leq \sup_{l \in \mathbf{H_P}} \{\sigma_{\mathbf{P}}(l)/\sigma_{\mathbf{Q}}(l)\} < \infty.$$

(Use the result of Problem 7.2.)

PROPOSITION 7.4. *Let $\mathbf{P} \in \mathcal{G}_0(\mathbf{X})$; let* $\operatorname{supp}(\mathbf{P})$ *and* $\mathbf{N_P}$ *be, respectively, the topological and the linear supports of* $\mathbf{P}$ *with respect to the weak topology. Then*

1) $\operatorname{supp}(\mathbf{P}) = \mathbf{N_P} = \bigcap_{x' \in \operatorname{Ker} I^*} \operatorname{Ker} x'$;

2) *if* $\mathbf{H_P^-}$ *is the closure of* $\mathbf{H_P}$ *in* $\mathbf{X}$, *then* $\mathbf{N_P} = \mathbf{H_P^-}$;

3) $\mathbf{P}\{\mathbf{N_P}\} = 1$.

PROOF. We write $U = \bigcap_{x' \in \operatorname{Ker} I^*} \operatorname{Ker} x'$. First, let us show that $\mathbf{H_P^-} = U$. If $x \in \mathbf{H_P}$, then $x = Iz$, $z \in \mathbf{X'_P}$, and

$$\langle x, x' \rangle = \langle Iz, x' \rangle = (z, I^* x') = 0$$

for $x' \in \operatorname{Ker} I^*$. Therefore, $x \in \operatorname{Ker} x'$ and $\mathbf{H_P} \subset U$. Since U is closed, we have $\mathbf{H_P^-} \subset U$. Conversely, if $x \notin \mathbf{H_P^-}$, then, by the Hahn–Banach theorem, there exists a vector $x' \in \mathbf{X}'$ such that $\langle x, x' \rangle = 1$ and $\langle \cdot, x' \rangle = 0$ on $\mathbf{H_P^-}$. Therefore, $x' \in \operatorname{Ker} I^*$, and hence, $x \notin U$. It follows that $\mathbf{H_P^-} \supset U$.

Thus, $\mathbf{H_P^-} = U$. If $\mathbf{X}$ is finite-dimensional, then the verification of statement 1) is straightforward. In the general case, only the inclusion $\operatorname{supp}(\mathbf{P}) \subset U$ is obvious. Let $x \in U$. Let us choose a basic neighborhood

$$V = \{y \in X \mid \max_{1 \leq i \leq m} |\langle y - x, x'_i \rangle| < \varepsilon\} \ni x$$

and verify that $\mathbf{P}\{V\} > 0$. We define a mapping $f \colon \mathbf{X} \to \mathbf{R}^m$ by setting $f(y) = \{\langle y, x'_i \rangle\}$. By applying statement 1) to the finite measure $\mathbf{P}f^{-1}$ and the point $f(x)$, we see that $f(x) \in \operatorname{supp}(\mathbf{P}f^{-1})$. Therefore,

$$\mathbf{P}\{V\} = \mathbf{P}f^{-1}\{f(V)\} = \mathbf{P}f^{-1}\{v \in \mathbf{R}^m \mid \max_{1 \leq i \leq m} |v_i - f(x)_i| < \varepsilon\} > 0.$$

Hence, $x \subset \operatorname{supp}(\mathbf{P})$, and consequently, $U \subset \operatorname{supp}(\mathbf{P})$. Thus, we have $U = \operatorname{supp}(\mathbf{P}) = \mathbf{H_P^-}$. Since $\mathbf{P}$ is a Radon measure and, by definition, U is a linear set, it follows that (see §1) $\mathbf{P}\{\operatorname{supp}(\mathbf{P})\} = 1$, the linear support $N_{\mathbf{P}}$ exists, and $N_{\mathbf{P}} = \operatorname{supp}(\mathbf{P})$. This implies the desired equations in 1)–3). □

REMARK. Since, by Proposition 7.3, 4) and Theorem 7.1, the space $\mathbf{H_P}$ is separable, it follows from Proposition 7.4 that the support of any measure $\mathbf{P} \in \mathcal{G}_0(\mathbf{X})$ is a linear separable subset of $\mathbf{X}$.

This statement remains valid if the Radon measure $\mathbf{P}$ is defined on a wider σ-algebra $\mathfrak{B}$ generated by some local convex topology compatible with the duality between $\mathbf{X}$ and $\mathbf{X}'$. However, if $\mathbf{P}$ is not a Radon measure, then the support need not be linear or separable.

Partitions into parallel lines. Let us study partitions of a space with Gaussian measure and calculate the corresponding conditional distributions. We use the notation introduced in §3.

Let $l \in \mathbf{H_P}$ and let Γ be the partition of $\mathbf{X}$ into lines parallel to l, that is, orbits of the group (7.13). On each line $\gamma = \{y \in \mathbf{X} \mid y = x + cl, c \in \mathbf{R}^1\} \in \mathbf{X}/\Gamma$, we define the Lebesgue measure λ_γ by setting

$$\lambda_\gamma\{y \mid y = x + cl, c_1 \leq c \leq c_2\} = c_2 - c_1.$$

By Theorem 7.1, there exists a linear measurable functional $z \in \mathbf{X}'_{\mathbf{P},0}$ such that $Iz = l$. On γ we introduce the one-dimensional Gaussian measure

$$\mathbf{P}_\gamma\,(dy) = (2\pi)^{-1/2}|z| \exp\Big\{-\frac{\langle y, z/|z|\rangle^2}{2}\Big\}\lambda_\gamma\,(dy). \tag{7.18}$$

Note that the functional z is defined and linear everywhere on some linear space $\mathbf{L} \in \mathfrak{C}$ with $\mathbf{P}\{\mathbf{L}\} = 1$. Since l is an admissible translation, we have $l \in L$. Therefore, the measure $\mathbf{P}_\gamma$ is well defined on the set $\pi(\mathbf{L}) = \{\gamma \mid \gamma \subset L\}$, and $\mathbf{P}_\Gamma\{\pi(\mathbf{L})\} = 1$.

THEOREM 7.2. *The system* $\{\mathbf{P}_\gamma\}$ *is the system of conditional measures for* $\mathbf{P}$ *with respect to* Γ.

PROOF. According to (3.1), for $C \in \mathfrak{C}$ and $A \in \mathfrak{C}_{\mathbf{X},\Gamma}$ we must verify the relation

$$\mathbf{P}\{C \cap \pi^{-1}(A)\} = \int_A \mathbf{P}_\gamma\{C\}\mathbf{P}_\Gamma\,(d\gamma). \tag{7.19}$$

We shall only verify (7.19) for some special sets. Let $f = \{f_i\}$, $1 \leqslant i \leqslant n$, and $g = \{g_j\}$, $1 \leqslant j \leqslant m$, be two systems of centered functionals, $f_i, g_j \in \mathbf{X}'_{\mathbf{P},0}$ with $(f_i, z) = (g_j, z) = 0$. We also choose three Borel sets $Z \in \mathfrak{B}(\mathbf{R}^1)$, $F \in \mathfrak{B}(\mathbf{R}^n)$, and $G \in \mathfrak{B}(\mathbf{R}^m)$, and set

$$C = \{x \in \mathbf{X} \mid \langle x, z\rangle \in Z,\ \{\langle x, f_i\rangle\} \in F\}, \qquad A = \pi(\{x \in \mathbf{X} \mid \{\langle x, g_j\rangle\} \in G\}).$$

Let us verify (7.19). In view of orthogonality, the functional z is independent of f and g. Therefore,

$$\begin{aligned}\mathbf{P}\{C \cap \pi^{-1}(A)\} &= \mathbf{P}\{x \in \mathbf{X} \mid \langle x, z\rangle \in Z,\ \{\langle x, f_i\rangle\} \in F,\ \{\langle x, g_j\rangle\} \in G\}\\ &= \mathbf{P}\{x \in \mathbf{X} \mid \langle x, z\rangle \in Z\}\mathbf{P}\{\{\langle x, f_i\rangle\} \in F,\ \{\langle x, g_j\rangle\} \in G\}\\ &= \mathcal{N}(0, |z|^2)\{Z\}\mathbf{P}\{\{\langle x, f_i\rangle\} \in F,\ \{\langle x, g_j\rangle\} \in G\}.\end{aligned}$$

On the other hand, the definition of $\mathbf{P}_\gamma$ yields

$$\mathbf{P}_\gamma\{C\} = \mathbf{1}_{\{\langle x, f_i\rangle\} \in F}\mathcal{N}(0, |z|^2)\{Z\}.$$

Thus, the right-hand side of (7.19) is equal to

$$\begin{aligned}&\mathbf{P}_\Gamma\{A \cap \{\gamma \mid \{\langle x, f_i\rangle\} \in F,\ x \in \gamma\}\}\mathcal{N}(0, |z|^2)\{Z\}\\ &\quad = \mathbf{P}\{x \in \Gamma \mid \{\langle x, f_i\rangle\} \in F,\ \{\langle x, g_j\rangle\} \in G\}\mathcal{N}(0, |z|^2)\{Z\}.\end{aligned}$$

We have verified (7.19) for sets A and C of special structure. The general verification of (7.19) is left to the reader as a simple exercise in measure theory. □

Examples of Gaussian measures.

PROBLEM 7.4. Let $\mathbf{X} = \mathbf{X}' = \mathbf{R}^1$ and $\mathbf{P} = \mathcal{N}(a, \sigma^2) \in \mathcal{G}(\mathbf{X})$. Construct the spaces $\mathbf{X}'_{\mathbf{P}}$, $\mathbf{X}'^{A}_{\mathbf{P}}$, and $\mathbf{X}'_{\mathbf{P},0}$. Define the operators I^* and I. Verify (7.14) by straightforward computation. Do the same for $\mathbf{X} = \mathbf{X}' = \mathbf{R}^n$, $n > 1$.

We present some examples of Gaussian measures on infinite-dimensional spaces.

EXAMPLE 7.1. Let T be an arbitrary set, $\mathbf{X} = \mathbf{R}^T$ the space of all functions on T, and $\mathbf{X}'$ the linear space spanned by the coordinate functionals $\pi_t(x)$, $t \in T$, given by $\pi_t(x) = x(t)$, $x \in \mathbf{X}$.

We treat $\mathbf{X}$ as the space of trajectories of a Gaussian process $\{x_t(\omega), t \in T\}$ and a Gaussian measure $\mathbf{P}$ on $\mathbf{X}$ as the distribution of this process. To specify $\mathbf{P}$, it suffices to specify the mean value (barycenter), that is, a function $a(\cdot) \in \mathbf{X}$ and a symmetric function $K\colon T \times T \to \mathbf{R}^1$ satisfying the positive semidefiniteness condition

$$\sum_{i,j=1}^{n} K(t_i, t_j) v_i v_j \geq 0$$

for all $n > 0$, $v \in \mathbf{R}^n$, and $t_1, \ldots, t_n \in T$.

Then it follows from the classical Kolmogorov theorem on consistent systems of finite-dimensional distributions ([**18**, Ch. 5, §5.1], [**55**, Ch. III, §4]) that there exists a unique measure $\mathbf{P}$ on $(\mathbf{X}, \mathfrak{C})$ such that the coordinate functional π_t has the distribution $\mathcal{N}(a(t), K(t,t))$ with respect to $\mathbf{P}$ and

$$\mathsf{E}_{\mathbf{P}}\big(\pi_s - a(s)\big)\big(\pi_t - a(t)\big) = K(s,t).$$

If T is not countable, then the Borel σ-algebra $\mathfrak{B}$ is wider than the cylinder σ-algebra $\mathfrak{C}$, and not each Gaussian measure is a Radon measure.

PROBLEM 7.5. Construct the operator I for Example 7.1. Verify that $II^*\pi_t = K(t, \cdot)$. By using this fact, prove that the function $K(t, \cdot)$ is an admissible translation for the measure $\mathbf{P}$.

EXAMPLE 7.2. Let us consider the most important special case of the previous example. Suppose that $\mathbf{X} = \mathbf{R}^\infty$ is the space of sequences and $\mathbf{X}'$ is the space of finite sequences. The duality between $\mathbf{X}$ and $\mathbf{X}'$ is defined in the standard way,

$$\langle x, x' \rangle = \sum_{t=1}^{\infty} x_t x'_t.$$

Let $a(t) \equiv 0$ and

$$K(t_1, t_2) = \begin{cases} 0 & \text{if } t_1 \neq t_2, \\ 1 & \text{if } t_1 = t_2. \end{cases}$$

Then the measure $\mathbf{P}$ is the distribution of a sequence of independent $\mathcal{N}(0,1)$-distributed random variables.

By applying Theorem 7.1, we obtain $\mathbf{H}_{\mathbf{P}} = \mathbf{l}^2$; that is, square summable sequences are admissible translations for $\mathbf{P}$.

EXAMPLE 7.3. Let $\mathbf{X}$ be a separable Hilbert space with inner product $\langle \cdot, \cdot \rangle$. We set $\mathbf{X}' = \mathbf{X}$ and use $\langle \cdot, \cdot \rangle$ to define the duality between $\mathbf{X}$ and $\mathbf{X}'$.

In this case $\mathfrak{C} = \mathfrak{B}$ (each open set of $\mathbf{X}$ can be represented as a countable union of open balls, an open ball as a countable union of closed balls, and a closed ball as a countable intersection of half-spaces belonging to $\mathfrak{C}$). To construct a Gaussian measure $\mathbf{P}$ on $(\mathbf{X}, \mathfrak{B})$, we choose several objects, namely, a vector $a \in \mathbf{X}$, a basis $\{e_j\}_{j=1}^{\infty}$ in $\mathbf{X}$, a sequence $\{\xi_j\}_{j=1}^{\infty}$ of independent random variables with distribution $\mathcal{N}(0,1)$, and a sequence $\{\sigma_j\}_{j=1}^{\infty}$ of nonnegative numbers such that $\sum_{j=1}^{\infty} \sigma_j^2 < \infty$.

We define $\mathbf{P}$ to be the distribution of the $\mathbf{X}$-valued random vector

$$\Xi = a + \sum_{j=1}^{\infty} \sigma_j \xi_j e_j.$$

This method permits one to obtain an *arbitrary* Gaussian measure on $\mathbf{X}$.

PROBLEM 7.6. Prove that $\mathbf{X}'_{\mathbf{P},0}$ is a Hilbert space and the functionals $f_j(\cdot) = \sigma_j^{-1}\langle \cdot - a, e_j\rangle$ form a basis in this space. Derive the following formulas for the operators I^*, I, and K:

$$I^*\Big(\sum_j x_j e_j\Big) = \sum_j \sigma_j x_j f_j, \qquad I\Big(\sum_j z_j f_j\Big) = \sum_j \sigma_j z_j e_j,$$
$$K\Big(\sum_j x_j e_j\Big) = \sum_j \sigma_j^2 x_j e_j.$$

Use the relation $\mathbf{H}_{\mathbf{P}} = I(\mathbf{X}'_{\mathbf{P},0})$ to obtain the following representation for the space of admissible translations of the measure $\mathbf{P}$:

$$\mathbf{H}_{\mathbf{P}} = \Big\{ l \in \mathbf{X} \,\Big|\, \sum_j |\langle l, e_j\rangle|^2 \sigma_j^{-2} < \infty \Big\} \subset \mathbf{X}.$$

The following theorem proves useful when one tries to find the space $\mathbf{H}_{\mathbf{P}}$ in various situations.

THEOREM 7.3. *Suppose that the correlation operator of the measure* $\mathbf{P}$ *admits a factorization* $K = JJ^*$, *where* J *is a linear operator that acts from a Hilbert space* $\mathbf{L}$ *into* $\mathbf{X}$ *and* J^* *is the adjoint of* J. *Then* $\mathbf{H}_{\mathbf{P}} = J(\mathbf{L})$.

REMARK. For $\mathbf{L} = \mathbf{X}'_{\mathbf{P},0}$ and $J = I$, where I is introduced in (7.10), we obtain just the first statement of Theorem 7.1.

PROOF. Let $\mathbf{L}_0$ be the closure of $J^*(\mathbf{X}')$ in $\mathbf{L}$, and let $\mathbf{L}_0^{\perp}$ be the orthogonal complement of $\mathbf{L}_0$. The mapping $\varkappa\colon I^*(x') \to J^*(x')$ is an isometry between $I^*(\mathbf{X}')$ and $J^*(\mathbf{X}')$; indeed,

$$(J^*x', J^*y')_{\mathbf{L}} = \langle JJ^*x', y'\rangle = \langle Kx', y'\rangle = \langle II^*x', y'\rangle = (I^*x', I^*y').$$

The isometry $\varkappa$ extends to an isometry between $\mathbf{X}'_{\mathbf{P},0}$ and $\mathbf{L}_0$. Let us show that $I = J\varkappa$. Indeed,

$$J\varkappa z = J\varkappa I^* x' = JJ^*x' = II^*x' = Iz$$

for $z = I^*x' \in \mathbf{X}'_{\mathbf{P},0}$. Hence, by Theorem 7.1 we have

$$\mathbf{H}_{\mathbf{P}} = I(\mathbf{X}'_{\mathbf{P},0}) = J\varkappa(\mathbf{X}'_{\mathbf{P},0}) = J(\mathbf{L}_0).$$

On the other hand, for $l \in \mathbf{L}_0^{\perp}$ and $x \in \mathbf{X}'$ we have $\langle Jl, x'\rangle = (l, J^*x')_{\mathbf{L}} = 0$. Therefore, $Jl = 0$ and $J(\mathbf{L}) = J(\mathbf{L}_0) = \mathbf{H}_{\mathbf{P}}$. □

We consider some examples to illustrate how this theorem can be used.

EXAMPLE 7.4. *A space with a reproducing kernel.* Let $\mathbf{P}$ be a Gaussian measure on $\mathbf{X} = \mathbf{R}^T$ with correlation function $K(s,t)$ (see Example 7.1). Let $\mathbf{L}$ be the Hilbert space with reproducing kernel K, that is, the space of functions on T obtained by completing the linear span of the functions $K_s = K(s,\cdot)$ with respect to the inner product $(K_s, K_t) = K(s,t)$.

We define an operator $J^*\colon \mathbf{X}' \to \mathbf{L}$ by setting $J^*(\pi_s) = K_s$. Then J is an embedding of $\mathbf{L}$ in $\mathbf{X}$, and $K = JJ^*$. By Theorem 7.3, the space $\mathbf{L}$ is just the space of admissible translations for $\mathbf{P}$ (cf. Problem 7.5).

EXAMPLE 7.5. Let $\mathbf{X}$ be the space of bounded measurable functions defined on a metric compactum T. We define the dual space $\mathbf{X}'$ as the space of finite signed measures $\mathbf{Z}(T)$. The duality is given by the formula

$$\langle f, \nu \rangle = \int_T f(t)\,\nu(dt), \qquad f \in \mathbf{X}, \quad \nu \in \mathbf{X}'.$$

Let $\mathbf{P} \in \mathcal{G}_0(\mathbf{X})$. The correlation operator $K\colon \mathbf{X}' \to \mathbf{X}$ is an integral operator of the form

$$K\nu(s) = \int_T K(s,t)\nu(dt), \tag{7.20}$$

where the kernel $K(s,t) = \int_{\mathbf{X}} f(s)f(t)\mathbf{P}(df)$ is a positive semidefinite function on T^2. Let $\mathbf{L}$ be the completion of the linear span of the functions

$$f_\nu(\cdot) = \int_T K(\cdot,t)\nu\,(dt), \qquad \nu \in X',$$

with respect to the inner product

$$(f_\nu, f_\mu)_L = \int_{T^2} K(s,t)\,\nu(ds)\mu(dt).$$

Then K can be treated as a linear operator from $\mathbf{X}'$ to $\mathbf{L}$, and the embedding $J\colon \mathbf{L} \to \mathbf{X}$ is the adjoint of K. Therefore, for $J^* = K$ we have $K = JJ^*$, and by Theorem 7.3, the admissible translations of the measure $\mathbf{P}$ are just the elements of $\mathbf{L}$.

EXAMPLE 7.6. *The Wiener field.* Let $\mathbf{P}$ be the centered Gaussian measure on $\mathbf{X} = \mathbf{C}(T)$, $T = [0,1]^d$, corresponding to the correlation function

$$K(s,t) = \prod_{i=1}^d \min(s_i, t_i), \qquad s,t \in [0,1]^d. \tag{7.21}$$

To describe admissible translations, we construct a factorization (7.20) of the operator K different from that considered in the preceding example. We set $\mathbf{L} = \mathbf{L}^2(T)$ and define operators $J\colon \mathbf{L} \to \mathbf{X}$ and $J^*\colon \mathbf{X}' \to \mathbf{L}$ by setting

$$\begin{aligned} Jh(s) &= \int_{[0,s]} h(t)\lambda^d\,(dt), \qquad [0,s] = \prod_{i=1}^d [0,s_i], \quad s \in T; \\ J^*\nu(t) &= \int_{T\setminus[0,t]} \nu\,(ds). \end{aligned} \tag{7.22}$$

PROBLEM 7.7. Verify that the operators J and J^* are adjoints of each other and $K = JJ^*$.

Thus, formula (7.22) provides the general form of admissible translations, and moreover, $\sigma_{\mathbf{P}}(Jh) = (\int_T |h|^2)^{-1}$ is the admissibility.

The one-parameter case ($d = 1$) is of particular importance. Here the measure $\mathbf{P}$ defined on $\mathbf{C}[0,1]$ is the distribution of the *Wiener process* (*Brownian motion*), that is, the Gaussian random process $w(t)$, $t \in [0,1]$, defined by the conditions $\mathsf{E}w(t) = 0$ and $\mathsf{E}w(s)w(t) = \min(s,t)$. This measure is called the *Wiener* measure. By applying formulas (7.22) and (7.14) to $\mathbf{P}$, we obtain the following classical result.

THEOREM 7.4. *Let $\mathbf{P}$ be the Wiener measure on $\mathbf{C}[0,1]$. For a function $l \in \mathbf{C}[0,1]$ to be an admissible translation of $\mathbf{P}$, it is necessary and sufficient that $l(0) = 0$, l is absolutely continuous, and $l' \in \mathbf{L}^2[0,1]$. If l is an admissible translation, then $\mathcal{I}(l) = \sigma_{\mathbf{P}}^{-2}(l) = \int_0^1 l'^2(s)\,ds$ and*

$$\frac{d\mathbf{P}^l}{d\mathbf{P}}(w) = \exp\left\{\int_0^1 l'(s)\,dw(s) - \frac{1}{2}\int_0^1 l'^2(s)\,ds\right\}. \tag{7.23}$$

EXAMPLE 7.7. *Semistable Gaussian processes* (*fractional Brownian motion*). Let us consider the centered Gaussian process $w_\alpha(t)$ on $\mathbf{R}^+$ with correlation function

$$K(s,t) = \frac{1}{2}[s^\alpha + t^\alpha - |s-t|^\alpha].$$

We can choose the parameter α in the interval $(0,2]$. Let $\mathbf{P}_\alpha$ be the Gaussian measure on $\mathbf{X} = \mathbf{C}[0,1]$ corresponding to the restriction of $w_\alpha(t)$ to the interval $[0,1]$. The correlation operator has the integral form (7.20). Straightforward calculations show that we have the factorization $K = JJ^*$, where $J\colon L = \mathbf{L}^2(\mathbf{R}^1) \to \mathbf{X}$ is the integral operator with kernel

$$j(s,t) = c_\beta\{(s-t)^{-\beta}\mathbf{1}_{[0,s]}(t) + [(s-t)^{-\beta} - (-t)^{-\beta}]\mathbf{1}_{(-\infty,0]}(t)\},$$

$$\beta = \frac{1-\alpha}{2}, \qquad c_\beta = \left\{\int_{-\infty}^0 [(1-r)^{-\beta} - (-r)^{-\beta}]^2\,dr + \int_0^1 (1-r)^{-2\beta}\,dr\right\}^{-1/2}.$$

Note that the operator $J^*\colon \mathbf{X}' \to \mathbf{L}^2(\mathbf{R}^1)$ is also an integral operator with kernel $j^*(s,t) = j(t,s)$.

Thus, the admissible translations for $\mathbf{P}_\alpha$ have the following general form

$$l(s) = c_\beta \int_{-\infty}^0 h(t)[(s-t)^{-\beta} - (-t)^{-\beta}]\,dt + c_\beta \int_0^s h(t)(s-t)^{-\beta}\,dt, \tag{7.24}$$

where $h \in \mathbf{L}^2(\mathbf{R}^1)$ is called the *fractional derivative of order* $1-\beta$ of the function l.

For $\alpha = 1$ and $\beta = 0$, we arrive at the distribution of the Wiener process, and (7.24) is reduced to the statement of Theorem 7.4.

Let us give (without proof) a very useful sufficient condition for the function l to be representable in the form (7.24). This is a differentiability condition in terms of the Fourier transform.

PROPOSITION 7.5. *Suppose that $l(0) = 0$ and the Fourier transform $\widehat{l}$ of the function l belongs to $\mathbf{L}^2(\mathbf{R}^1, |u|^{1+\alpha}\lambda(du))$. Then l can be represented in the form* (7.24) *and*

$$\int_{\mathbf{R}^1} |h|^2\,dt \le \int_{\mathbf{R}^1} |\widehat{l}(u)|^2 |u|^{1+\alpha}\,du.$$

EXAMPLE 7.8. *Stationary Gaussian processes.* Let us consider a Gaussian process (ξ_t), $t \in \mathbf{R}^1$, with zero mean, correlation function $K(s,t) = K(t-s)$, and spectral measure B,

$$K(s,t) = \int_{\mathbf{R}^1} e^{i(t-s)u}\,B(du).$$

Let $\mathbf{P}$ be the measure on $\mathbf{L}^2[0,1]$ corresponding to the restriction of ξ to $[0,1]$ (in this example, we deal with $\mathbf{L}^2$-spaces consisting of *complex-valued* functions). The correlation operator K of the measure $\mathbf{P}$, which is the integral operator with kernel $K(\cdot,\cdot)$, can be factored as follows: $K = JJ^*$, where (verify this!)

$$J\colon \mathbf{L}^2(\mathbf{R}^1, B) \to \mathbf{L}^2[0,1],$$
$$Jh(s) = \int_{\mathbf{R}^1} e^{-isu} h(u)\,B(du), \quad s \in [0,1],$$
$$J^*g(u) = \int_0^1 e^{isu} g(s)\,ds, \quad u \in \mathbf{R}^1.$$

Therefore, l is an admissible translation for $\mathbf{P}$ if and only if

$$l(s) = \int_{\mathbf{R}^1} e^{-ist} h(u)\,B(du) \tag{7.25}$$

for some $h \in \mathbf{L}^2(\mathbf{R}^1, B)$.

Admissible translations are of the simplest form if the process ξ has a spectral density, $B(du) = b(u)\,du$, and if

$$c_1(1+|u|)^{-2n} \le b(u) \le c_2(1+|u|)^{-2n}, \tag{7.26}$$

where n is an integer. In this case, equation (7.25) has a solution if and only if $l, l^{(1)}, \dots, l^{(n-1)}$ are absolutely continuous and $l^{(n)} \in \mathbf{L}^2[0,1]$, that is, l belongs to the Sobolev space $\mathbf{W}_2^n$. The measure of admissibility is equivalent to the reciprocal of the Sobolev norm:

$$\sigma_{\mathbf{P}}(l) \asymp \|l\|^{-1}_{\mathbf{W}_2^n[0,1]} = \left(\sum_{k=0}^n \|l^{(k)}\|^2_{\mathbf{L}^2[0,1]} \right)^{-1/2}.$$

For the proof of these statements, the reader is referred to [**88**, Ch. II, §3]. Note that the description of admissible translations remains the same if condition (7.26) is satisfied only at infinity rather than on the entire line. This follows from Problem 7.3 and the statement that the measures corresponding to stationary processes whose spectral densities coincide outside some neighborhood of zero are equivalent [**52**].

EXAMPLE 7.9. *Gaussian white noise.* Let T be a set, $\mathfrak{T}$ an algebra of subsets of T, $\mathbf{X}$ the linear space of simple finitely additive signed measures defined on $\mathfrak{T}$, and $\mathbf{X}'$ the space of simple $\mathfrak{T}$-measurable functions of the form

$$x'(t) = \sum_{i=1}^n c_i \mathbf{1}_{A_i}(t); \qquad c_i \in \mathbf{R}^1, \quad A_i \in \mathfrak{T}.$$

The duality between $\mathbf{X}$ and $\mathbf{X}'$ is defined by the relation

$$\langle x, x' \rangle = \sum_{i=1}^{n} c_i x\{A_i\} = \int_T x'(t)\, x(dt).$$

Let m be a countably additive measure on $(T, \mathfrak{T})$. The centered Gaussian measure $\mathbf{P}$ on $\mathbf{X}$ with correlation operator

$$\langle Kx', y' \rangle = \int_T x'(t) y'(t)\, m(dt)$$

is called the *Gaussian white noise* with variance m.

The signed measure of a given set A is $\mathcal{N}(0, m(A))$-distributed with respect to $\mathbf{P}$. The signed measures of disjoint sets are independent. Therefore, the white noise can be considered as the distribution of a random finitely additive signed measure with independent values.

PROBLEM 7.8. Prove that the core $\mathbf{H_P} \subset \mathbf{X}$ has the form

$$\mathbf{H_P} = \{r(t)\, m(dt) \in \mathbf{X} \mid r(\cdot) \in \mathbf{L}^2(T, \mathfrak{T}, m)\}.$$

§8. Smooth functionals

Spaces of smooth functionals and differential operators. Suppose that $\langle \mathbf{X}, \mathbf{X}' \rangle$ is a pair of spaces in duality, $\mathbf{P} \in \mathcal{G}_0(\mathbf{X})$ a centered Gaussian measure, $\mathbf{H_P} \subset \mathbf{X}$ the core of $\mathbf{P}$, and $f \colon \mathbf{X} \to \mathbf{R}^1$ a stochastic functional. We define the notion of smoothness of f with respect to $\mathbf{P}$, examine the smoothness of $\mathbf{P}$ itself, and derive a useful formula of integration by parts. Then we use these results to study the smoothness of distributions of functionals.

Recall that $\mathbf{H_P}$ is a separable Hilbert space with norm $|l|^2 = \mathcal{I}(l)$ defined in (7.17). Let $\mathbf{H} \subset \mathbf{H_P}$ be a finite-dimensional subspace. A functional f is said to be m times continuously differentiable along $\mathbf{H}$ (we write $f \in \mathbf{C}^m_{\mathbf{H}}(\mathbf{X})$) if for any $x \in \mathbf{X}$ the function $h \mapsto f(x+h)$, $h \in \mathbf{H}$, is m times continuously Fréchet differentiable on $\mathbf{H}$ with respect to the Hilbert structure inherited from $\mathbf{H_P}$. By $f^{(i)}_{\mathbf{H}}$ we denote the ith derivative at a point $x \in \mathbf{X}$. We treat this derivative as a polylinear form, $f^{(i)}_{\mathbf{H}} \in \mathcal{L}^i(\mathbf{H}, \mathbf{R}^1)$, and use the explicit notation

$$f^{(i)}_{\mathbf{H}}(x)[h_1, \ldots, h_i], \qquad x \in \mathbf{X}, \quad h_j \in \mathbf{H},$$

whenever necessary.

For the derivative, we use the Hilbert–Schmidt norm

$$\big|f^{(i)}_{\mathbf{H}}(x)\big|^2 = \sum_{\alpha} \big|f^{(i)}_{\mathbf{H}}[e_{\alpha_1}, \ldots, e_{\alpha_i}]\big|^2,$$

where $\{e_j\}$ is an orthonormal basis in $\mathbf{H}$.

Let us choose a multi-index $q = (q(0), \ldots, q(m))$, $q(j) \geq 1$. A functional $f \in \mathbf{L}^{q(0)}(\mathbf{X}, \mathbf{P})$ is said to be $\mathbf{H_P}$-*differentiable of order* q if for any increasing system of finite-dimensional subspaces $\mathbf{H}_n \nearrow \mathbf{H}$ and the corresponding orthogonal projections $Q_n \colon \mathbf{H_P} \to \mathbf{H}_n$, there exists a sequence of functionals $f_n \in \mathbf{C}^m_{\mathbf{H}_n}$ convergent to f in $\mathbf{L}^{q(0)}(\mathbf{X}, \mathbf{P})$ such that the sequence of mappings

$$x \to f^{(j)}_n(x)[Q_n(\cdot), \ldots, Q_n(\cdot)]$$

converges in $\mathbf{L}^{q(j)}(\mathbf{X}, \mathbf{P}, \mathcal{L}_2^j(\mathbf{H_P}, \mathbf{R}^1))$ for each $j = 1, 2, \ldots, m$.

The limits of these sequences are called the *derivatives of f along* $\mathbf{H_P}$ and denoted by $f^{(j)}$. These derivatives are Hilbert–Schmidt polylinear forms, $f^{(j)}(x) \in \mathcal{L}_2^j(\mathbf{H_P}, \mathbf{R}^1)$, and are independent of the choice of the sequences $\mathbf{H}_n$ and f_n (the proof is left to the reader). In particular, the first derivative $f^{(1)}$ can be treated as an element of the space $\mathbf{H_P} = \mathcal{L}_2^1(\mathbf{H_P}, \mathbf{R}^1)$.

By $\mathcal{W}^q(\mathbf{X}, \mathbf{P})$ we denote the space of q times $\mathbf{H_P}$-differentiable functionals. In some problems, only the *local* integrability of the derivatives is required. A functional f is said to belong to the class $\mathcal{W}^q_{\text{loc}}(\mathbf{X}, \mathbf{P})$ if for some sequences of open sets $A_n \nearrow \mathbf{X}$ and functionals $f_n \in \mathcal{W}^q(\mathbf{X}, \mathbf{P})$, we have $f_n = f$ on A_n for all n. One can verify that in this case the sequences of derivatives $f_n^{(j)}$ converge $\mathbf{P}$-almost everywhere. Just as above, we denote the corresponding limits by $f^{(j)}$.

In quite a similar way, one can define the spaces $\mathbf{C_H}(\mathbf{X}, B)$, $\mathcal{W}^q(\mathbf{X}, \mathbf{P}, B)$, and $\mathcal{W}^q_{\text{loc}}(\mathbf{X}, \mathbf{P}, B)$ of differentiable functionals with values in an arbitrary Hilbert space B.

PROBLEM 8.1. Suppose that $\mathbf{X}$ is equipped with a norm that makes $\mathbf{X}$ a Banach space and a functional f is m times Fréchet differentiable with respect to this norm. Prove that if $\|f^{(j)}(\cdot)\| \in \mathbf{L}^{q(j)}(\mathbf{X}, \mathbf{P})$, $0 \le j \le m$, then $f \in \mathcal{W}^q(\mathbf{X}, \mathbf{P})$.

Let us consider a measurable mapping $V\colon \mathbf{X} \to \mathbf{H_P}$, which will be called a *vector field* (there is a vector $V(x)$ assigned to each point $x \in \mathbf{X}$).

We define a differential operator D on the space $\mathcal{W}^{1,1}_{\text{loc}}(\mathbf{X}, \mathbf{P})$ by setting

$$Dg(x) = (g^{(1)}(x), V(x)), \qquad x \in \mathbf{X}. \tag{8.1}$$

Let us show how the operator D acts in the scale $\{\mathcal{W}^q\}$.

By $d \times m$ we denote the multi-index $(d, \ldots, d)$ of length m.

PROPOSITION 8.1. *Suppose that $d > 1$, $q = (q_0, \ldots, q_m)$ is a multi-index, and $s_i = (d^{-1} + q_{i+1}^{-1})^{-1} \ge 1$. Consider the multi-indices $s = (s_0, \ldots, s_{m-1})$ and $d \times m = (d, \ldots, d)$ of length m. Suppose that $V \in \mathcal{W}^{d\times m}(\mathbf{X}, \mathbf{P}, \mathbf{H_P})$. Then the following assertions hold:*

1) $D\colon \mathcal{W}^q(\mathbf{X}, \mathbf{P}) \to \mathcal{W}^s(\mathbf{X}, \mathbf{P})$;

2) *if $q_i \ge d/(d-i)$, $1 \le i \le m$, then the iterated differentiation operator*

$$D^m\colon \mathcal{W}^q(\mathbf{X}, \mathbf{P}) \to \mathbf{L}^1(\mathbf{X}, \mathbf{P})$$

is well defined;

3) *similar statements are valid for the spaces $\mathcal{W}^q_{\text{loc}}$.*

PROOF. Let us estimate Dg for $g \in \mathcal{W}^q(\mathbf{X}, \mathbf{P})$ in terms of moments. Using the Cauchy and Hölder inequalities, we obtain

$$\begin{aligned}\int_{\mathbf{X}} |Dg(x)|^{s_0}\, \mathbf{P}(dx) &\le \int_{\mathbf{X}} |g^{(1)}(x)|^{s_0} |V(x)|^{s_0}\, \mathbf{P}(dx) \\ &\le \Big(\int_{\mathbf{X}} |g^{(1)}(x)|^{q_1}\, \mathbf{P}(dx)\Big)^{(d-s_0)/d} \Big(\int_{\mathbf{X}} |V(x)|^{d}\, \mathbf{P}(dx)\Big)^{s_0/d}.\end{aligned}$$

Similar estimates can be obtained for the derivatives of the functional Dg; to this end, instead of the inequality $|Dg| \le |g^{(1)}|\,|V|$ we use the inequality

$$|Dg^{(i)}(x)| \le 2^i \big(\max_{1\le j\le i+1} |g^{(j)}(x)|\big)\big(\max_{0\le j\le i} |V^{(j)}(x)|\big).$$

As a result, we obtain $(Dg)^{(i)} \in \mathbf{L}^{s_i}(\mathbf{X},\mathbf{P})$, $0 \le i \le m-1$, and hence, $Dg \in \mathcal{W}^s(\mathbf{X},\mathbf{P})$.

The second statement is proved by applying the first statement m times. □

Integration by parts and smoothness of the Gaussian measure. Our immediate goal is to define the derivative of $\mathbf{P}$ with respect to D and derive the corresponding formula of integration by parts. First, we obtain a finite-dimensional result.

LEMMA 8.1. *Suppose that $V\colon \mathbf{R}^n \to \mathbf{R}^n$ is a $\mathbf{C}^1$ vector field, $\mathbf{P}$ is a standard Gaussian measure on $\mathbf{R}^n$, and $q \ge 1$. If $|V|, |V^{(1)}| \in \mathbf{L}^q(\mathbf{R}^n,\mathbf{P})$, then*

$$\int_{\mathbf{R}^n} \left((x,V(x)) - \operatorname{tr} V^{(1)}(x)\right) \mathbf{P}(dx) = 0. \tag{8.2}$$

PROOF. Let $n=1$. Integrating by parts, we obtain

$$\begin{aligned}\int_{\mathbf{R}^1} (x,V(x))\,\mathbf{P}(dx) &= \frac{1}{\sqrt{2\pi}}\int_{-\infty}^{\infty} xV(x)e^{-x^2/2}\,dx \\ &= \frac{1}{\sqrt{2\pi}}\int_{-\infty}^{\infty} V'(x)e^{-x^2/2}\,dx = \int_{\mathbf{R}^1} \operatorname{tr} V^{(1)}(x)\,\mathbf{P}(dx).\end{aligned}$$

By applying the one-dimensional result to the restrictions of the scalar projections of V on lines parallel to the coordinate axes, we obtain the multidimensional result. □

LEMMA 8.2. *Let $\mathbf{H} \subset \mathbf{H_P}$ be the n-dimensional subspace spanned by the orthonormal system $\{l_j = Kx'_j,\ 1 \le j \le n, x'_j \in \mathbf{X}'\}$, where K is the correlation operator of the measure $\mathbf{P}$. Let $V\colon \mathbf{X} \to \mathbf{H}$ be a vector field such that $V \in \mathbf{C}^1_{\mathbf{H}}(\mathbf{X},\mathbf{H}) \cap \mathcal{W}^{q,q}(\mathbf{X},\mathbf{P},\mathbf{H})$, $q > 1$. Then*

$$\int_X \left(\langle x, K^{-1}V(x)\rangle - \operatorname{tr} V^{(1)}(x)\right) \mathbf{P}(dx) = 0. \tag{8.3}$$

PROOF. First we note that the integrand is well defined for all $x \in \mathbf{X}$; if $V(x) = \sum_j v_j(x) l_j$, then

$$\langle x, K^{-1}V(x)\rangle = \sum_j v_j(x)\langle x, K^{-1}l_j\rangle = \sum_j v_j(x)\langle x, x'_j\rangle.$$

Moreover, for $x \in \mathbf{H}$ we have

$$\langle x, K^{-1}V(x)\rangle = (x, V(x))_{\mathbf{H}},$$

so that (8.3) is a generalization of (8.2). By applying (8.2) to each n-dimensional plane of the form $y + \mathbf{H}$ and then integrating with respect to y, we get (8.3). □

LEMMA 8.3. *If the assumptions of Lemma 8.2 are satisfied, then*

$$\int_{\mathbf{X}} \left(g^{(1)}(x), V(x)\right) \mathbf{P}(dx) = \int_{\mathbf{X}} g(x)\left[\langle x, K^{-1}V(x)\rangle - \operatorname{tr} V^{(1)}(x)\right] \mathbf{P}(dx) \tag{8.4}$$

for each functional $g \in \mathbf{C}^1_{\mathbf{H}}(\mathbf{X}) \cap \mathcal{W}^{q',q'}(\mathbf{X},\mathbf{P})$, $q' = q/(q-1)$.

PROOF. First, assume that g and $g^{(1)}$ are uniformly bounded on $\mathbf{X}$. Then the vector field $V_1(x) = g(x)V(x)$ satisfies the assumptions of Lemma 8.2. We apply (8.3) to V_1 and use the identity (verify it!)

$$\operatorname{tr} V_1^{(1)}(x) = g(x) \operatorname{tr} V^{(1)}(x) + (g^{(1)}(x), V(x)).$$

By substituting this expression into (8.3), we obtain (8.4). To prove the statement in the general case, one approximates the functional g by bounded functionals and then passes to the limit in (8.4). □

We can treat (8.4) as a formula of integration by parts. We set

$$\delta_V(x) = \operatorname{tr} V^{(1)}(x) - \langle x, K^{-1}V(x)\rangle \tag{8.5}$$

and define the derivative of $\mathbf{P}$ with respect to D by the formula

$$D\mathbf{P}(dx) = \delta_V(x)\,\mathbf{P}(dx). \tag{8.6}$$

Then we can rewrite (8.4) in the canonical form

$$\int_{\mathbf{X}} (Dg)(x)\,\mathbf{P}(dx) = -\int_{\mathbf{X}} g(x)(D\mathbf{P})\,(dx). \tag{8.7}$$

So far, this formula has been proved only for a finite-dimensional field V. In the general case it also holds, but the expression (8.5) for the density of the derivative of $\mathbf{P}$ cannot be interpreted literally, since each of the two terms on the right-hand side in (8.5) taken separately need not be well defined. A generalization to the infinite-dimensional case is given by the following theorem.

THEOREM 8.1. *Let d be an even number, $q = (d, d)$, $q' = \big(d/(d-1), d/(d-1)\big)$, and $V \in \mathcal{W}^q(\mathbf{X}, \mathbf{P}, \mathbf{H_P})$. Then there exists a function $\delta_V \in \mathbf{L}^d(\mathbf{X}, \mathbf{P})$ such that* (8.7) *holds for each functional $g \in \mathcal{W}^{q'}(\mathbf{X}, \mathbf{P})$ and the measure* (8.6).

PROOF. Let g_n and V_n be approximating sequences for g and V, respectively. By (8.4), we have

$$\int_{\mathbf{X}} (g_n^{(1)}(x), V_n(x))\,\mathbf{P}(dx) = -\int_{\mathbf{X}} g_n(x)\delta_n(x)\,\mathbf{P}(dx), \tag{8.8}$$

where $\delta_n(x) = \operatorname{tr} V_n^{(1)}(x) - \langle x, K^{-1}V_n(x)\rangle$. Obviously, the left-hand side of (8.8) converges to the left-hand side of (8.4). Let us show that δ_n is a Cauchy sequence in $\mathbf{L}^d(\mathbf{X}, \mathbf{P})$. To this end, we need the following estimate [**1**].

LEMMA 8.4. *Suppose that d is even, $V\colon \mathbf{R}^n \to \mathbf{R}^n$ is a continuously differentiable field, and $\mathbf{P}$ is the standard Gaussian measure on $\mathbf{R}^n$. Then there exists a constant $c(d)$, independent of n and V, such that*

$$\int_{\mathbf{R}^n} \big((x, V(x)) - \operatorname{tr} V^{(1)}(x)\big)^d\,\mathbf{P}(dx) \le c(d)\int_{\mathbf{R}^n} \big(|V(x)|^d + |V^{(1)}(x)|^d\big)\,\mathbf{P}(dx). \tag{8.9}$$

Since $\delta_n(x) - \delta_k(x) = \operatorname{tr} U(x) - (x, U(x))$ and $U = V_n - V_k$, it follows that δ_n is a Cauchy sequence in $\mathbf{L}^d(\mathbf{X}, \mathbf{P})$. We write $\delta_V = \lim_{n\to\infty} \delta_n$. Using the Hölder inequality, we can easily justify the passage to the limit on the right-hand side in (8.8). Thus, (8.8) yields (8.7) as $n \to \infty$. □

Let us generalize Theorem 8.1 to the case of iterated differentiation. Suppose that $m \geq 1$ and $d > m$ is even. For a vector field V, we define functions $a_0, \ldots, a_m : \mathbf{X} \to \mathbf{R}^1$ by setting

$$a_0 \equiv 1, \qquad a_{i+1} = (a_i^{(1)}, V) + a_i \delta_V, \tag{8.10}$$

where δ_V is defined in (8.5). The reason for introducing these functions is that

$$D^i\, \mathbf{P}(dx) = a_i(x)\, \mathbf{P}(dx), \tag{8.11}$$

which is a generalization of (8.6). Let us justify (8.11). Needless to say, the sequence a_i is not well defined by (8.10) for all fields V. But it is well defined at least for m times continuously differentiable fields. To state the theorem, we need spaces of measures differentiable with respect to D on some definite function classes S_j. These spaces, $\overline{\mathbf{W}}(D \mid S_0, \ldots, S_m)$, were defined in §6. For example, the statement of Theorem 8.1 means that

$$\mathbf{P} \in \overline{\mathbf{W}}^1\big(D \mid \mathbf{L}^1(\mathbf{X}, \mathbf{P}), \mathcal{W}^{\frac{d}{d-1} \times 2}(\mathbf{X}, \mathbf{P})\big).$$

THEOREM 8.2. *Let $m \geq 1$, and let $d > m$ be even. Consider the multi-indices $q = d \times (m+1)$ and $q_i = (d/(d-i)) \times (i+1)$. If $V \in \mathcal{W}^q(\mathbf{X}, \mathbf{P}, \mathbf{H_P})$, then $\mathbf{P} \in \overline{\mathbf{W}}^m\big(D \mid \mathbf{L}^1(\mathbf{X}, \mathbf{P}), \mathcal{W}^{q_1}(\mathbf{X}, \mathbf{P}), \ldots, \mathcal{W}^{q_m}(\mathbf{X}, \mathbf{P})\big)$. For each $i = 1, \ldots, m$, we can choose a representing measure for the functional $D^i\mathbf{P}$ in the form $a_i(x)\mathbf{P}(dx)$, where $a_i \in \mathbf{L}^{d/i}(\mathbf{X}, \mathbf{P})$.*

PROOF. For $i = 1$, everything has been proved in Theorem 8.1. For $i > 1$, the proof is similar. Instead of (8.8), we pass to the limit in the chain of relations

$$\begin{aligned}\int_{\mathbf{X}} D_n^i g_n(x)\, \mathbf{P}(dx) &= \cdots = (-1)^j \int_{\mathbf{X}} D_n^{i-j} g_n(x) a_{j,n}(x)\, \mathbf{P}(dx) \\ &= \cdots = (-1)^i \int_{\mathbf{X}} g_n(x) a_{i,n}(x)\, \mathbf{P}(dx).\end{aligned}$$

Here D_n is the differential operator generated by a finite-dimensional field V_n that approximates V, $a_{1,n}, \ldots, a_{m,n}$ is the sequence of the form (8.10) corresponding to the field V_n, and $g_n(x)$ is a smooth functional that approximates $g \in \mathcal{W}^{q_i}(\mathbf{X}, \mathbf{P})$.

To justify passing to the limit, we again use Lemma 8.4.

In the limit, we obtain the following formula of integration by parts:

$$\int_{\mathbf{X}} D^i g(x)\, \mathbf{P}(dx) = (-1)^i \int_{\mathbf{X}} g(x) a_i(x)\, \mathbf{P}(dx). \tag{8.12}$$

This relation just means that $a_i(x)\, \mathbf{P}(dx)$ is a representing measure for $D^i\mathbf{P}$.

Sometimes it is useful to express the density of the highest derivative of $\mathbf{P}$, that is, the function a_i, $i > 1$, in terms of the density $a_1 = \delta_V$ of the first derivative. It follows from (8.10) that a_i can be represented as a linear combination of the expressions

$$(\delta_V)^{\beta_1} (D\delta_V)^{\beta_2} \ldots (D^{i-1}\delta_V)^{\beta_i}, \qquad \beta_j \geq 0, \qquad \sum_{j=1}^{i} j\beta_j = i. \tag{8.13}$$

For example, $a_2 = \delta_V^2 + D\delta_V$ and $a_3 = \delta_V^3 + 3\delta_V(D\delta_V) + D^2\delta_V$. □

§9. Distributions of smooth functionals

Absolute continuity of distributions. Theorem 4.6 provides a general criterion for absolute continuity of the distribution of a stochastic functional in terms of admissible semigroups. Since a Gaussian measure has a large supply of admissible linear groups (7.13), we can specify the general theory developed in §4 for the Gaussian case.

Suppose that $\langle \mathbf{X}, \mathbf{X}' \rangle$ is a pair of spaces in duality, $\mathbf{P} \in \mathcal{G}(\mathbf{X})$, $\mathbf{H_P} \subset \mathbf{X}$ is the core of the Gaussian measure $\mathbf{P}$, and $\tau_{\mathbf{H}}$ is the discrete topology in $\mathbf{H_P}$. Let $B \in \mathfrak{B}_{\mathbf{X}}$. By (4.21), a functional $f \colon \mathbf{X} \to \mathbf{R}^1$ belongs to the class $\mathcal{M}(B, \mathbf{H_P})$ if for each $x \in B$ there exist a direction $l \in \mathbf{H_P}$ and a neighborhood $V \subset \mathbf{X}$ such that the function $y \to D_l f(y)$ is continuous on $V \cap B$ and $D_l f(x) \neq 0$.

PROPOSITION 9.1. *Suppose that $B \in \mathfrak{B}_{\mathbf{X}}$, $f \colon \mathbf{X} \to \mathbf{R}^1$ is a measurable functional, and $f \in \mathcal{M}(B, \mathbf{H_P})$. Then $\mathbf{P}_B f^{-1} \ll \lambda$. In particular, if $\mathbf{P}\{B\} = 1$, then the distribution of f is absolutely continuous.*

PROOF. Let us verify the assumptions of Theorem 4.6. Obviously, the admissible semigroups $\{G^l, l \in \mathbf{H_P}\}$ form a $(B, \mathbf{H_P})$-sufficient class of admissible mappings. Thus, the first assumption of the theorem is satisfied. By virtue of the remark following Theorem 4.6, the second assumption is satisfied and we need not verify it. The third assumption of Theorem 4.6 is included in Proposition 9.1. Finally, the fourth assumption is also satisfied, since $\mathbf{P}$ is a Radon measure. □

By way of example, let us consider the distribution of an integral functional. Suppose that (T, μ) is a standard Borel space with measure, $\mathbf{X} = \mathbf{L}^2(T, \mu)$, and $f \colon \mathbf{X} \to \mathbf{R}^1$ is an integral functional of the form

$$f(x) = \int_T q\big(x(t)\big)\, \mu(dt). \tag{9.1}$$

Suppose that the kernel q satisfies the local Lipschitz condition and the derivative q' exists, is continuous, and does not vanish on some set $S \subset \mathbf{R}^1$ with $\lambda\{\mathbf{R}^1 \setminus S\} = 0$.

Let $\mathbf{P} \in \mathcal{G}(\mathbf{X})$ satisfy the nondegeneracy condition

$$|I^* y'|^2 = \mathbf{D_P}\langle \cdot, y' \rangle > 0 \tag{9.2}$$

for any nonzero $y' \in \mathbf{X}' = \mathbf{X}$. We also assume that the measure $\mathbf{P}$ is concentrated on bounded functions, that is, $\mathbf{P}\{\mathbf{L}^\infty(T, \mu)\} = 1$.

PROPOSITION 9.2. *Under the above assumptions, $\mathbf{P} f^{-1} \ll \lambda$.*

PROOF. We set $B_0 = \big\{x \in \mathbf{X} \mid \mu\{t \mid x(t) \in S\} = \mu(T)\big\} \subset \mathbf{X}$. For each $M > 0$, we introduce the set $B = B(M) = \big\{x \in \mathbf{X} \mid \ \|x\|_\infty \le M\big\} \cap B_0$. We intend to apply Proposition 9.1 to B, f, and $\mathbf{P}$. To this end, we must verify that $f \in \mathcal{M}(B, \mathbf{H_P})$. Indeed, for $x \in B_0$ and $l \in \mathbf{H_P} \subset \mathbf{L}^\infty(T, \mu)$, we have

$$D_l f(x) = \int_T q'(x(t)) l(t)\, \mu(dt). \tag{9.3}$$

Using this expression and Lebesgue's dominated convergence theorem, we see that $D_l f(\cdot)$ is continuous. On the other hand, for $x \in B_0$ we have $q'(x(t)) \neq 0$ for μ-almost all $t \in T$. Therefore, the expression (9.3) is nonzero for some $l \in \mathbf{X}' = \mathbf{X}$. By Proposition 7.4, condition (9.2) implies that $\mathbf{H_P}$ is dense in $\mathbf{X}$; therefore, one can choose an $l \in \mathbf{H}_\mathbf{P}$ so that $D_l f(x) \neq 0$. Hence, $f \in \mathcal{M}(B, \mathbf{H}_\mathbf{P})$, and by Proposition 9.1, we have $\mathbf{P}_{B(M)} f^{-1} \ll \lambda$.

It remains to note that, by assumption,

$$\lim_{M\to\infty} \mathbf{P}\{B(M)\} = \mathbf{P}\{\mathbf{L}^\infty(T,\mu) \cap B_0\} = \mathbf{P}\{B_0\}.$$

But it follows from (9.2) that $\mathbf{P}\{B_0\} = 1$. Therefore, letting $M \to \infty$, we obtain $\mathbf{P} f^{-1} \ll \lambda$. □

Similar assertions can be proved for integral functionals on the space of continuous functions, for the integral functional with the kernel $q\big(x(t), t\big)$, and for an integral functional whose kernel is a smooth function only on a small part of the real axis (in this case, it is necessary to require that **P**-almost all sample functions pass through this part). For the corresponding results see [**34, 40, 58**].

PROBLEM 9.1. Let $\mathbf{X} = \mathbf{C}[0,1]$, and let $\mathbf{P}$ be a Wiener measure on $\mathbf{X}$. We define an integral functional f by setting

$$f(x) = \int_0^1 q\big(x(t)\big)\, dt.$$

Suppose that the derivative $q'(\cdot)$ exists, is continuous, and does not vanish on some open set whose closure contains zero. Prove that $\mathbf{P} f^{-1} \ll \lambda$.

Estimates for the distribution density. In Proposition 9.1, the assumption that the derivative of the functional is nondegenerate, that is, $D_l f(x) \neq 0$ for an appropriate l, is most important for the absolute continuity of the distribution. We develop this approach and obtain some estimates for the distribution density in terms of the probability of the sets on which the differential $D.f$ is close to zero. In particular, these bounds prove useful in finding conditions for the density to be bounded.

Suppose that $\mathbf{P} \in \mathcal{G}(\mathbf{X})$, $B \in \mathfrak{C}_\mathbf{X}$, $\{B_n\}$, $\{E_n\} \subset \mathfrak{C}_\mathbf{X}$ are two covers of the set B, $\{l_n\} \subset \mathbf{H}_\mathbf{P}$ is a sequence of admissible translations, $\sigma(l_n)$ is the admissibility (7.16) of l_n, and $f\colon \mathbf{X} \to \mathbf{R}^1$ is a measurable functional. We set

$$\delta_n = \inf_{x\in E_n} \big[\sup\{c \in \mathbf{R}^1 \mid x + cl_n \in E_n\} - \inf\{c \in \mathbf{R}^1 \mid x + cl_n \in E_n\}\big],$$

which means that δ_n is the minimum length of the interval cut by the set E_n on a straight line parallel to l_n.

Suppose that

1) $B_n \subset E_n$;

2) E_n cuts a convex set on each line parallel to l_n;

3) the functional f is monotone on B_n along l_n and its derivative satisfies the estimate

$$\inf_{x\in B_n} |D_{l_n} f(x)| \geq d_n. \tag{9.4}$$

THEOREM 9.1. *Under the above assumptions, we have*

$$\operatorname*{ess\,sup}_{v\in\mathbf{R}^1} \frac{d\mathbf{P}_B f^{-1}}{d\lambda}(v) \leq 2(\sqrt{e}+1) \sum_n d_n^{-1} \mathbf{P}\{E_n\} \times \max\big\{\delta_n^{-1}, \sigma(l_n)^{-1}(1 + |2\ln \mathbf{P}\{E_n\}|^{1/2})\big\}. \tag{9.5}$$

PROOF. Let Γ_n be the partition of $\mathbf{X}$ into lines parallel to l_n. By $\mathbf{P}_n$ and $\{\mathbf{P}^n_\gamma\}$ we denote the corresponding quotient measures and families of conditional measures. Since $B \subset \bigcup_n B_n$, it follows that

$$(9.6) \qquad \sup \frac{d\mathbf{P}_B f^{-1}}{d\lambda} \le \sum_n \sup \frac{d\mathbf{P}_{B_n} f^{-1}}{d\lambda} \le \sum_n \int_{\mathbf{X}/\Gamma_n} \sup \frac{d\mathbf{P}^n_{\gamma, B_n} f^{-1}}{d\lambda} \mathbf{P}_n(d\gamma).$$

By using formula (4.18) and the estimate (9.4), we see that the nth term in (9.6) does not exceed

$$(9.7) \qquad d_n^{-1} \int_{\mathbf{X}/\Gamma_n} \sup_{B_n \cap \pi^{-1}(\gamma)} \frac{d\mathbf{P}^n_\gamma}{d\lambda^n_\gamma} \mathbf{P}_n(d\gamma).$$

Here λ^n_γ is the invariant measure (4.4) on the straight line $\pi^{-1}(\gamma)$:

$$\lambda^n_\gamma\{y \mid y = x + cl_n,\ c \in [c_1, c_2]\} = c_2 - c_1.$$

To estimate the density of the conditional measure, we use the following elementary result.

LEMMA 9.1. *Suppose that $\mathcal{N} = \mathcal{N}(a, \sigma^2)$ is a Gaussian measure in $\mathbf{R}^1$, $p = d\mathcal{N}/d\lambda$, U is an interval in $\mathbf{R}^1$, and $\lambda(U) \ge \delta$. Set $\rho = \inf_{c \in U} |c - a|$. Then*

$$(9.8) \qquad \mathcal{N}\{U\} \le \exp\{-\rho^2/2\sigma^2\},$$

$$(9.9) \qquad \sup_{c \in U} p(c) \le 2\sqrt{e} \max\{\delta^{-1}, \sigma^{-1}, \rho\sigma^{-2}\} \mathcal{N}\{U\}.$$

We use this lemma in the following situation "embedded" in the space $\mathbf{X}$: $\mathcal{N} = \mathbf{P}^n_\gamma$, a_γ is the barycenter of the measure $\mathcal{N}$, and

$$U = \pi^{-1}(\gamma) \cap E_n, \qquad \sigma = \sigma(l_n), \qquad \delta = \delta_n,$$
$$\rho = \rho_\gamma = \inf\big\{|c|, c \mid a_\gamma + cl_n \in E_n\big\}.$$

By using (9.9), we obtain the upper bound

$$2\sqrt{e} d_n^{-1} \int_{\mathbf{X}/\Gamma_n} \max\{\delta_n^{-1}, \sigma(l_n)^{-1}, \rho_\gamma \sigma(l_n)^{-2}\} \mathbf{P}^n_\gamma\{E_n\}\, \mathbf{P}_n(d\gamma)$$

for (9.7).

Let us split the interval into two parts according to the conditions

$$\rho_\gamma \sigma(l_n)^{-1} \le |2 \ln \mathbf{P}\{E_n\}|^{1/2} + 1, \qquad \rho_\gamma \sigma(l_n)^{-1} > |2 \ln \mathbf{P}\{E_n\}|^{1/2} + 1.$$

In the first domain of integration, we use the total probability formula for $\mathbf{P}\{E_n\}$ and obtain the majorant

$$2\sqrt{e} d_n^{-1} \max\big\{\delta_n^{-1}, \sigma(l_n)^{-1}(|2 \ln \mathbf{P}\{E_n\}|^{1/2} + 1)\big\} \mathbf{P}\{E_n\}.$$

In the second domain of integration, we use the estimate (9.8) for $\mathbf{P}^n_\gamma\{E_n\}$ in the form

$$\mathbf{P}^n_\gamma\{E_n\} \le \exp\{-\rho_\gamma^2/2\sigma(l_n)^2\}.$$

We substitute this estimate into (9.7) and, since the function $v \to v\exp\{-v^2/2\}$ is monotone for $v \geq 1$, obtain the following majorant for (9.7):

$$2\sqrt{e}d_n^{-1} \max\{\delta_n^{-1}, \sigma(l_n)^{-1}(|2\ln \mathbf{P}\{E_n\}|^{1/2}+1)\}e^{-(|2\ln \mathbf{P}\{E_n\}|+1)/2}$$
$$= 2d_n^{-1} \max\{\delta_n^{-1}, \sigma(l_n)^{-1}(|2\ln \mathbf{P}\{E_n\}|^{1/2}+1)\}\mathbf{P}\{E_n\}.$$

The sum of these two estimates yields (9.5). □

To use Theorem 9.1, one must choose the cover B_n and the translations l_n so that the estimates of d_n in (9.4) and of $\sigma(l_n)$ are not too small. Then one must extend B_n so that condition 2) preceding Theorem 9.1 is satisfied. This extension is taken as E_n.

Numerous examples of calculating the admissibility of measures, given in §7, allow one to apply this scheme to a wide class of Gaussian measures.

PROBLEM 9.2. Let $\mathbf{X}$, $\mathbf{P}$, f, and q be the same as in Problem 9.1. Suppose that the kernel q is monotone on some intervals $(-a, 0)$ and $(0, a)$ and satisfies the estimate $|q'(r)| \geq b(r)$ for almost all $r \in (-a, a)$. What conditions on b guarantee the boundedness of the density of the distribution $\mathbf{P}f^{-1}$?

For the history of this problem, see the bibliographical notes in the end of the book. Sakhanenko [**93**] proved that the function $b(r) = |r|^{-11}\exp\{-\pi^2 r^{-2}/8\}$ guarantees the boundedness of the density, whereas this is not the case for $b(r) = |r|^{-9}\exp\{-\pi^2 r^{-2}/8\}o(1)$.

Smoothness of distributions. Suppose that $f\colon \mathbf{X} \to \mathbf{R}^1$ is a stochastic functional defined on a space $\mathbf{X}$ with a Gaussian measure $\mathbf{P} \in \mathcal{G}_0(\mathbf{X})$. We study the smoothness of the distribution of f with respect to $\mathbf{P}$. More precisely, we obtain the conditions under which the distribution $\mathbf{P}f^{-1}$ belongs to the class of measures $\overline{\mathbf{W}}_1^m$ introduced in §6. To verify whether a distribution is smooth, we need to apply the general Theorem 6.1 in the Gaussian situation. In §8, we introduced the tools, namely, the spaces $\mathcal{W}^q(\mathbf{X}, \mathbf{P})$ of smooth functionals and $\mathcal{W}^q(\mathbf{X}, \mathbf{P}, \mathbf{H_P})$ of smooth vector fields.

Thus, suppose that $V\colon \mathbf{X} \to \mathbf{H_P}$ is a vector field, D the differential operator (8.1) corresponding to V, and $\delta_V(x)$ the density of the measure $D\mathbf{P}$, defined in Theorem 8.1.

We use the notation $I(l, i)$ introduced in Theorem 6.1 for the set of nonnegative integer multi-indices $k = (k_1, \ldots, k_l)$ such that $\sum_{j=1}^{l} jk_j = l - i$. We write $|k| = k_1 + \cdots + k_l$.

THEOREM 9.2. *Suppose that m is a positive integer, $d > m+1$ is even, and the following three conditions are satisfied:*

1) *$V \in \mathcal{W}^{d\times(m+1)}(\mathbf{X}, \mathbf{P}, \mathbf{H_P})$ and $f \in \mathcal{W}^s(\mathbf{X}, \mathbf{P})$, where $s = (s_0, \ldots, s_{m+1})$ and $s_i = d/(d-i)$;*

2)

$$\mathbf{P}\{x \in \mathbf{X} \mid Df(x) = 0\} = 0; \tag{9.10}$$

3) *the integral*

$$\int_{\mathbf{X}} \big[|Df|^{-m-|k|} \times |D^2 f|^{k_1} \times \cdots \times |D^{m+1} f|^{k_m} \times |\delta_V|^{\beta_1} \times \cdots \times |D^{i-1}\delta_V|^{\beta_i}\big](x)\,\mathbf{P}(dx) \tag{9.11}$$

is finite for all $i = 0, \dots, m$ *and any multi-indices* $k \in I(m,i)$ *and* $\beta \in I(i,0)$.

Then $\mathbf{P}f^{-1} \in \overline{\mathbf{W}}_1^m$.

PROOF. Let us more precisely describe the sets of functions on which we consider the operator D. We introduce the multi-indices $q_i = (d/(d-i)) \times (i+1)$, $1 \le i \le m$, and we set $S_0 = \mathbf{L}^1(\mathbf{X}, \mathbf{P})$ and $S_i = \mathcal{W}^{q_i}(\mathbf{X}, \mathbf{P})$, $1 \le i \le m$. Then, by Theorem 8.2, the measure $\mathbf{P}$ is smooth, that is,

$$\mathbf{P} \in \overline{\mathbf{W}}^m(D \mid S_0, S_1, \dots, S_m).$$

This smoothness is one of the conditions under which Theorem 6.1 can be used. Next, we must verify that the integrals (6.2) are finite. This is a direct consequence of the fact that the integrals (9.11) are finite and that the measure $\mu^i = D^i\mathbf{P} = a_i(x)\mathbf{P}(dx)$ can be represented as a sum of the products (8.13). Obviously, the other assumptions of Theorem 6.1 are also satisfied. Therefore, $\mathbf{P}f^{-1} \in \overline{\mathbf{W}}_1^1$. □

The first assumption of Theorem 9.2 comprises the requirements imposed on the smoothness of f and V. The second assumption means that the derivative of f is nondegenerate along V. The third assumption combines quantitative requirements imposed on smoothness (the terms $|D^j f|$, $2 \le j \le m+1$), nondegeneracy (the factor $|Df|$ raised to a negative power), and compatibility of V with $\mathbf{P}$ (the factors including δ_V).

Note that, by Proposition 8.1 and Theorem 8.1, the first assumption of the theorem provides that all the expressions in (9.11) are well defined. Therefore, the third assumption is purely a moment condition.

To illustrate the above arguments, we consider the simplest special case $m = 1$.

COROLLARY. *Suppose* $V \in \mathcal{W}^{d,d}(\mathbf{X}, \mathbf{P}, \mathbf{H_P})$, $f \in \mathcal{W}^{1,d/(d-1),d/(d-2)}(\mathbf{X}, \mathbf{P})$, (9.10) *holds, and the integrals*

$$\int_{\mathbf{X}} \left[|Df|^{-2}|D^2 f|\right](x)\,\mathbf{P}(dx), \qquad \int_{\mathbf{X}} \left[|Df|^{-1}|\delta_V|\right](x)\,\mathbf{P}(dx)$$

are finite.

Then $\mathbf{P}f^{-1} \in \overline{\mathbf{W}}_1^1$. *Thus, the distribution density of the functional* f *is of bounded variation; in particular, the density is bounded.*

It follows from the corollary that, to apply Theorem 9.2, we must require the functional f to be at least two times differentiable along V.

To use Theorem 9.2 for the proof that the distribution of f is smooth, we must be able to construct an appropriate vector field V. If all moments of the $(m+1)$st derivative of f are defined, then V can be readily constructed. This is shown in the following theorem.

THEOREM 9.3. *Suppose that* $m \ge 1$, $f \in \bigcap_{d>1} \mathcal{W}^{d \times (m+2)}(\mathbf{X}, \mathbf{P})$, *and*

$$\mathsf{E}|f^{(1)}|^{-2m-\delta} = \int_{\mathbf{X}} |f^{(1)}(x)|^{-2m-\delta}\,\mathbf{P}(dx) < \infty \tag{9.12}$$

for some $\delta > 0$.

Then $\mathbf{P}f^{-1} \in \overline{\mathbf{W}}_1^m$.

PROOF. Let us verify that the assumptions of Theorem 9.2 are satisfied for the functional f and the vector field $V(x) = f^{(1)}(x)$. We see that the derivatives of f and V have the required smoothness and integrability. Furthermore, in view of the identities

$$Df(x) = \big(f^{(1)}(x), V(x)\big) = \big(f^{(1)}(x), f^{(1)}(x)\big) = |f^{(1)}(x)|^2,$$

(9.10) follows from (9.12).

Let us verify that the integrals (9.11) are finite. Under the assumptions of our theorem, it is convenient to couple all factors of the form $|D^j f|$, $2 \le j \le m+1$, with the compensating factors $|Df|^{-1}$, since the functionals $|D^j f|\,|Df|^{-1}$ have all moments. For example,

$$|D^2 f|\,|Df|^{-1} = 2\Big|f^{(2)}(x)\big[f^{(1)}(x), f^{(1)}(x)\big]\Big|\,|Df|^{-1} \le 2|f^{(2)}(x)|.$$

The factors $|D^j \delta_V|$, $0 \le j \le m-1$, also have all moments. It remains to consider the first factor. After the compensating factors have been canceled out, the first factor acquires the form

$$|Df|^{-m-|k|+\sum_j k_j} = |Df|^{-m} = |f^{(1)}|^{-2m}.$$

Therefore, the Hölder inequality and condition (9.12) ensure that (9.11) is finite. □

§10. Convexity and the isoperimetric property of the Gaussian measure

The main goal of this and the next section is to obtain some information about the distribution of a convex functional defined on a space with Gaussian measure. This problem is simplified by the fact that the Gaussian measure is, in a sense, convex. This property is closely related to the famous isoperimetric theorem stating that *of all sets of the same measure, a half-space has the least "surface area."* Following Ehrhard, who proposed a unified approach to all these problems, we first consider the properties of the standard Gaussian measure on $\mathbf{R}^m$, then extend the results to arbitrary Gaussian measures, and finally arrive at a qualitative description of the distribution of a convex functional.

Symmetrization of the Gaussian measure on $\mathbf{R}^m$. In this subsection, $\mathbf{P}_m$ stands for the *standard* Gaussian measure on $\mathbf{R}^m$, that is, the Gaussian measure with respect to which the functionals of the form $(\cdot, e)$, where $e \in \mathbf{R}^m$ is a unit vector, have the distribution $\mathcal{N}(0,1)$. By $\mathbf{P}_k$ we also denote the projection of the measure $\mathbf{P}_m$ on a k-dimensional linear subspace $\mathbf{L} \subset \mathbf{R}^m$ $(k \le m)$. By $\Pi(e,r) = \{x \in \mathbf{R}^m \mid (x,e) > r\}$, we denote half-spaces.

Suppose that $1 \le k \le m$, $\mathbf{L}$ is an $(n-k)$-dimensional subspace of $\mathbf{R}^m$ and $e \in \mathbf{R}^m$ is a unit vector orthogonal to $\mathbf{L}$. Let us define the *Gaussian k-symmetrization with respect to* $\mathbf{L}$ *in the direction* e. This is a mapping that assigns a set A' to each open or closed set $A \subset \mathbf{R}^m$. The set A' is defined as follows. Let $x \in \mathbf{L}$. Then the following properties hold:

1) if $\mathbf{P}_k\{A \cap (x + \mathbf{L}^\perp)\} = 0$, then $A' \cap (x + \mathbf{L}^\perp) = \varnothing$;

2) if $\mathbf{P}_k\{A \cap (x + \mathbf{L}^\perp)\} = 1$, then $A' \cap (x + \mathbf{L}^\perp) = x + \mathbf{L}^\perp$;

3) if $0 < \mathbf{P}_k\{A \cap (x + \mathbf{L}^\perp)\} < 1$, then either

$$A' \cap (x + \mathbf{L}^\perp) = \Pi(e, a) \cap (x + \mathbf{L}^\perp)$$

(if A is open), or

$$A' \cap (x + \mathbf{L}^\perp) = \Pi^-(e, a) \cap (x + \mathbf{L}^\perp)$$

(if A is closed).

Here the number a is determined from the equation

$$\mathbf{P}_k\{A \cap (x + \mathbf{L}^\perp)\} = \mathbf{P}_k\{\Pi(e, a) \cap (x + \mathbf{L}^\perp)\}.$$

We write $A' = S[A] = S(\mathbf{L}, e)[A]$ for the symmetrization.

The symmetrization of a set A can be described as follows. We take a plane $x + \mathbf{L}^\perp$ parallel to the k-dimensional space $\mathbf{L}^\perp$, calculate the Gaussian measure of the trace of A on $x + \mathbf{L}^\perp$, and replace this trace by the k-dimensional half-space of the same measure with boundary hyperplane orthogonal to e. The union of all the half-spaces thus constructed gives $S[A]$.

EXAMPLE 10.1. If $m = k = 1$, $\mathbf{L} = \{0\}$, and $e = -1$, then for each set A, its symmetrization is the ray $S(\mathbf{L}, e)[A] = \big(-\infty,\ \Phi^{-1}(\mathbf{P}_1\{A\})\big)$ of the same measure, where $\Phi(r) = \mathcal{N}(0,1)\{(-\infty, r]\}$.

If $e = 1$, then the symmetrization yields the ray

$$S(\mathbf{L}, e)[A] = \big(\Phi^{-1}(1 - \mathbf{P}_1\{A\}), +\infty\big)$$

directed rightward.

For $k = m > 1$ and $\mathbf{L} = \{0\}$, we obtain almost the same result: the symmetrization of an open set A is the half-space

$$S(\{0\}, e)[A] = \Pi\big(e, \Phi^{-1}(1 - \mathbf{P}_m\{A\})\big),$$

also directed rightward.

The reader will have no difficulty in verifying the simplest properties of symmetrization.

PROPOSITION 10.1. *Let $S = S(\mathbf{L}, e)$ be a Gaussian symmetrization. Then the following properties hold.*

(i) *If $A \subset B$, then $S[A] \subset S[B]$ (monotonicity).*

(ii) *If A_i is an increasing sequence of open sets and $A = \bigcup A_i$, then $S[A] = \bigcup_i S[A_i]$ (lower continuity).*

(iii) *$S(\mathbf{L}, e)[A^\complement] = S(\mathbf{L}, -e)[A]^\complement$ (compatibility with the operation of taking the complement).*

(iv) *$S[A]$ is invariant along the space $(\mathbf{L} + \mathcal{L}(e))^\perp$, and $S[A] + ce \subset S[A]$ for $c \geq 0$ (the invariance along $\mathbf{L}^\perp$).*

(v) *If $h \in \mathbf{L}$, then $S[A + h] = S[A] + h$. In particular, if $\mathbf{M} \subset \mathbf{L}$ is a linear subspace and A is invariant along $\mathbf{M}$, then $S[A]$ is also invariant along $\mathbf{M}$ (the invariance along $\mathbf{L}$).*

(vi) *If B is invariant under $\mathbf{L}^\perp$, that is, $B = B + \mathbf{L}^\perp$, then*

$$\mathbf{P}_m\{B \cap A\} = \mathbf{P}_m\{B \cap S[A]\}.$$

In particular, $\mathbf{P}_m\{A\} = \mathbf{P}_m\{S[A]\}$ (measure preservation).

The most remarkable property of symmetrization is that it reduces the surface of a set.

Let $\mathcal{D}_r$ be the closed ball of radius r centered at the origin, and let $A^r = A + \mathcal{D}_r$ be the r-neighborhood of a set A.

THEOREM 10.1. *Let $S = S(\mathbf{L}, e)$ be a k-symmetrization in $\mathbf{R}^m$. Then*

$$S[A^r] \supset S[A]^r \tag{10.1}$$

for any closed set A and any $r > 0$.

The inclusion (10.1) is called the *reduction of the surface* and S is said to *reduce the surface*. This terminology becomes clear if we pass to the measure $\mathbf{P}_m$ in (10.1) and use the property of measure preservation under symmetrization. Then we obtain

$$\mathbf{P}_m\{A^r\} = \mathbf{P}_m\{S[A^r]\} \geq \mathbf{P}_m\{S[A]^r\}, \qquad \mathbf{P}_m\{A\} = \mathbf{P}_m\{S[A]\}. \tag{10.2}$$

Hence,

$$\mathbf{P}_m\{A^r \setminus A\} \geq \mathbf{P}_m\{S[A]^r \setminus S[A]\},$$

that is, the Gaussian measure of the layer of width r around A is greater than that of the layer of the same width around $S[A]$.

Let us outline the basic steps in the proof of Theorem 10.1.

First, we prove (10.1) for the simplest symmetrizations ($k = 1$).

Then we represent a k-symmetrization via superpositions of 1-symmetrizations. The inclusion (10.1) readily extends from the components of the superposition to its result.

In particular, we cover the case of m-symmetrization, which transforms each set into a half-space.

By applying inequality (10.2) to m-symmetrization, we obtain the famous *isoperimetric property* of a half-space with respect to the Gaussian measure: *the measure of the r-neighborhood of a half-space is less than or equal to the measure of the r-neighborhood of any measurable set of the same measure as the half-space.*

We present the properties of Gaussian symmetrizations according to the scheme shown in Figure 1, on the next page.

PROOF OF THEOREM 10.1. We start with the case in which $m = 1$, $L = \{0\}$, $e = 1$, and A is a closed interval in $\mathbf{R}^1$. Let $q = \mathbf{P}_1\{A\}$ be the measure of A. Closed intervals of measure q form the one-parameter family $A_t = [t, s(t)]$, $t \in [-\infty, \Phi^{-1}(1-q)]$, where the function $t \to s(t)$ is determined by the condition $\mathbf{P}_1\{A_t\} = q$. Let us minimize the measure $q_r(t) = \mathbf{P}_1\{(A_t)^r\}$ with respect to t. Inequality (10.1) implies that q_r attains its minimum values at the extreme values of t, namely, $t_1 = -\infty$ and $t_2 = \Phi^{-1}(1-q)$, for which A_t becomes a ray. Note that for any t we have

$$S(L, -1)[A_t] = A_{t_1}, \qquad S(L, 1)[A_t] = A_{t_2}.$$

By differentiating q_r with respect to t, we obtain

$$q_r'(t) = p(t)\left[\frac{p(s(t)+r)}{p(s(t))} - \frac{p(t-r)}{p(t)}\right], \tag{10.3}$$

where p is the standard density (1.6) with $a = 0$ and $\sigma = 1$.

A simple analysis of (10.3) shows that $q_r(t)$ is a unimodal function with maximum at the point $t_0 = -s(t_0)$, which corresponds to the symmetric position of the

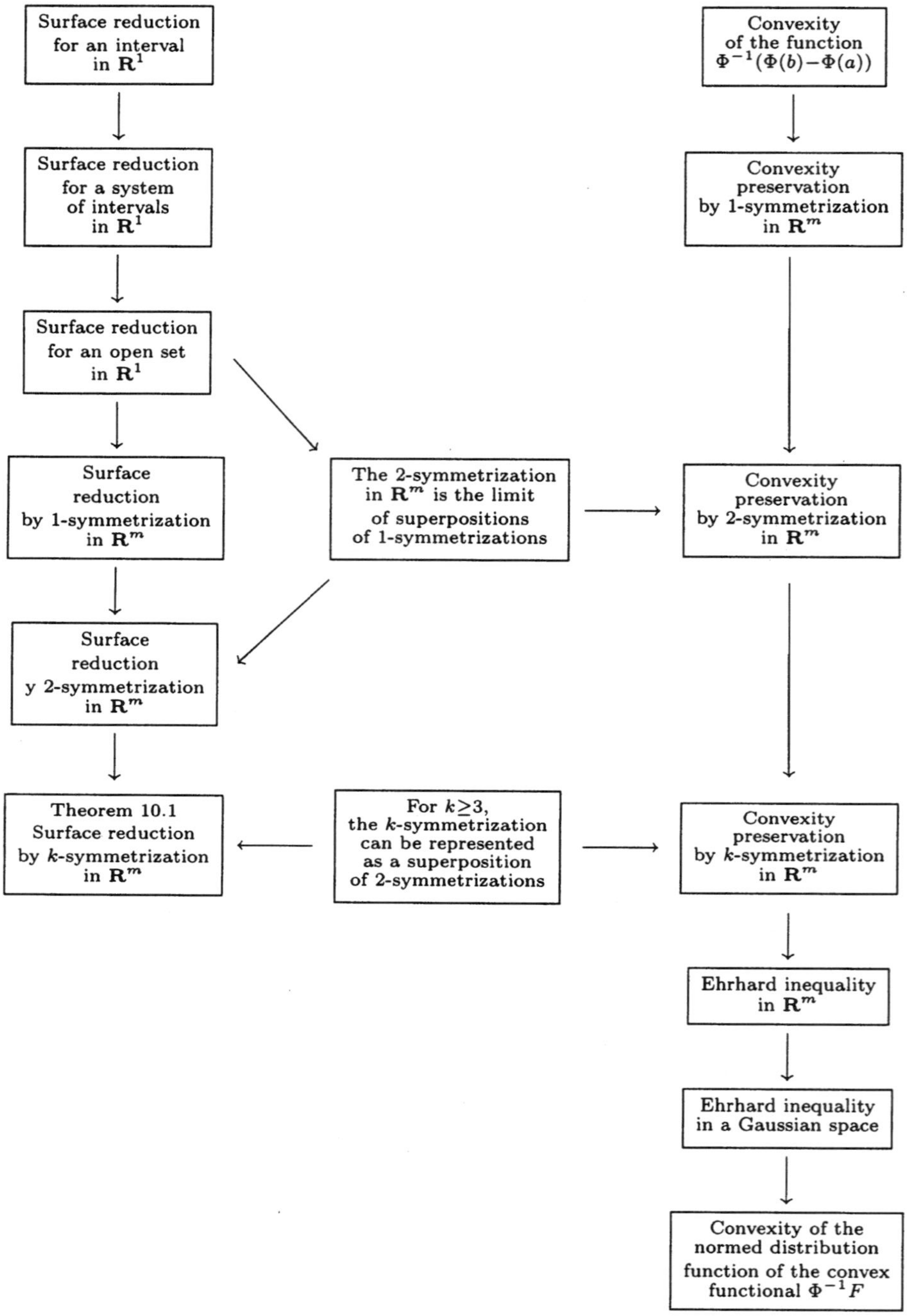
Surface reduction for an interval in $\mathbf{R}^1$
Surface reduction for a system of intervals in $\mathbf{R}^1$
Surface reduction for an open set in $\mathbf{R}^1$
Surface reduction by 1-symmetrization in $\mathbf{R}^m$
Surface reduction y 2-symmetrization in $\mathbf{R}^m$
Theorem 10.1 Surface reduction by k-symmetrization in $\mathbf{R}^m$
The 2-symmetrization in $\mathbf{R}^m$ is the limit of superpositions of 1-symmetrizations
For $k \geq 3$, the k-symmetrization can be represented as a superposition of 2-symmetrizations
Convexity of the function $\Phi^{-1}(\Phi(b)-\Phi(a))$
Convexity preservation by 1-symmetrization in $\mathbf{R}^m$
Convexity preservation by 2-symmetrization in $\mathbf{R}^m$
Convexity preservation by k-symmetrization in $\mathbf{R}^m$
Ehrhard inequality in $\mathbf{R}^m$
Ehrhard inequality in a Gaussian space
Convexity of the normed distribution function of the convex functional $\Phi^{-1}F$

FIGURE 1

segment A_t, and with minima at t_1 and t_2. Thus, we have proved formula (10.1) for a closed interval.

Now let $\{A_i,\ 1 \le i \le n+1\}$ be a family of disjoint segments numbered from the left to the right. By induction over n, we shall prove (10.1) for the set $A = \bigcup_{i=1}^{n+1} A_i$ and for both symmetrizations $S_+ = S(\{0\}, 1)$ and $S_- = S(\{0\}, -1)$ in $\mathbf{R}^1$. We have already proved the base clause ($n = 0$).

Without loss of generality, we can assume that $(A_j)^r \cap (A_k)^r = \varnothing$ for $j \neq k$; otherwise, we can prove the inclusion (10.1) for shorter intervals, for which the left-hand side of (10.1) remains unchanged and the right-hand side is smaller.

We set $J = \bigcup_{i=2}^{n} A_i$ and replace the extreme intervals A_1 and A_{n+1} by the rays $S_-[A_1]$ and $S_+[A_{n+1}]$ of the same measure so that

$$(10.4)\qquad \begin{aligned} S_-[A] &= S_-\Big[\bigcup_{i=1}^{n+1} A_i\Big] = S_-\big[S_-[A_1] \cup J \cup S_+[A_{n+1}]\big], \\ S_-[A^r] &= S_-\big[S_-[(A_1)^r] \cup J^r \cup S_+[(A_{n+1})^r]\big]. \end{aligned}$$

Then we substitute the already known inequalities

$$S_-[(A_1)^r] \supset S_-[A_1]^r, \qquad S_+[(A_{n+1})^r] \supset S_+[A_{n+1}]^r$$

into (10.4), thus obtaining

$$(10.5)\qquad S_-[A^r] \supset S_-\big[S_-[A_1]^r \cup J^r \cup S_+[A_{m+1}]^r\big].$$

Let $I = S_-[A_1]^r \cup J^r \cup S_+[A_{m+1}]^r$. The complement of I is a union of n closed intervals; therefore, by the induction assumption, the inclusion (10.1) is satisfied for $I^{\complement}$, and we obtain

$$\begin{aligned} S_-\big[S_-[A_1] \cup J \cup S_+[A_{m+1}]\big] &= S_-\big[((I^{\complement})^r)^{\complement}\big] \\ &= S_+\big[(I^{\complement})^r\big]^{\complement} \subset \big(S_+[I^{\complement}]^r\big)^{\complement} = \big(S_-[I]\big)^{-r}. \end{aligned}$$

Thus,

$$S_-\big[S_-[A_1] \cup J \cup S_+[A_{m+1}]\big]^r \subset S_-[I].$$

By comparing this inclusion with (10.4) and (10.5), we obtain

$$S_-[A^r] \supset S_-[I] \supset S_-\big[S_-[A_1] \cup J \cup S_+[A_{m+1}]\big]^r = S_-[A]^r.$$

Thus, we have proved (10.1) for unions of finitely many intervals in $\mathbf{R}^1$ (obviously, it makes no difference whether the intervals are open or closed). Since any open set in $\mathbf{R}^1$ is a union of a countably many intervals, we can use the fact that the symmetrization is continuous (Proposition 10.1 (ii)) to pass in (10.1) from finite systems of open intervals to any open set in $\mathbf{R}^1$.

To consider an arbitrary closed set A, we use the following standard argument. Let $\mathcal{D}^0_\varepsilon$ be the *open* ball of radius ε. Then

$$(10.6)\qquad \begin{aligned} &A = \bigcap_{\varepsilon>0} (A + \mathcal{D}^0_\varepsilon), \qquad A^r = \bigcap_{\varepsilon>0} (A + \mathcal{D}^0_\varepsilon)^r, \\ &S[A^r] = \bigcap_{\varepsilon>0} S[(A + \mathcal{D}^0_\varepsilon)^r] \supset \bigcap_{\varepsilon>0} S[A + \mathcal{D}^0_\varepsilon]^r = S[A]^r. \end{aligned}$$

Now let us prove (10.1) for 1-symmetrization in $\mathbf{R}^m$. Let $e \in \mathbf{R}^m$ be a unit vector, $\mathbf{L} = \mathcal{L}\{e\}^\perp$ a hyperplane, and $S = S(\mathbf{L}, e)$ the corresponding symmetrization.

For $x \in \mathbf{L}$ by $R_x = \{y : y = x + \rho e,\ \rho \in \mathbf{R}^1\}$ we denote the one-dimensional affine subspace in which the symmetrization is performed. We verify (10.1) in these subspaces, that is, for each x we prove the inclusion

$$S[A^r] \cap R_x \supset S[A]^r \cap R_x. \tag{10.7}$$

We rewrite the left-hand side as $S[A^r \cap R_x]$ and the right-hand side as

$$S[A]^r \cap R_x = \bigcup_{y \in \mathbf{L}} \left(S[A] \cap R_y\right)^r \cap R_x.$$

Each term in the union can be represented as the one-dimensional neighborhood

$$\left(S[A] \cap R_y\right)^r \cap R_x = S[A \cap R_y] + (\mathcal{D}_r \cap R_{x-y}). \tag{10.8}$$

By applying the one-dimensional inclusion (10.1) to the set $A \cap R_y + x - y \subset R_x$, we obtain

$$S[(A + R_y)^r \cap R_x] \supset S[A \cap R_y] + (\mathcal{D}_r \cap R_{x-y}).$$

Comparing this expression with (10.8), we get

$$S[(A \cap R_y)^r \cap R_x] \supset (S[A] \cap R_y)^r \cap R_x.$$

By taking the union over all $y \in \mathbf{L}$, we obtain

$$S[A^r \cap R_x] \supset S[A]^r \cap R_x,$$

which proves (10.7).

Thus, we have shown that 1-symmetrization reduces surfaces in $\mathbf{R}^m$.

PROBLEM 10.1. Let $k < m$. Assume that k-symmetrization reduces surfaces in $\mathbf{R}^k$. Prove that then k-symmetrization also reduces surfaces in $\mathbf{R}^m$. (For $k = 1$, this statement has just been proved; the argument can be transferred literally to the case $k > 1$.)

Now we must study 2-symmetrizations. We represent them as a limit of superpositions of 1-symmetrizations. We start with the space $\mathbf{R}^2$.

Let $\{e_j\}$, $j = 0, 1, 2, \dots$, be the sequence of unit vectors in $\mathbf{R}^2$ determined by the following relations:

$$e_0 = (0,1); \qquad e_1 = (1,0); \qquad (e_j,\, e_{j-1} + e_0) = 0; \qquad (e_j,\, e_0) < 0.$$

Then $e_n \to -e_0$ and for $j > 1$, the angle between e_j and $-e_0$ is half the angle between e_{j-1} and e_0.

Let $S_j = S(\mathcal{L}\{e_j\}^\perp, e_j)$ be the 1-symmetrization in $\mathbf{R}^2$ corresponding to e_j. We set $Q_j = S_j \circ S_{j-1} \circ \dots \circ S_0$.

LEMMA 10.1. *For any $j \geq 0$, $t, t' \in \mathbf{R}_+$, a closed set $A \subset \mathbf{R}^2$, and $x \in Q_j[A]$, we have*

$$x + te_0 + t'e_j \in Q_j[A]. \tag{10.9}$$

PROOF. We proceed by induction on j. For $j = 0$, the statement of the lemma follows from the invariance of symmetrization (see Proposition 10.1 (iv)). Let us pass from j to $j + 1$. We write $h_j = e_j + e_0$. For $r > 0$, we construct the segment

$$\Delta_{j,r} = \{y : y = \lambda r e_0 + (1 - \lambda) r e_j,\ 0 \le \lambda \le 1\}.$$

Consider the line $R_\alpha = \{y \mid y = \alpha h_j + \rho e_{j+1},\ \rho \in \mathbf{R}^1\}$, which is one of the subspaces in which the symmetrization S_{j+1} is performed. The line R_α contains the set $Q_j[A] \cap R_\alpha$ obtained as a result of the previous symmetrizations. From all points of this set, we draw cones with generators e_0, e_j, thus obtaining the new set

$$B = \{y : y = x + te_0 + t'e_j;\ x \in Q_j[A] \cap R_\alpha;\ t, t' \in \mathbf{R}_+\} \subset Q_j[A].$$

Let us find the traces of B on the lines R_β with $\beta > \alpha$. The vectors e_0 and e_j form the same angles with e_{j+1}. Therefore, $B \cap R_\beta$ is a translated neighborhood of the original set $Q_j[A] \cap R_\alpha$:

$$B \cap R_\beta = (Q_j[A] \cap R_\alpha) + \Delta_{j,r} + (\beta - \alpha) h_j.$$

By applying the inclusion (10.1) to the one-dimensional set $B \cap R_\beta$ inside R_β, we obtain

$$\begin{aligned} S_{j+1}[B \cap R_\beta] &\supset S_{j+1}\big[Q_j[A] \cap R_\alpha + (\beta - \alpha) h_j\big] + \Delta_{j,r} \\ &= \{u : u = x + te_0 + t'e_j;\ x \in S_{j+1}\big[Q_j[A] \cap R_\alpha\big];\ t, t' \in \mathbf{R}_+\} \cap R_\beta. \end{aligned}$$

Now we take the union over $\beta \ge \alpha$, thus obtaining the inclusion

$$S_{j+1}[B] \supset \{y \mid y = x + te_0 + t'e_j;\ x \in S\big[Q_j[A] \cap R_\alpha\big];\ t, t' \in \mathbf{R}_+\}.$$

In particular, for $x \in S_{j+1}\big[Q_j[A] \cap R_\alpha\big] = Q_{j+1}[A] \cap R_\alpha$ we have

$$y = x + te_0 + t'e_j \in S_{j+1}[B] \subset S_{j+1} \circ Q_j[A] = Q_{j+1}[A]. \quad \square$$

Lemma 10.1 has the following geometric interpretation: for each $x \in Q_j[A]$, the set $Q_j[A]$ contains the cone with vertex x and generators e_0 and e_j. Since the angle between e_0 and e_j tends to π, we see that for large j the set $Q_j[A]$ looks like a half-plane whose boundary is orthogonal to e_1. This is just the half-plane obtained by applying the 2-symmetrization $Q = S(\{0\}, e_1)$ to A. The relation

$$Q[A] = \lim_{j \to \infty} Q_j[A]$$

suggests that it is possible to pass from 1-symmetrization to 2-symmetrization. The following lemma contains the precise statement.

LEMMA 10.2. *Let Q_j, Q be the above-defined transformations of subsets of $\mathbf{R}^2$. Then for any $\varepsilon \in (0, 1/2)$ there exists a large number $\rho(\varepsilon)$ such that*

$$\lim_{j \to \infty} Q_j[A] \cap \mathcal{D}_r = Q[A] \cap \mathcal{D}_r \tag{10.10}$$

for $r > \rho(\varepsilon)$ uniformly with respect to a closed set A with $\varepsilon \le \mathbf{P}_2\{A\} \le 1 - \varepsilon$.

The convergence of sets is understood in the Hausdorff metric

$$d(A, B) = \sup_{c \in \mathbf{R}^2} \left| \inf_{b \in B} |c - b| - \inf_{a \in A} |c - a| \right|.$$

PROBLEM 10.2. Derive this lemma from the previous lemma.

By using the inclusion (10.1), which has already been proved for S_j, we obtain

$$Q_j[A]^r = S_j\big[Q_{j-1}[A]\big]^r \subset S_j\big[Q_{j-1}[A]^r\big] \subset \cdots \subset S_j \circ S_{j-1} \circ \cdots \circ S_0[A^r] = Q_j[A^r].$$

By passing to the limit with respect to j and using (10.10), we arrive at the inclusion

$$Q[A]^r \subset Q[A^r].$$

This proves (10.1) for 2-symmetrizations in $\mathbf{R}^2$.

PROBLEM 10.3. Prove (10.1) for an arbitrary 2-symmetrization in $\mathbf{R}^m$. Use the result of Problem 10.1.

The following two lemmas allow us to pass from low- to arbitrary-order symmetrizations.

LEMMA 10.3. *Let $\mathbf{M}_1$, $\mathbf{M}_2$, and $\mathbf{M}_3$ be pairwise orthogonal subspaces of $\mathbf{R}^m$. Set $\mathbf{L}_1 = \mathbf{M}_1 + \mathbf{M}_2$ and $\mathbf{L}_2 = \mathbf{M}_2 + \mathbf{M}_3$. Then the Gaussian symmetrizations $S_1 = S(\mathbf{L}_1, e)$ and $S_2 = S(\mathbf{L}_2, e)$ satisfy*

$$S_1 \circ S_2 = S_2 \circ S_1 = S(\mathbf{M}_2, e).$$

PROOF. Let $\mathbf{H} = (\mathbf{M}_1 + \mathbf{M}_2 + \mathbf{M}_3 + \mathcal{L}\{e\})^\perp$. The invariance of symmetrizations (Proposition 10.1 (iv)) implies that

$$S_1[A] = S_1[A] + (\mathbf{L}_1 + \mathcal{L}\{e\})^\perp = S_1[A] + \mathbf{M}_3 + \mathbf{H},$$
$$S_2[A] = S_2[A] + \mathbf{M}_1 + \mathbf{H}$$

for any closed set A. Since $\mathbf{M}_1 \subset \mathbf{L}_1$, it follows from Proposition 10.1 (v) that

$$S_1 \circ S_2[A] = S_1 \circ S_2[A] + \mathbf{M}_1.$$

Consequently,

$$\begin{aligned} S_1 \circ S_2[A] &= S_1 \circ S_2[A] + \mathbf{M}_3 + \mathbf{H} = \big(S_1 \circ S_2[A] + \mathbf{M}_1\big) + \mathbf{M}_3 + \mathbf{H} \\ &= S_1 \circ S_2[A] + (\mathbf{M}_1 + \mathbf{M}_3 + \mathbf{H}) = S_1 \circ S_2[A] + (\mathbf{M}_2 + \mathcal{L}\{e\})^\perp. \end{aligned}$$

Thus, the set $S_1 \circ S_2[A]$ is invariant with respect to $(\mathbf{M}_2 + \mathcal{L}\{e\})^\perp$. By definition, the same is true of $S(\mathbf{M}_2, e)[A]$. Moreover, both sets are invariant under translations along e (see Proposition 10.1 (iv)). Therefore, the intersection of either set with the affine subspace $\mathbf{N}_x = x + \mathbf{M}_2^\perp$ is a half-space. Let us compare the measures of these half-spaces. By Proposition 10.1 (vi), we have

$$\begin{aligned} \mathbf{P}_k\{S_1 \circ S_2[A] \cap \mathbf{N}_x\} &= \mathbf{P}_k\{S_1[A] \cap \mathbf{N}_x\} = \mathbf{P}_k\{A \cap \mathbf{N}_x\} \\ &= \mathbf{P}_k\{S(\mathbf{M}_2, e)[A] \cap \mathbf{N}_x\}, \qquad k = \dim \mathbf{N}_x. \end{aligned}$$

Since the measures coincide, it follows that the half-spaces also coincide,

$$S_1 \circ S_2[A] \cap \mathbf{N}_x = S(\mathbf{M}_2, e)[A] \cap \mathbf{N}_x.$$

By taking the union over $x \in \mathbf{M}_2$, we obtain

$$S_1 \circ S_2[A] = S(\mathbf{M}_2, e)[A]. \quad \square$$

LEMMA 10.4. *Let $Q = S(\mathbf{L}, e)$ be a k-symmetrization in $\mathbf{R}^m$ for $m \geq 3$ and $k \geq 2$. Then there exist 2-symmetrizations $Q_1, Q_2, \dots, Q_{k-1}$ such that*

$$Q = Q_1 \circ Q_2 \circ \cdots \circ Q_{k-1}. \tag{10.11}$$

PROOF. Let $h \in (\mathbf{L} + \mathcal{L}\{e\})^{\perp}$ be a unit vector. We set $\mathbf{L}_1 = (\mathcal{L}\{h\} + \mathcal{L}\{e\})^{\perp}$ and $\mathbf{L}_2 = \mathbf{L} + \mathcal{L}\{h\}$. Then the assumptions of Lemma 10.3 are satisfied for $\mathbf{M}_1 = (\mathbf{L} + \mathcal{L}\{h\} + \mathcal{L}\{e\})^{\perp}$, $\mathbf{M}_2 = \mathbf{L}$, and $\mathbf{M}_3 = \mathcal{L}\{h\}$. Hence,

$$Q = S(\mathbf{L}_1, e) \circ S(\mathbf{L}_2, e).$$

Note that $Q_1 = S(\mathbf{L}_1, e)$ is a 2-symmetrization, whereas $S(\mathbf{L}_2, e)$ is a $(k-1)$-symmetrization.

In a similar way, we can represent $S(\mathbf{L}_2, e)$ as a superposition, etc. At each stage, we extract a new factor Q_j so that the order of the remaining symmetrization decreases until it becomes equal to 2. □

Now we can prove Theorem 10.1. We have already proved this theorem for all symmetrizations in $\mathbf{R}^1$ and $\mathbf{R}^2$, as well as for k-symmetrizations in $\mathbf{R}^m$, $k = 1, 2$ (see Problem 10.3). Let us consider the symmetrization Q of order $k \geq 3$ and represent it in the form (10.11). Then we have

$$\begin{aligned} Q[A^r] &= Q_1 \circ Q_2 \circ \cdots \circ Q_{k-1}[A^r] \supset Q_1 \circ \cdots \circ Q_{k-2}\big[Q_{k-1}[A]^r\big] \\ &\supset \cdots \supset Q_1\big[(Q_2 \circ \cdots \circ Q_{k-1}[A])^r\big] \supset Q_1\big[Q_2 \circ \cdots \circ Q_{k-1}[A]\big]^r = \big(Q[A]\big)^r. \end{aligned}$$

The proof of Theorem 10.1 is complete. □

Convexity and the isoperimetric property of the Gaussian measure on $\mathbf{R}^m$.

THEOREM 10.2 (isoperimetric inequality). *Let $\mathbf{P}_m$ be the standard Gaussian measure on $\mathbf{R}^m$, and let $A \subset \mathbf{R}^m$ be a Borel subset. Then*

$$\Phi^{-1}\big(\mathbf{P}_m\{A^r\}\big) \geq \Phi^{-1}\big(\mathbf{P}_m\{A\}\big) + r \tag{10.12}$$

for any $r > 0$.

PROOF. First, suppose that A is closed. We take any m-symmetrization S and apply Theorem 10.1 in the form (10.2). Then the sets $S[A]$ and $S[A]^r$ are half-spaces, and since the symmetrization preserves the measure, we have

$$\mathbf{P}_m\{S[A]\} = \mathbf{P}_m\{A\}.$$

Hence,

$$\mathbf{P}_m\{S[A]^r\} = \Phi\big(\Phi^{-1}(\mathbf{P}_m\{A\}) + r\big).$$

We substitute this relation into (10.2) and obtain (10.12), since Φ is monotone.

To prove (10.12) for an arbitrary Borel subset A, we use the regularity of the measure $\mathbf{P}_m$ and approximate A by closed sets $A_i \subset A$, $\mathbf{P}_m\{A_i\} \to \mathbf{P}_m\{A\}$. By applying (10.12) to A_i and taking the limit as $i \to \infty$, we prove the theorem. □

THEOREM 10.3 (convexity is preserved under symmetrization). *Let A be an open or closed subset of $\mathbf{R}^m$, and let S be a Gaussian symmetrization in $\mathbf{R}^m$. If A is convex, then so is $S[A]$.*

PROOF. First, we consider the 1-symmetrization $S = S(\mathbf{L}, e)$, where e is a unit vector in $\mathbf{R}^m$ and $\mathbf{L} = \mathcal{L}\{e\}^\perp$. For $x \in \mathbf{L}$, we set $R_x = x + \mathcal{L}\{e\}$. The set $A \cap R_x$ is a ray or a segment, and the set $S[A] \cap R_x$ is a ray. We must verify that

$$S[A] \cap R_{\gamma x_1 + (1-\gamma)x_2} \supset \gamma\big(S[A] \cap R_{x_1}\big) + (1-\gamma)\big(S[A] \cap R_{x_2}\big) \tag{10.13}$$

for $x_1, x_2 \in \mathbf{L}$ and $\gamma \in [0,1]$. This inclusion is equivalent to the convexity of $S[A]$. Let us introduce some notation for the related segments and rays. Let

$$\begin{aligned} A \cap R_{x_j} &= x_j + [a_j, b_j]e, \qquad j = 1,2,\\ S[A] \cap R_{x_j} &= x_j + [c_j, \infty]e, \qquad j = 1,2. \end{aligned}$$

By the definition of symmetrization, we have $\Phi(b_j) - \Phi(a_j) = 1 - \Phi(c_j)$.

In this notation, (10.13) is reduced to the following inequality for the function Φ:

$$\begin{aligned} \Phi^{-1}\big(\Phi(\gamma b_1 + (1-\gamma)b_2) - \Phi(\gamma a_1 + (1-\gamma)a_2)\big) &\geq -\gamma_1 c_1 - (1-\gamma)c_2 \\ &= \gamma\Phi^{-1}\big(\Phi(b_1) - \Phi(a_1)\big) + (1-\gamma)\Phi^{-1}\big(\Phi(b_2) - \Phi(a_2)\big). \end{aligned} \tag{10.14}$$

In turn, (10.14) is an immediate consequence of the following simple lemma, which we present without proof.

LEMMA 10.5. *The function $g(a,b) = \Phi^{-1}\big(\Phi(b) - \Phi(a)\big)$ is concave in the domain $\{a < b\}$.*

Thus, we have proved Theorem 10.3 for 1-symmetrizations. For 2-symmetrizations, we prove the theorem by passing to the limit, as described in Lemma 10.2. Finally, for higher-order symmetrizations, the convexity preservation follows from Lemma 10.4. □

The following theorem gives precise meaning to the statement that the Gaussian measure is concave.

THEOREM 10.4 (the Ehrhard inequality). *Suppose that $\mathbf{P}_m$ is the standard Gaussian measure on $\mathbf{R}^m$, $A, B \subset \mathbf{R}^m$ are nonempty convex subsets, and $\gamma \in [0,1]$. Then*

$$\Phi^{-1}\big(\mathbf{P}_m\{\gamma A + (1-\gamma)B\}\big) \geq \gamma\Phi^{-1}\big(\mathbf{P}_m\{A\}\big) + (1-\gamma)\Phi^{-1}\big(\mathbf{P}_m\{B\}\big). \tag{10.15}$$

PROOF. In $\mathbf{R}^{m+1}$, consider the subsets $A' = A \times \{1\}$ and $B' = B \times \{0\}$. The set

$$C = \{y \in \mathbf{R}^{m+1} \mid y = \beta a + (1-\beta)b;\ a \in A', b \in B', \beta \in [0,1]\}$$

is concave, and then

$$C \cap \big(\mathbf{R}^m \times \{\gamma\}\big) = \big(\gamma A + (1-\gamma)B\big) \times \{\gamma\}.$$

First, assume that A and B are open.

Consider the m-symmetrization $S(\mathcal{L}\{e\}, h)$, where $e = (0, 0, \dots, 0, 1) \in \mathbf{R}^{m+1}$ and h is a unit vector orthogonal to e. By Theorem 10.3, the convexity of the set $S[C]$ is equivalent to the concavity of the function

$$f(\gamma) = \Phi^{-1}\big(\mathbf{P}_m\{C \cap (\mathbf{R}^m \times \{\gamma\})\}\big) = \Phi^{-1}\big(\mathbf{P}_m\{\gamma A + (1-\gamma)B\}\big)$$

on the interval $[0,1]$. Hence,

$$f(\gamma) \geq \gamma f(1) + (1-\gamma) f(0),$$

which coincides with the desired inequality (10.15).

Now let A and B be arbitrary convex sets. Then we apply our theorem to their interiors A^0 and B^0 and use the fact that $\mathbf{P}_m\{A^0\} = \mathbf{P}_m\{A\}$ and $\mathbf{P}_m\{B^0\} = \mathbf{P}_m\{B\}$. An arbitrary convex set need not be a Borel set, but it necessarily belongs to the complement of $\mathfrak{B}$ with respect to the measure $\mathbf{P}_m$, so that all measures involved in (10.15) are well defined. □

Let us discuss how one can extend the statements of Theorems 10.2 and 10.4 to the case of an arbitrary Gaussian measure on $\mathbf{R}^m$.

PROBLEM 10.4. Let $\mathbf{P}$ be a Gaussian measure on $\mathbf{R}^m$ with mean value a and correlation operator $K\colon \mathbf{R}^m \to \mathbf{R}^m$. Then $\mathbf{P}$ is the image of the standard Gaussian measure $\mathbf{P}_m$ under the affine transformation

$$G(x) = a + K^{1/2}x.$$

To generalize the isoperimetric inequality (10.12), we give a new interpretation of the notion of a neighborhood of a set. Whereas for the standard Gaussian measure we have $A^r = A + \mathcal{D}_r$, for an arbitrary measure $\mathbf{P}$ we introduce the *ellipsoid of concentration* $\mathrm{Ell}_r = K^{1/2}\mathcal{D}_r$ and set $A^r = A + \mathrm{Ell}_r$. The new definition of a neighborhood is related to the transformation G by

$$(G(A))^r = G(A) + \mathrm{Ell}_r = G(A) + K^{1/2}\mathcal{D}_r = a + K^{1/2}(A + \mathcal{D}_r) = G(A^r).$$

With regard to the result of Problem 10.4, we can reformulate the isoperimetric inequality as follows.

THEOREM 10.2′. *Suppose that* $\mathbf{P}$ *is a Gaussian measure on* $\mathbf{R}^m$, $A \in \mathfrak{B}(\mathbf{R}^m)$, *and* $r > 0$. *Then*

$$\Phi^{-1}\big(\mathbf{P}\{A + \mathrm{Ell}_r\}\big) \geq \Phi^{-1}\big(\mathbf{P}\{A\}\big) + r. \tag{10.16}$$

Since G preserves convexity and commutes with the summation of sets, we obtain the following modification of the Ehrhard inequality.

THEOREM 10.4′. *Suppose that* $\mathbf{P}$ *is a Gaussian measure on* $\mathbf{R}^m$, $A, B \subset \mathbf{R}^m$ *are convex subsets, and* $\gamma \in [0,1]$. *Then*

$$\Phi^{-1}\big(\mathbf{P}\{\gamma A + (1-\gamma)B\}\big) \geq \gamma\Phi^{-1}\big(\mathbf{P}\{A\}\big) + (1-\gamma)\Phi^{-1}\big(\mathbf{P}\{B\}\big). \tag{10.17}$$

Convexity and the isoperimetric property of a Gaussian measure on an infinite-dimensional space. To extend the above results to Gaussian measures on an arbitrary linear space $\mathbf{X}$, we shall use the following construction.

Let $\{f_i\}_{i=1}^{\infty}$ be a countable set of functionals from $\mathbf{X}'$. We define a projection $F_m\colon \mathbf{X} \to \mathbf{R}^m$ by setting $F_m(x) = \big\{\langle x, f_1\rangle, \dots, \langle x, f_m\rangle\big\}$.

If $A \subset \mathbf{X}$ is a compact set, then $F_m(A)$ is compact in $\mathbf{R}^m$, and hence, the set $A_m = F_m^{-1}F_m(A) \subset \mathbf{X}$ is a cylinder, $A_m \in \mathfrak{C}_0$.

We set $A_\infty = \bigcap_{m=1}^{\infty} A_m$.

PROPOSITION 10.2. *Let* $\Sigma = \sigma\{f_i\}$ *be the minimal* σ*-algebra with respect to which all functionals* f_i *are measurable. Then* A_∞ *is the smallest element of* Σ *that contains* A. *Moreover,* $A_\infty = A + \mathcal{K}$, *where* $\mathcal{K} = \bigcap_{i=1}^{\infty} \operatorname{Ker} f_i$.

PROOF. First, note that the sets A_m form a decreasing sequence and each of them contains A. Therefore, $A_\infty \supset A$. Each element of Σ is invariant under the addition of $\mathcal{K}$. Thus, $A_\infty = A_\infty + \mathcal{K} \supset A + \mathcal{K}$. Let us prove the opposite inclusion.

Let $x \in A_\infty$. By the definition of A_m and A_∞, there exist elements $x_m \in A$ such that $F_m(x_m) = F_m(x)$. In other words, for $i \le m$ we have $\langle x_m, f_i\rangle = \langle x, f_i\rangle$. Consider the limit points of the sequence x_m. Since A is compact, it follows that there exists at least one limit point, l. Then for any i and $\varepsilon > 0$, there exists an $m \ge i$ such that

$$x_m \in \{y \in \mathbf{X} \mid |\langle y, f_i\rangle - \langle l, f_i\rangle| < \varepsilon\},$$

whence it follows that

$$|\langle x - l, f_i\rangle| = |\langle x_m - l, f_i\rangle| \le \varepsilon.$$

Therefore,

$$\langle x, f_i\rangle = \langle l, f_i\rangle, \qquad x - l \in \mathcal{K}, \quad x = l + (x - l) \in A + \mathcal{K}.$$

We see that $A_\infty = A + \mathcal{K}$.

Finally, if $C \in \Sigma$ and $A \subset C$, then

$$C = C + \mathcal{K} \supset A + \mathcal{K} = A_\infty. \quad \square$$

In this proposition, we did not consider the measure. It turns out that by an appropriate choice of the system of functionals, one can ensure that the measures of A and A_∞ coincide.

PROPOSITION 10.3. *Suppose that $A \subset \mathbf{X}$ is a compact set and $\mathbf{P}$ is a measure defined on the cylinder σ-algebra $\mathfrak{C}$. Then $\mathbf{P}^*\{A\} = \mathbf{P}\{A_\infty\}$ for some countable set $\{f_i\} \subset \mathbf{X}'$. If $\mathbf{P}$ can be extended to a Radon measure on $\mathfrak{B}_\mathbf{X}$, then $\mathbf{P}\{A\} = \mathbf{P}\{A_\infty\}$.*

PROOF. By the definition of the outer measure $\mathbf{P}^*$, we have

$$\mathbf{P}^*\{A\} = \inf\{\mathbf{P}\{E\},\, E \in \mathfrak{C}_0, A \subset E\}.$$

Let us choose E_j so that $A \subset E_j$ and

$$\mathbf{P}^*\{A\} \le \mathbf{P}\{E_j\} + 1/j.$$

We set

$$E = \bigcap_{j=1}^\infty E_j.$$

Then $A \subset E$ and $\mathbf{P}^*\{A\} = \mathbf{P}^*\{E\} = \mathbf{P}\{E\}$. Since $E \in \mathfrak{C}$, we have $E \in \Sigma = \sigma\{f_i\}$ for some countable set of functionals $\{f_i\}$. By Proposition 10.2, we have $A_\infty \subset E$. Therefore,

$$\mathbf{P}^*\{A\} \le \mathbf{P}^*\{A_\infty\} = \mathbf{P}\{A_\infty\} \le \mathbf{P}\{E\} = \mathbf{P}^*\{A\}.$$

It follows that $\mathbf{P}^*\{A\} = \mathbf{P}\{A_\infty\}$.

If $\mathbf{P}$ admits a Radon extension, then $\mathbf{P} = \mathbf{P}^*$ on compact sets. Thus, $\mathbf{P}\{A\} = \mathbf{P}^*\{A\} = \mathbf{P}\{A_\infty\}$. $\square$

Now we are in a position to state and prove infinite-dimensional analogs of Theorems 10.2 and 10.4. In what follows, $\mathbf{P} \in \mathcal{G}_0(\mathbf{X})$ is a centered Radon Gaussian measure defined on $\mathfrak{C} = \mathfrak{C}(\mathbf{X})$.

To generalize the isoperimetric inequality (10.16), we must generalize the ellipsoid of concentration in Theorem 10.2′ to the infinite-dimensional case. We present two definitions and prove that they are equivalent.

DEFINITION 1. The set

$$\mathrm{Ell} = \bigcap_{\{x'\in \mathbf{X}' \mid \mathsf{E}_{\mathbf{P}}\langle\cdot,x'\rangle^2\le 1\}} \bigl\{x\in\mathbf{X} \mid |\langle x,x'\rangle|\le 1\bigr\}\subset \mathbf{X}$$

is called the *ellipsoid of concentration of the measure* $\mathbf{P}$.

DEFINITION 2. The set $\mathrm{Ell}_I = I(\mathcal{D}_1)$, where the operator $I\colon \mathbf{X}'_{\mathbf{P}}\to\mathbf{X}$ is defined in Proposition 7.3 and $\mathcal{D}_1$ is the unit ball in the Hilbert space $\mathbf{X}'_{\mathbf{P}}$, is called the *ellipsoid of concentration of the measure* $\mathbf{P}$.

PROPOSITION 10.4. *The ellipsoid* Ell_I *is compact and* $\mathrm{Ell}_I = \mathrm{Ell}$.

PROOF. Since $\mathbf{P}$ is a Radon measure, it follows that $\mathbf{X}_{\mathbf{P}}$ is separable. Hence, $\mathcal{D}_1$ is compact in the weak topology of $\mathbf{X}'_{\mathbf{P}}$. The operator I is weakly continuous; hence $\mathrm{Ell}_I = I(\mathcal{D}_1)$ is compact in $\mathbf{X}$.

Let us show that $\mathrm{Ell}_I = \mathrm{Ell}$. Let $x = Iz$, $z\in\mathcal{D}_1$. Then

$$|\langle x,x'\rangle| = |\langle Iz,x'\rangle| = |(z,I^*x')|\le |z|\,|I^*x'|\le \mathsf{E}_{\mathbf{P}}^{1/2}\langle\cdot,x'\rangle^2$$

for any $x'\in\mathbf{X}$, and therefore, $x\in\mathrm{Ell}$. Thus, we have shown that $\mathrm{Ell}_I\subset\mathrm{Ell}$. Conversely, let $x\in\mathbf{X}\setminus\mathrm{Ell}_I$. We show that $x\in\mathbf{X}\setminus\mathrm{Ell}$. We separate the compact set Ell_I from the point x by a continuous functional $x'\in\mathbf{X}'$ so that $\langle x,x'\rangle>1$. However, for $z\in\mathcal{D}_1$ we have $Iz\in\mathrm{Ell}_I$ and $\langle Iz,x'\rangle\le 1$. Set $z = I^*x'/|I^*x'|\in\mathcal{D}_1$. Then

$$|I^*x'| = (I^*x',I^*x')/|I^*x'| = (z,I^*x') = \langle Iz,x'\rangle\le 1.$$

Hence, $x\in\bigl\{y \mid |\langle y,x'\rangle|>1\bigr\}\subset\mathbf{X}\setminus\mathrm{Ell}$. The proof is complete. □

For $A\in\mathfrak{B}$, we set $A^r = A + r\,\mathrm{Ell}$.

THEOREM 10.5 (the infinite-dimensional isoperimetric inequality). *Let* $\mathbf{P}\in\mathcal{G}(\mathbf{X})$, $A\in\mathfrak{B}(\mathbf{X})$, *and* $r>0$. *Then*

$$\Phi^{-1}(\mathbf{P}_*\{A^r\})\ge\Phi^{-1}(\mathbf{P}\{A\}) + r. \tag{10.18}$$

PROOF. We shall successively verify (10.18) for the cases in which A is a cylinder, a compact set, and an arbitrary Borel set.

The proof is based on the fact that a finite-dimensional projection takes the ellipsoid of concentration of the measure $\mathbf{P}$ to the ellipsoid of concentration of the projection of $\mathbf{P}$. Let $\mathcal{F} = \{f_1,\dots,f_n,\dots,f_m\}\subset X'$ be a set of functionals such that $f_1,\dots,f_n$ are orthonormal and $f_{n+1},\dots,f_m$ are degenerate with respect to $\mathbf{P}$; furthermore, let $F\colon\mathbf{X}\to\mathbf{R}^m$ be the projection generated by $\mathcal{F}$. Next, let $\overline{\mathbf{P}} = \overline{\mathbf{P}}F^{-1}$ be the projection of $\mathbf{P}$, and let $\overline{\mathrm{Ell}}\subset\mathbf{R}^m$ be the ellipsoid of concentration of $\overline{\mathbf{P}}$:

$$\overline{\mathrm{Ell}} = \Bigl\{c\in\mathbf{R}^m \mid \sum_{i=1}^n c_i^2\le 1;\ c_{n+1}=\dots=c_m=0\Bigr\}.$$

LEMMA 10.6. *Let $\bar{A} \in \mathfrak{B}(\mathbf{R}^m)$, and let $A = F^{-1}(\bar{A}) \in \mathfrak{C}_0(\mathbf{X})$ be a cylinder. Then*

$$F(r\,\mathrm{Ell}) = r\,\overline{\mathrm{Ell}}, \tag{10.19}$$

$$F(A + r\,\mathrm{Ell}) = \bar{A} + r\,\overline{\mathrm{Ell}}, \tag{10.20}$$

$$A + r\,\mathrm{Ell} = F^{-1}(\bar{A} + r\,\overline{\mathrm{Ell}}). \tag{10.21}$$

PROOF. Suppose that $x \in r\,\mathrm{Ell}$; then $\langle x, f_{n+j}\rangle = \langle Iz, f_{n+j}\rangle = (z, I^* f_{n+j}) = 0$, since the f_{n+j} are degenerate for $j \geq 1$. Furthermore,

$$\begin{aligned}\sum_{i=1}^{n} |\langle x, f_i\rangle|^2 &= \Big\langle x, \sum_{i=1}^{n} \langle x, f_i\rangle f_i \Big\rangle = \Big(z, \sum_{i=1}^{n} \langle x, f_i\rangle I^* f_i\Big) \\ &\leq |z| \cdot \Big|\sum_{i=1}^{n} \langle x, f_i\rangle f_i\Big| \leq r\Big(\sum_{i=1}^{n} |\langle x, f_i\rangle|^2\Big)^{1/2}.\end{aligned}$$

It follows that $\sum_{i=1}^{n} |\langle x, f_i\rangle|^2 \leq r^2$, and hence, $F(x) \in r\,\overline{\mathrm{Ell}}$ and $F(r\,\mathrm{Ell}) \subset r\,\overline{\mathrm{Ell}}$. Conversely, let $c \in r\,\overline{\mathrm{Ell}}$. We set $x = \sum_{i=1}^{n} c_i I I^* f_i \in \mathbf{X}$. One can readily verify that $x \in r\,\mathrm{Ell}$ and $F(x) = c$. Thus, $r\,\overline{\mathrm{Ell}} \subset F(r\,\mathrm{Ell})$, which completes the proof of (10.19).

Since the projection F is linear, we obtain (10.20) as follows:

$$F(A + r\,\mathrm{Ell}) = F(A) + F(r\,\mathrm{Ell}) = \bar{A} + r\,\overline{\mathrm{Ell}}.$$

It follows that

$$A + r\,\mathrm{Ell} \subset F^{-1}(\bar{A} + r\,\overline{\mathrm{Ell}}).$$

Let us prove the opposite inclusion. Let $x \in F^{-1}(\bar{A} + r\,\overline{\mathrm{Ell}})$. Then for some $l \in A$ and $d \in r\,\mathrm{Ell}$, we have $F(x) = F(b) + F(d)$. Thus, $F(x - d) = F(l) \in \bar{A}$. Therefore, $x - d \in A$ and $x = (x - d) + d \in A + r\,\mathrm{Ell}$. The proof of Lemma 10.6 is complete. □

Now let us derive inequality (10.18) for A being the cylinder from Lemma 10.6. By applying Theorem 10.2′ to $\overline{\mathbf{P}}$ and $\bar{A}$, we obtain

$$\Phi^{-1}\big(\overline{\mathbf{P}}\{\bar{A} + r\,\overline{\mathrm{Ell}}\}\big) \geq \Phi^{-1}(\overline{\mathbf{P}}\{\bar{A}\}) + r.$$

Now by (10.21) we have the estimates

$$\begin{aligned}\mathbf{P}\{A + r\,\mathrm{Ell}\} &= \mathbf{P}F^{-1}\{\bar{A} + r\,\overline{\mathrm{Ell}}\} = \overline{\mathbf{P}}\{\bar{A} + r\,\overline{\mathrm{Ell}}\} \geq \Phi\big(\Phi^{-1}(\overline{\mathbf{P}}\{\bar{A}\}) + r\big) \\ &= \Phi\big(\Phi^{-1}(\mathbf{P}F^{-1}\{\bar{A}\}) + r\big) = \Phi\big(\Phi^{-1}(\mathbf{P}\{A\}) + r\big).\end{aligned}$$

Since Φ is monotone, this inequality is equivalent to the statement of the theorem.

Now let A be a compact set. We shall use the construction of Propositions 10.2 and 10.3, that is, construct a system of functionals such that $\mathbf{P}\{A_\infty\} = \mathbf{P}\{A\}$ and $\mathbf{P}\{(A + r\,\mathrm{Ell})_\infty\} = \mathbf{P}\{A + r\,\mathrm{Ell}\}$. (The latter relation can be ensured, since the set Ell is compact by Proposition 10.4, and hence, so is $A + r\,\mathrm{Ell}$.)

By Proposition 10.2, we have

$$A_\infty + r\,\mathrm{Ell} = (A + \mathcal{K}) + r\,\mathrm{Ell} = (A + r\,\mathrm{Ell}) + \mathcal{K} = (A + r\,\mathrm{Ell})_\infty = \bigcap_m (A + r\,\mathrm{Ell})_m.$$

Obviously, $(A + r\,\mathrm{Ell})_m = A_m + r\,\mathrm{Ell}$. An application of inequality (10.18) to the cylinder A_m yields

$$\Phi^{-1}\big(\mathbf{P}\{(A + r\,\mathrm{Ell})_m\}\big) = \Phi^{-1}\big(\mathbf{P}\{A_m + r\,\mathrm{Ell}\}\big) \geq \Phi^{-1}\big(\mathbf{P}\{A_m\}\big) + r.$$

By letting $m \to \infty$, we obtain

$$\Phi^{-1}\big(\mathbf{P}\{(A + r\,\mathrm{Ell})_\infty\}\big) \geq \Phi^{-1}\big(\mathbf{P}\{A_\infty\}\big) + r.$$

Now the theorem (inequality (10.18)) follows from Proposition 10.3.

Finally, let A be an arbitrary Borel set. We choose compact sets $A_n \subset A$ so that $\mathbf{P}\{A_n\} \to \mathbf{P}\{A\}$. By applying the isoperimetric inequality to A_n and then passing to the limit, we prove the theorem for A. □

THEOREM 10.6 (the infinite-dimensional Ehrhard inequality). *Suppose that* $\mathbf{P} \in \mathcal{G}(\mathbf{X})$, $A, B \in \mathfrak{B}(\mathbf{X})$, $\gamma \in [0,1]$, *and* A *and* B *are convex sets. Then*

$$\Phi^{-1}\big(\mathbf{P}_*\{\gamma A + (1-\gamma)B\}\big) \geq \gamma\Phi^{-1}\big(\mathbf{P}\{A\}\big) + (1-\gamma)\Phi^{-1}\big(\mathbf{P}\{B\}\big). \tag{10.22}$$

PROOF. Let $C = \gamma A + (1-\gamma)B$. We choose compact sets $A' \subset A$ and $B' \subset B$ so that $\mathbf{P}\{A'\} \geq \mathbf{P}\{A\} - \varepsilon$ and $\mathbf{P}\{B'\} \geq \mathbf{P}\{B\} - \varepsilon$. Then $C' = \gamma A' + (1-\gamma)B'$ is also a compact set.

We use Proposition 10.3 and choose a system of functionals $\mathcal{F} = \{f_i\} \subset \mathbf{X}'$ such that $\mathbf{P}\{A'\} = \mathbf{P}\{A'_\infty\}$, $\mathbf{P}\{B'\} = \mathbf{P}\{B'_\infty\}$ and $\mathbf{P}\{C'\} = \mathbf{P}\{C'_\infty\}$. Without loss of generality, we can assume that the functionals f_i are $\mathbf{P}$-orthogonal and their variances are either 0 or 1 (this can be achieved by a linear change of the f_i).

By applying Theorem 10.4 to the finite-dimensional compact sets $F_m(A')$ and $F_m(B')$ and the Gaussian measure $\mathbf{P}F_m^{-1}$, we obtain

$$\begin{aligned}
\Phi^{-1}\big(\mathbf{P}\{C'_m\}\big) &= \Phi^{-1}\big(\mathbf{P}F_m^{-1}\{F_m(C')\}\big) \\
&= \Phi^{-1}\big(\mathbf{P}F_m^{-1}\{\gamma F_m(A) + (1-\gamma)F_m(B')\}\big) \\
&\geq \gamma\Phi^{-1}\big(\mathbf{P}F_m^{-1}\{F_m(A')\}\big) + (1-\gamma)\Phi^{-1}\big(\mathbf{P}F_m^{-1}\{F_m(B')\}\big) \\
&= \gamma\Phi^{-1}\big(\mathbf{P}\{A'_m\}\big) + (1-\gamma)\Phi^{-1}\big(\mathbf{P}\{B'_m\}\big).
\end{aligned}$$

By letting $m \to \infty$, we obtain

$$\Phi^{-1}\big(\mathbf{P}\{C'_\infty\}\big) \geq \gamma\Phi^{-1}\big(\mathbf{P}\{A'_\infty\}\big) + (1-\gamma)\Phi^{-1}\big(\mathbf{P}\{B'_\infty\}\big).$$

Taking into account the fact that A'_∞, B'_∞, and C'_∞ have the same measure as A', B', and C', respectively, we obtain

$$\Phi^{-1}\big(\mathbf{P}\{C'\}\big) \geq \gamma\Phi^{-1}\big(\mathbf{P}\{A'\}\big) + (1-\gamma)\Phi^{-1}\big(\mathbf{P}\{B'\}\big).$$

The substitution of the definitions of A', B', and C' yields

$$\Phi^{-1}\big(\mathbf{P}_*\{C\}\big) \geq \Phi^{-1}\big(\mathbf{P}\{C'\}\big) \geq \gamma\Phi^{-1}\big(\mathbf{P}\{A\} - \varepsilon\big) + (1-\gamma)\Phi^{-1}\big(\mathbf{P}\{B\} - \varepsilon\big).$$

Letting $\varepsilon \to 0$, we arrive at (10.22). □

The following corollary of Theorem 10.6 is often used in applications.

COROLLARY (Anderson's inequality). *Let $C \in \mathfrak{B}(\mathbf{X})$ be a convex centrally symmetric set, and let $\mathbf{P} \in \mathcal{G}_0(\mathbf{X})$. Then*

$$\mathbf{P}\{C\} \geq \mathbf{P}\{C + l\}$$

for any $l \in \mathbf{X}$.

PROOF. We set $A = C + l$, $B = C - l$, and $\gamma = 0.5$ and use Theorem 10.6. Then the set

$$\gamma A + (1-\gamma)B = \gamma C + (1-\gamma)C + l(2\gamma - 1) = (C + C)/2 = C$$

satisfies the estimate

$$\Phi^{-1}(\mathbf{P}\{C\}) \geq (\Phi^{-1}(\mathbf{P}\{A\}) + \Phi^{-1}(\mathbf{P}\{B\}))/2.$$

Since C is symmetric and $\mathbf{P}$ is centered, we have

$$B = -A, \qquad \mathbf{P}\{B\} = \mathbf{P}\{A\}, \qquad \Phi^{-1}(\mathbf{P}\{C\}) \geq \Phi^{-1}(\mathbf{P}\{A\}). \quad \square$$

§11. Convex functionals and their distributions

A functional $f\colon \mathbf{X} \to (-\infty, +\infty]$ is said to be *convex* if

$$f(\beta x + (1-\beta)y) \leq \beta f(x) + (1-\beta) f(y)$$

for any $x, y \in \mathbf{X}$ and $\beta \in [0,1]$.

If the opposite inequality holds and the range of f is $[-\infty, +\infty)$, then we obtain the definition of a *concave* functional. The multiplication by -1 takes convex functionals to concave functionals and vice versa; therefore, in what follows we consider only convex functionals.

EXAMPLE 11.1. Let $\{f_t\}_{t\in T}$ be a family of linear functionals defined on $\mathbf{X}$, and let $\{e_t\}_{t \in T}$ be a family of real numbers. Then the functional

$$f(x) = \sup_{t\in T}\{f_t(x) + e_t\} \tag{11.1}$$

is convex.

Let us show that under very general topological assumptions on $\mathbf{X}$ and f, a convex functional f can be represented in the form (11.1). Recall that a functional $f\colon \mathbf{X} \to \overline{\mathbf{R}}$ is said to be *lower semicontinuous* if

$$f(x) \leq \varliminf_{y \to x} f(y)$$

for all $x \in \mathbf{X}$.

Equivalently, one can require all the sets $\{x \in \mathbf{X} \mid f(x) \leq r\}$ to be closed. We denote the class of convex lower semicontinuous functionals on $\mathbf{X}$ by $\mathcal{V}(\mathbf{X})$.

PROPOSITION 11.1. *Let $\langle \mathbf{X}, \mathbf{X}' \rangle$ be a pair of spaces in duality, and let $f \in \mathcal{V}(\mathbf{X})$. Then there exists a family $\{\langle\cdot, f_t\rangle,\ f_t \in \mathbf{X}'\}$ of continuous linear functionals and a family of numbers $\{e_t\}$ such that*

$$f(x) = \sup_{t\in T}\big(\langle x, f_t\rangle + e_t\big). \tag{11.2}$$

PROOF. The set $F = \{(x, r) \mid f(x) \le r\} \subset \mathbf{X} \times \mathbf{R}^1$ is convex and closed, since f is convex and lower semicontinuous. Let $T = \mathbf{X} \times \mathbf{R}^1 \setminus F$. By the Hahn–Banach theorem, for any point $t = (x_t, r_t) \in T$, there exists a functional $(f_t^0, e_t^0) \in \mathbf{X}' \times \mathbf{R}^1$ that separates t from F. This means that $e_t^0 < 0$ and

$$1 < \langle x_t, f_t^0 \rangle + r_t e_t^0, \tag{11.3}$$

but

$$1 \ge \langle x_t, f_t^0 \rangle + f(x) e_t^0$$

for all $x \in \mathbf{X}$.

We set

$$f_t = -f_t^0 / e_t^0, \qquad e_t = (e_t^0)^{-1}.$$

Then

$$\langle x, f_t \rangle + e_t \le f(x)$$

for all $x \in \mathbf{X}$. Hence,

$$f(x) \ge \sup_{t \in T} (\langle x, f_t \rangle + e_t).$$

On the other hand, let us choose an $x \in \mathbf{X}$ and consider a sequence $t_n = (x, r_n)$, where $r_n \to f(x)$. Then (11.3) yields

$$\langle x, f_{t_n} \rangle + e_{t_n} > r_n.$$

Therefore,

$$\sup_{t \in T} (\langle x, f_t \rangle + e_t) \ge \overline{\lim_{n \to \infty}} (\langle x, f_{t_n} \rangle + e_{t_n}) \ge f(x). \quad \square$$

One can readily verify the converse statement: if a functional f admits the representation (11.2), then $f \in \mathcal{V}(\mathbf{X})$.

Note that, generally speaking, the class $\mathcal{V}(\mathbf{X})$ depends on the topology with respect to which the semicontinuity of functionals is defined. However, (11.2) shows that for all topologies compatible with the duality $\langle \mathbf{X}, \mathbf{X}' \rangle$, the class $\mathcal{V}(\mathbf{X})$ is the same.

It is often convenient to have a representation (11.2) with a countable indexing set T. The following result shows that this is possible if we neglect a set of measure zero. If a representation (11.2) is given, then we write

$$f(T'; x) = \sup_{t \in T'} (\langle x, f_t \rangle + e_t)$$

for $T' \subset T$.

PROPOSITION 11.2. *Suppose that $\mathbf{P} \in \mathcal{G}_0(\mathbf{X})$, $f \in \mathcal{V}(\mathbf{X})$, and a representation (11.2) is given. Then there exists a countable subset $T_1 \subset T$ such that $f(x) = f(T_1; x)$ for $\mathbf{P}$-almost all $x \in \mathbf{X}$.*

PROOF. First, let us consider the case when $\inf_{t \in T} e_t > -\infty$, and assume that $\mathbf{P}\{f < \infty\} > 0$. Then

$$\mathbf{P}\Big\{ \sup_{t \in T} \langle x, f_t \rangle < \infty \Big\} > 0. \tag{11.4}$$

It is well known that under condition (11.4) there exist a set $\mathbf{X}_0 \subset \mathbf{X}$ of full measure and a metric d on the set T such that the space (T, d) is separable and the function $\langle x, f_\cdot \rangle \colon T \to \mathbf{R}^1$ is continuous on (T, d) for $x \in \mathbf{X}_0$ [**114, 115, 155**]. We replace d by the stronger measure $d_1(t_1, t_2) = d(t_1, t_2) + |e_{t_1} - e_{t_2}|$. Then the space

(T, d_1) remains separable, but the functions $\langle x, f_{\cdot}\rangle + e_{\cdot}$ are continuous with respect to d_1.

Let T_1 be a countable d_1-dense subset of T. We choose $x \in \mathbf{X}_0$, $t \in T$, and a sequence $\{t_n\} \subset T_1$ such that $d_1(t_n, t) \to 0$. Then

$$f(T_1; x) \geq \lim_{n\to\infty} (\langle x, f_{t_n}\rangle + e_{t_n}) = \langle x, f_t\rangle + e_t.$$

By taking the upper bound with respect to t, we obtain the desired inequality

$$f(T_1; x) \geq f(T, x) = f(x).$$

Now let us consider the general case, allowing $\inf_{t\in T} e_t = -\infty$. Let $T^n = \{t \in T \mid e_t \geq -n\}$ and $f^{(n)}(x) = f(T^n; x)$. In T^n, we choose countable subsets T_1^n such that $f^{(n)}(\cdot) = f(T_1^n; \cdot)$, $\mathbf{P}$-almost surely. Let $T_1 = \bigcup_n T_1^n$.

Then

$$f(x) = \sup_n f^{(n)}(x), \qquad f(T_1; x) = \sup_n f(T_1^n; x)$$

for any $x \in \mathbf{X}$, which implies that $f(x) = f(T_1; x)$, $\mathbf{P}$-almost surely.

It remains to consider the degenerate case $\mathbf{P}\{f < \infty\} = 0$. Since $\mathbf{P}$ is a Radon measure, it follows that for any $M > 0$ and $\varepsilon > 0$ there exists a compact set $\Delta \subset \{x \mid f(x) > M\}$ such that $\mathbf{P}\{\Delta\} \geq 1 - \varepsilon$. By choosing a finite subcover of the cover of Δ by the open sets $\{x \mid \langle x, f_t\rangle + e_t > M\}$, we obtain a finite set $T' \subset T$ such that

$$\mathbf{P}\{\max_{t\in T'} (\langle x, f_t\rangle + e_t) \geq M\} \geq \mathbf{P}\{\Delta\} \geq 1 - \varepsilon.$$

This proves our proposition. □

The distribution of a convex functional. Let $\mathbf{P} \in \mathcal{G}(\mathbf{X})$ and $f \in \mathcal{V}(\mathbf{X})$. The following results characterize the distribution of f with respect to the measure $\mathbf{P}$. Consider the distribution function

$$F(r) = \mathbf{P}\{x \mid f(x) \leq r\}.$$

PROPOSITION 11.3. *The function* $\Psi(r) = \Phi^{-1}(F(r))$ *is concave.*

PROOF. Let us choose $r_1, r_2 \in \mathbf{R}^1$ and $\gamma \in [0, 1]$. Set $r_3 = \gamma r_1 + (1-\gamma) r_2$ and $A_i = \{x \in \mathbf{X} \mid f(x) \leq r_i\}$, $i = 1, 2, 3$. Then the sets A_i are closed and satisfy

$$A_3 \supset \gamma A_1 + (1-\gamma) A_2,$$

since f is convex. By Theorem 10.6, we obtain

$$\Phi^{-1}(\mathbf{P}\{A_3\}) \geq \gamma \Phi^{-1}(\mathbf{P}\{A_1\}) + (1-\gamma)\Phi^{-1}(\mathbf{P}\{A_2\}),$$

which is equivalent to the desired inequality

$$\Psi\big(\gamma r_1 + (1-\gamma) r_2\big) \geq \gamma \Psi(r_1) + (1-\gamma)\Psi(r_2). \quad \square$$

THEOREM 11.1. *The distribution function F has the following properties.*

(i) *F is everywhere continuous except possibly at the separation point*

$$r_0 = \inf\{r \mid F(r) > 0\} = \inf\{r \mid \mathbf{P}\{f \le r\} > 0\}. \tag{11.5}$$

(ii) *F is absolutely continuous on the interval $(r_0, +\infty)$.*

(iii) *F is differentiable everywhere on $(r_0, +\infty)$ except on an at most countable exceptional set $\Delta \subset (r_0, +\infty)$.*

(iv) *The derivative F' is positive and continuous everywhere on $(r_0, +\infty) \setminus \Delta$; it has one-sided limits and jumps downwards at each point $r \in \Delta$.*

(v) *F' is the distribution density of the functional f, that is,*

$$\mathbf{P}\{x \mid f(x) \in A\} = \int_A F'(r)\,\lambda(dr)$$

for any $A \in \mathfrak{B}(\mathbf{R}^1)$.

(vi) *For any $r > r_0$, the variation of F' on $[r, \infty)$ is bounded. In particular, F' is bounded on $[r, \infty)$.*

PROOF. The assertion of the theorem follows from the representation $F = \Phi \circ \Psi$, where Ψ is the concave function defined in Proposition 11.3. The derivation of (i)–(vi) from elementary properties of convex functions on $\mathbf{R}^1$ is left to the reader. □

The following examples show that the unpleasant "singularities" allowed by Theorem 11.1 can indeed occur. Specifically, F can have a jump at r_0; F' can have jumps on a countable set Δ; F' can be unbounded in any neighborhood of r_0 and hence need not be a function of bounded variation on $(r_0, +\infty)$.

EXAMPLE 11.2. Let η_n be independent $\mathcal{N}(0,1)$-distributed random variables, $\sigma_n = (2\ln n + 2\ln\ln n)^{-1/2}$, and $\xi_n = \sigma_n \eta_n$. Then

$$\mathbf{P}\{\sup \xi_n < \infty\} = 1, \tag{11.6}$$

$$\mathbf{P}\{\sup \xi_n = 1\} > 0. \tag{11.7}$$

Indeed, simple calculations show that the sum

$$S(M) = \sum_{n=1}^{\infty} \mathbf{P}\{\xi_n \ge M\} = \sum_{n=1}^{\infty} \left(1 - \Phi\Big(\frac{M}{\sigma_n}\Big)\right) \sim \sum_{n=1}^{\infty} (\ln n)^{-1/2} (n \ln n)^{-M^2}$$

is finite for $M \ge 1$ and infinite for $0 < M < 1$.

Now it follows from the Borel–Cantelli lemma that the event $\{\xi_n \ge 1\}$ almost surely occurs only finitely many times. Thus, (11.6) holds. It also follows from the Borel–Cantelli lemma that $\mathbf{P}\{\sup \xi_n \ge 1\} = 1$.

Note that for sufficiently large $n \ge N$ we have

$$\ln \mathbf{P}\{\xi_n \le 1\} \ge -\mathbf{P}\{\xi_n \ge 1\}/2.$$

Thus,

$$\mathbf{P}\{\sup \xi_n \le 1\} = \exp\left\{\sum_{n=1}^{\infty} \ln \mathbf{P}\{\xi_n \le 1\}\right\} \ge \exp\left\{\sum_{n=1}^{N} \ln \mathbf{P}\{\xi_n \le 1\} - \frac{S(1)}{2}\right\} > 0,$$

which proves (11.7).

Let $\mathbf{P}$ be the Gaussian measure on $\mathbf{R}^\infty$ corresponding to the distribution of ξ_n, and let $f\colon \mathbf{R}^\infty \to \overline{\mathbf{R}}$ be the functional defined by the formula $f(x) = \sup_n x_n$. Then $\mathbf{P}$ and f satisfy the assumptions of Theorem 11.1. The distribution $\mathbf{P}f^{-1}$ has the separation point $r_0 = 1$ and an atom at the point r_0.

The same reasoning allows us to study the following situation.

EXAMPLE 11.3. Let η_n and σ_n be taken from the previous example. Set $\zeta_n = \eta_n - \sigma_n^{-1} = \eta_n - (2\ln n + 2\ln\ln n)^{1/2}$. Then

$$\mathbf{P}\{0 < \sup \zeta_n < \infty\} = 1, \qquad \mathbf{P}\{\sup \zeta_n = 0\} > 0.$$

Thus, the distribution of the functional $f = \sup \zeta_n$ has an atom and, in contrast with Example 11.2, all variables have the same variances.

PROBLEM 11.1. Let $\mathbf{X} = \mathbf{R}^1$, $\mathbf{P} = \mathcal{N}(0,1)$, and $r_0 \in \mathbf{R}^1$, and let $\Delta \subset (r_0, +\infty)$ be a countable set. Construct a convex function $f\colon \mathbf{X} \to \mathbf{R}^1$ such that r_0 is a separation point of the corresponding distribution function F and F' jumps downwards at each point of Δ and is unbounded in any neighborhood of r_0.

THEOREM 11.2. *Suppose that* $\mathbf{P} \in \mathcal{G}_0(\mathbf{X})$, $f \in \mathcal{V}(\mathbf{X})$, *and* $\inf_{t\in T}\{e_t\} > -\infty$ *in the representation* (11.2). *Moreover, let* $\mathsf{E}_\mathbf{P}|\langle\cdot, f_t\rangle|^2 = |I^* f_t|^2 = \sigma^2$ *for some* $\sigma > 0$ *and for all* $t \in T$. *Then the distribution* $F = \mathbf{P}f^{-1}$ *is absolutely continuous and has the density*

$$F'(r) = q(r)\exp\{-r^2/2\sigma^2\}, \tag{11.8}$$

where $q(\cdot)$ *is a nondecreasing function on* $\mathbf{R}^1$ *continuous everywhere except possibly for the separation point* r_0 *defined in* (11.5).

PROOF. First, let us we prove the theorem for the case in which the indexing set T in the representation (11.2) is finite. Let $T = \{1, 2, \ldots, n\}$. Then we have the explicit expression

$$F'(r) = \sum_{t=1}^{n} \mathbf{P}\Big\{ \max_{s\neq t}(\langle\cdot, f_s\rangle + e_s) \le r \mid \langle\cdot, f_t\rangle + e_t = r\Big\} p(r - e_t),$$

where p is the Gaussian density (1.6) with parameters $(0, \sigma^2)$. Let us show that each term in this sum has the form (11.3). Let t be fixed. Let $k_{st} = \mathsf{E}_\mathbf{P}\langle\cdot, f_s\rangle\langle\cdot, f_t\rangle\sigma^{-2}$ be the correlation coefficient. Then for each s we have the representation

$$f_s = k_{st} f_t + n_s,$$

where $I^* f_t$ and $I^* n_s$ are orthogonal and, by virtue of the Gaussian property, independent. Therefore,

$$\begin{aligned}
&\mathbf{P}\Big\{ \max_{s\neq t}(\langle\cdot, f_s\rangle + e_s) \le r \mid \langle\cdot, f_t\rangle + e_t = r\Big\} \\
&\qquad = \mathbf{P}\Big\{ \max_{s\neq t}(k_{st}\langle\cdot, f_s\rangle + n_s + e_s) \le r \mid \langle\cdot, f_t\rangle = r - e_t\Big\} \\
&\qquad = \mathbf{P}\Big\{ \max_{s\neq t}(n_s + e_s + k_{st}(r - e_t))\Big\} \le r.
\end{aligned}$$

Since $k_{st} \le 1$, it follows that the last probability is monotone with respect to r, whence we obtain the representation (11.8) for a finite T. In particular, (11.8) implies the inequality

$$q(r) \le \left(\sqrt{2\pi}\sigma^{-1}\right)^{-1}\left(1-\Phi(r/\sigma)\right)^{-1} < \infty. \tag{11.9}$$

If $\Delta_1 = [\delta_1, \delta_2]$ and $\Delta_2 = [\delta_3, \delta_4]$ are intervals in $\mathbf{R}^1$ and $\delta_2 \le \delta_3$, then from (11.8) we obtain the estimate

$$\frac{F\{\Delta_1\}}{\mathcal{N}(0,\sigma^2)\{\Delta_1\}} \le \frac{F\{\Delta_2\}}{\mathcal{N}(0,\sigma^2)\{\Delta_2\}}. \tag{11.10}$$

Now suppose that the set T is countable. We represent it as the union of an expanding sequence of finite sets T^n. Then $f(T^n; x) \to f(x)$ for each $x \in \mathbf{X}$. Hence, we have the weak convergence of distributions

$$F_n = \mathbf{P}f^{-1}(T^n;\cdot) \Longrightarrow \mathbf{P}f^{-1} = F.$$

We have already proved the desired estimates for the distributions F_n. By (11.9), they are uniformly absolutely continuous on any ray $(-\infty, r]$. Hence, F is also an absolutely continuous distribution. By passing to the limit in the estimate (11.10) for F_n, we obtain the estimate of the same form for F. By letting the lengths of the intervals Δ_i tend to zero, we see that the density F has the form (11.8).

Finally, let us consider the general case. By Proposition (11.2), it can be reduced to the case of a finite T, which has already been considered.

It follows from (11.8) that F' cannot jump downwards. Theorem 11.1 (iv) prohibits other types of jumps for F'. Thus, F' is continuous.

The proof of Theorem 11.2 is complete. □

Let $\mathbf{X} = \mathbf{R}^1$, $\mathbf{P} = \mathcal{N}(0,1)$, and $f(x) = x - 1$. Then

$$F'(r) = (2\pi)^{-1/2}\exp\{-(x+1)^2/2\}.$$

Obviously, the conclusion of the theorem (about the monotonicity of q) does not hold, since the condition $\inf\{e_t\} \ge 0$ is violated. Theorem 11.2 can be generalized to the case $e_- = \inf\{e_t\} > -\infty$. Instead of (11.8), here we can guarantee the representation

$$F'(r) = q(r)\exp\big\{-(r-e_-)^2/2\sigma^2\big\}.$$

Let $\mathbf{X} = \mathbf{R}^1$, $\mathbf{P} = \mathcal{N}(0,1)$, and $f(x) = \sup(x, -x) = |x|$. Then $F'(r) = \mathbf{1}_{\{r\ge 0\}}(2/\pi)^{1/2}\exp\{-r^2/2\}$. This example shows that F' may have a jump at the separation point r_0.

It follows from Example 11.3 that we cannot generalize Theorem 11.2 to the case $\inf_{t\in T}\{e_t\} = -\infty$.

Uniqueness of the extremal functional. Let $f \in \mathcal{V}(\mathbf{X})$ and let a representation (11.2) be given. For $x \in \mathbf{X}$ we construct the set of extremal indices

$$M_x = \{t \in T \mid f(x) = \langle x, f_t\rangle + e_t\}.$$

The set M_x may be empty, but more often M_x is a singleton if we identify $\mathbf{P}$-a.s. coinciding functionals. More precisely, we have the following statement.

THEOREM 11.3. *Let $\mathbf{P} \in \mathcal{G}_0(\mathbf{X})$ and $f \in \mathcal{V}(\mathbf{X})$. Then there exists an $X_1 \subset \mathbf{X}$ with $\mathbf{P}\{X_1\} = 1$ such that $I^*f_s = I^*f_t$, $\langle x, f_s\rangle = \langle x, f_t\rangle$, and $e_s = e_t$ for any $x \in X_1$ and $s, t \in M_x$.*

PROOF. Suppose that $s, t \in T$ and $I^* f_s \neq I^* f_t$. Then $(z, I^* f_s) < (z, I^* f_t)$ for some $z \in \mathbf{X}'_{\mathbf{P}}$. In $\mathbf{X}'_{\mathbf{P}}$, we choose small neighborhoods $V_s \ni I^* f_s$ and $V_t \ni I^* f_t$ such that

$$\sup_{V_s}(z, \cdot) < \inf_{V_t}(z, \cdot).$$

We need the following auxiliary lemma. □

LEMMA 11.1. *There exists a set $X_{s,t} \subset \mathbf{X}$ of full measure such that for each $x \in X_{s,t}$ one of the sets $M_x \cap I^{*-1}(V_s)$ or $M_x \cap I^{*-1}(V_t)$ is empty.*

PROOF. Consider the functionals

$$g_s(x) = f\big(\{u \in T \mid I^* f_u \in V_s\}; x\big), \quad g_t(x) = f\big(\{u \in T \mid I^* f_u \in V_t\}; x\big).$$

Note that if for some x both sets mentioned in the lemma are not empty, then $f(x) = g_s(x) = g_t(x)$. Therefore, we can set $X_{s,t} = \{x \mid g_s(x) \neq g_t(x)\}$. Let us prove that $\mathbf{P}\{X_{s,t}\} = 1$.

We set $l = Iz$ and consider the partition of $\mathbf{X}$ into lines parallel to the vector l. It suffices to verify that for any line $\gamma = \{x \mid x = x_0 + cl, c \in \mathbf{R}^1\}$ the set $\gamma \cap (\mathbf{X} \setminus X_{s,t})$ consists of at most one point. Let us choose γ and a point $x_1 \in \gamma \cap (\mathbf{X} \setminus X_{s,t})$. Let us verify that $x + cl \in X_{s,t}$ for $c > 0$. Indeed, by the definition of x_1 we have $g_s(x_1) = g_t(x_1)$. Using the definitions of z, l, g_s, and g_t, we obtain

$$\begin{aligned} g_s(x_1 + cl) &\leq g_s(x_1) + c \sup_{I^* f_u \in V_s} (\langle l, f_u \rangle) \leq g_s(x_1) + c \sup_{V_s}(z, \cdot) < g_t(x_1) + c \sup_{V_t}(z, \cdot) \\ &\leq g_t(x_1) + c \inf_{I^* f_u \in V_t} (\langle l, f_u \rangle) \leq g_t(x_1 + cl). \end{aligned}$$

Thus, $g_t(x_1 + cl) > g_s(x_1 + cl)$, and our lemma is thereby proved. □

Let us continue the proof of Theorem 11.3. We have an open cover of the set of pairs $\{(I^* f_s, I^* f_t) \mid s, t \in T, I^* f_s \neq I^* f_t\}$ by sets of the form $V_s \times V_t$. Since $\mathbf{X}'_{\mathbf{P}}$ is separable, it follows that this cover has a countable subcover; we denote the corresponding indices by s_n and t_n. Then we apply Lemma 11.1 and set $X_1 = N_{\mathbf{P}} \cap \bigcap_n X_{s_n, t_n}$. Let us prove that X_1 is the desired set. By Lemma 11.1 and Proposition 7.4 (iii), we have $\mathbf{P}\{X_1\} = 1$. If $x \in X_1$, $s, t \in M_x$, and $I^* f_s \neq I^* f_t$, then $I^* f_s \in V_{s_n}$ and $I^* f_t \in V_{t_n}$ for some n, and it turns out that $s \in M_x \cap I^{*-1}(V_{s_n})$ and $t \in M_x \cap I^{*-1}(V_{t_n})$. Thus, $x \in X_{s_n, t_n}$. Moreover, $x \in X_1$. This contradiction shows that $I^* f_s = I^* f_t$, that is, $f_s - f_t \in \operatorname{Ker} I^*$. Since $x \in \mathbf{N}_{\mathbf{P}}$, it follows from Proposition 7.4 (i) that $\langle x, f_s \rangle = \langle x, f_t \rangle$. Finally, by the definition of the set M_x we have $e_s = f(x) - \langle x, f_s \rangle = f(x) - \langle x, f_t \rangle = e_t$. The proof of the theorem is complete. □

COROLLARY. *Suppose that $\mathbf{X} \subset \mathbf{R}^T$ is some space of trajectories, $\mathbf{X}' = \mathcal{L}\{\pi_t\}$, $t \in T$ (see Example 7.1), $\mathbf{P} \in \mathcal{G}_0(\mathbf{X})$, and $f(x) = \sup_{t \in T} x(t)$. Furthermore, suppose that $\mathbf{P}\{x(t) = x(s)\} = 0$ for $s \neq t \in T$. Then almost surely the equality $x(t^*) = \sup_{t \in T} s(t)$ is attained at no more than one point $t^* \in T$.*

We can strengthen Theorem 11.3 by considering a wider set of extremal indices $M_x^0 \supset M_x$, namely,

$$M_x^0 = \big\{t \in T \mid \inf f\big(\{u \mid |I^* f_u - I^* f_t| < \varepsilon\}, x\big) = f(x)\big\}.$$

Then almost surely $I^* f_s = I^* f_t$ and $\langle x, f_s \rangle = \langle x, f_t \rangle$ for $s, t \in M_x^0$, but we need not have $e_s = e_t$.

PROBLEM 11.2. State and prove an analog of Theorem 11.3 for $\mathbf{P} \in \mathcal{G}(\mathbf{X})$.

The separation point and absolute continuity of distributions. Let $\mathbf{P} \in \mathcal{G}(\mathbf{X})$ and $f \in \mathcal{V}(\mathbf{X})$. Let us study the absolute continuity of the distribution $F = \mathbf{P}f^{-1}$. Theorem 11.1 shows that $F \ll \lambda$ if and only if F does not have an atom at the separation point r_0 defined in (11.5). In particular, if $r_0 = -\infty$, then $F \ll \lambda$. Therefore, the location of the separation point is of particular interest.

If we assume that f is a *continuous* functional, then we can give a descriptive geometric interpretation of the separation point.

PROBLEM 11.3. Let $\mathbf{P}$ be a Radon (not necessarily Gaussian) measure on a topological space $\mathbf{X}$, and let $f\colon \mathbf{X} \to \mathbf{R}^1$ be a functional continuous on $\operatorname{supp}(\mathbf{P})$. Then $r_0 = \inf_{x \in \operatorname{supp}(\mathbf{P})} f(x)$.

There is a natural question: how to verify the continuity of a convex functional? The following problem shows that continuity is equivalent to local boundedness.

PROBLEM 11.4. Suppose that $\mathbf{P} \in \mathcal{G}(\mathbf{X})$, $f\colon \mathbf{N_P} \to (-\infty, +\infty]$ is a convex functional, $N_0 \subset \mathbf{N_P}$ is the set of points of continuity of f, and $N = N_0 \cap \{x \mid f(x) < \infty\}$. Prove that

1. if $V \subset \mathbf{X}$ is open and $\sup\limits_{V \cap \mathbf{N_P}} f < \infty$, then $V \cap N_{\mathbf{P}} \subset N$;
2. the set N is convex and open in $\mathbf{N_P}$;
3. if $y \in \mathbf{N_P} \setminus N$, then $\inf\limits_{2y-N} f = +\infty$;
4. if $\mathbf{P}\{x \mid f(x) < \infty\} = 1$ and N is not empty, then $f < \infty$ on $N_{\mathbf{P}}$ and $N = N_{\mathbf{P}}$.

In connection with these problems, let us again consider the functional f and the Gaussian measure $\mathbf{P} \in \mathcal{G}_0(\mathbf{R}^\infty)$ from Example 11.2. In this case, we have $\mathbf{N_P} = \mathbf{R}^\infty$ and $\inf_{\mathbf{N_P}} f = -\infty$, but $r_0 = 1$. Therefore, in the statement of Problem 11.3 we cannot omit the assumption that f is continuous. The same $\mathbf{P}$ and f can be considered in the Banach space of bounded sequences equipped with the norm $\|x\| = \sup_n |x_n|$. Then the functional f is continuous, but $\mathbf{P}$ is not a Radon measure.

There is one more way to find the points at which the distribution $\mathbf{P}f^{-1}$ may have an atom. Let $f \in \mathcal{V}(\mathbf{X})$ and let a representation (11.2) be given. We additionally assume that for $\mathbf{P}$-almost all $x \in \mathbf{X}$ there exists a $t \in T$ such that

$$f(x) = \langle x, f_t \rangle + e_t. \tag{11.11}$$

A typical example of this situation is given by the functional $f(x) = \sup_{t \in T} x(t)$ defined on the space of continuous functions $\mathbf{C}(T)$, where T is a compact set.

Set

$$\begin{aligned} T_0 &= \{t \in T \mid |I^* f_t|^2 = \mathsf{E}_{\mathbf{P}}|\langle \cdot, f_t\rangle|^2 - \mathsf{E}^2_{\mathbf{P}}\langle \cdot, f_t\rangle = 0\}, \\ e^0 &= \sup_{t \in T_0}(e_t + \mathsf{E}_{\mathbf{P}}\langle \cdot, f_t\rangle). \end{aligned} \tag{11.12}$$

The following theorem shows that if there exists an atom of the distribution $\mathbf{P}f^{-1}$, then this atom is at the point e^0.

THEOREM 11.4. *Suppose that $\mathbf{P} \in \mathcal{G}(\mathbf{X})$, $f \in \mathcal{V}(\mathbf{X})$, and f satisfies condition (11.11). Then the distribution $\mathbf{P}f^{-1}$ is concentrated on $[e^0, +\infty]$ and is absolutely continuous on $(e^0, +\infty)$.*

PROOF. We assume that $\mathbf{P} \in \mathcal{G}_0(\mathbf{X})$. For each $t \in T_0$, we have

$$\mathbf{P}f^{-1}\{(-\infty, e_t)\} = \mathbf{P}\{x \mid f(x) < e_t\} \leq \mathbf{P}\{x \mid \langle x, f_t\rangle + e_t < e_t\} = 0.$$

Therefore,

$$\mathbf{P}f^{-1}\{(-\infty, e^0)\} = \mathbf{P}f^{-1}\Big\{\bigcup_{t\in T_0}(-\infty, e_t)\Big\} = 0.$$

Let us prove the absolute continuity. Let $\{z_n\}$ be a basis in $\mathbf{X}'_{\mathbf{P}}$. We set $T(n,m) = \{t \in T \mid |(z_n, I^*f_t)| > m^{-1}\}$. Obviously, $T \setminus T_0 \subset \bigcup_{n,m} T(n,m)$. By the assumptions of the theorem, we have

$$\{x \in \mathbf{N_P} \mid f(x) \in \Delta\} = \{x \in \mathbf{N_P} \mid f(T\setminus T_0; x) \in \Delta\} \subset \bigcup_{n,m}\{x \mid f(T(n,m); x) \in \Delta\}$$

for any Borel set $\Delta \subset (e^0, +\infty)$. Therefore, it suffices to verify the absolute continuity of the distribution of the functional $f(T(n,m);\cdot)$. For brevity, we assume that $T = T(n,m)$ for some n and m. Set $l = Iz_n$. Let us estimate the partial derivative

$$D_l f(x) = \lim_{c\to 0} c^{-1}\big(f(x+cl) - f(x)\big)$$

for any $x \in \mathbf{X}$. Let us choose a t satisfying condition (11.11). Then

$$f(x+cl) \geq \langle x + cl, f_t\rangle + e_t = f(x) + c\langle l, f_t\rangle = f(x) + c(z_n, I^*f_t).$$

Similarly, $f(x - cl) \geq f(x) - c(z_n, I^*f_t)$. Therefore, if $D_l f(x)$ is defined, then

$$|D_l f(x)| > |(z_n, I^*f_t)| > m^{-1}. \tag{11.13}$$

Now let us consider the partition of $\mathbf{X}$ into lines γ parallel to l. The restriction of f to each γ is a convex function. By inequality (11.13), the situation in which this function has a closed interval of points of minimum rather than one point is impossible. Taking into account the absolute continuity of the conditional measures $\mathbf{P}_\gamma$ (see Theorem 7.2), we see that Proposition 4.4 can be applied to these measures, and therefore, $\mathbf{P}_\gamma f^{-1} \ll \lambda$. By (3.4), it follows that $\mathbf{P}f^{-1} \ll \lambda$. □

COROLLARY. *Suppose that f satisfies condition* (11.11) *and $e^0 < r_0$. Then $\mathbf{P}f^{-1} \ll \lambda$.*

To prove this, it suffices to compare Theorem 11.1 (i), (ii) with Theorem 11.4.

Let us again consider Example 11.2, in which $-\infty = e^0 < r_0 = 1$ and the assertions of Theorem 11.4 and its corollary are not satisfied since condition (11.11) is violated.

PROBLEM 11.5. Let $\mathbf{X} = \mathbf{C}[0,1]$ and $f(x) = \sup_{t\in[0,1]} x(t)$. Construct a measure $\mathbf{P} \in \mathcal{G}_0(\mathbf{X})$ for which $e^0 = r_0 = 0$ and $\mathbf{P}f^{-1}\{0\} > 0$.

Theorem 11.4 suggests that it is degenerate linear functionals f_t, $t \in T$, that generate atoms in the distribution $\mathbf{P}f^{-1}$. It is natural to assume that if T_0 is empty, then $\mathbf{P}f^{-1} \ll \lambda$. Although this is not the case (see Examples 11.2 and 11.3), we have the following weaker result.

PROPOSITION 11.4. *Let $\mathbf{P} \in \mathcal{G}_0(\mathbf{X})$ and $f \in \mathcal{V}(\mathbf{X})$. Suppose that we have $\sigma = \inf_{t\in T}|I^*f_t| > 0$ and $e_- = \inf_{t\in T} e_t > -\infty$ in the representation* (11.2). *Then $\mathbf{P}f^{-1} \ll \lambda$, and the density $F' = d\mathbf{P}f^{-1}/d\lambda$ is bounded.*

PROOF. Let us choose an $r \in \mathbf{R}^1$ and set

$$\rho = \inf_{t \in T}(e_t - r)/\sigma_t > (-|e_-| - r)/\sigma > -\infty.$$

Then the functional

$$g(x) = \sup_{t \in T}\big(\langle x, f_t/\sigma_t\rangle + (e_t - r)/\sigma_t - \rho\big)$$

satisfies the assumptions of Theorem 11.2. Moreover,

$$\begin{aligned}\{x \mid f(x) \in [r-\varepsilon, r+\varepsilon]\} &= \{x \mid \sup(\langle x, f_t\rangle + e_t - r)/\sigma \in [-\varepsilon/\sigma, \varepsilon/\sigma]\} \\ &\subset \{x \mid g(x) \in [-\rho - \varepsilon/\sigma, -\rho + \varepsilon/\sigma]\}.\end{aligned}$$

Taking (11.8) and (11.9) into account, we obtain the estimate

$$\begin{aligned}\mathbf{P}f^{-1}\{[r-\varepsilon, r+\varepsilon]\} &\le \mathbf{P}g^{-1}\{[-\rho-\varepsilon/\sigma, -\rho+\varepsilon/\sigma]\} \\ &\le 2\varepsilon\sigma^{-1}(\sqrt{2\pi})^{-1}\big(1 - \Phi(-\rho+\varepsilon/\sigma)\big)^{-1}\exp\{-(|\rho| - \varepsilon/\sigma)^2/2\}\end{aligned}$$

for $\varepsilon < \sigma|\rho|$. Therefore, the distribution $F = \mathbf{P}f^{-1}$ is absolutely continuous, and

$$F'(r) \le \Big[\sqrt{2\pi}\sigma\Big(1 - \Phi\Big(\frac{r + |e_-|}{\sigma}\Big)\Big)\Big]^{-1}\exp\Big\{-\frac{(r+|e_-|)^2}{2\sigma^2}\Big\}$$

for almost all $r \in \mathbf{R}^1$. This estimate is uniform on all rays $(-\infty, r_1]$. By Theorem 11.1 (vi), F' is bounded on the entire axis. □

Example 11.3 shows that one cannot omit the condition $e_- > -\infty$ in the statement of the theorem.

Probabilities of large values of a convex functional. In many applications, it is important to know the behavior of the tail of the distribution of the stochastic functional $\mathbf{P}\{f \ge r\}$ as $r \to \infty$. For a wide class of convex functionals that "grow linearly at infinity," the tails of the corresponding distributions are similar to the "tails" of a Gaussian measure in $\mathbf{R}^1$.

THEOREM 11.5. *Let $\mathbf{P} \in \mathcal{G}_0(\mathbf{X})$ and $f \in \mathcal{V}(\mathbf{X})$. Let $e_- = \inf_{t\in T} e_t > -\infty$ and $\sigma^2 = \sup_{t \in T}|I^* f_t|^2 > 0$ in the representation* (11.2). *Suppose that $\mathbf{P}\{f < \infty\} = 1$. Then*

$$\mathbf{P}\{x \mid f(x) \ge r\} = \exp\{-(r+d)^2/2\sigma^2 + o(r)\} \tag{11.14}$$

as $r \to \infty$ for some $d \in (-\infty, -e_-]$.

Let us clarify the statement of the theorem. The condition $e_- > -\infty$ excludes functionals of "superlinear growth," for example, $f(x) = x^2$ for $\mathbf{X} = \mathbf{R}^1$. The condition $\sigma > 0$ excludes the degenerate case $f = \text{const}$.

The asymptotics (11.14) means that, neglecting a constant translation by d, the tail of the distribution of f corresponds to the tail of the distribution $\mathcal{N}(0, \sigma^2)$, that is, to the tail of the distribution of the linear functional whose variance is the maximum of the variances of all functionals on the right-hand side in (11.2).

LEMMA 11.2. *Let $\Psi(r) = \Phi^{-1}(F(r))$. Then there exists a finite limit $d = \lim_{r\to\infty}(\sigma\Psi(r) - r)$.*

PROOF. By Proposition 11.3, the function $\sigma\Psi(r)-r$ is concave. Therefore, the limit d exists. It suffices to prove that $d \neq \pm\infty$. First, let us obtain a lower bound. Suppose that $\rho_0 \in \mathbf{R}^1$ is chosen so that $F(\rho_0) > 0$, and hence, $\Psi(\rho_0) > -\infty$. We choose an arbitrary $r > 0$ and apply Theorem 10.5 to the set $A = \{x \mid f(x) \leq \rho_0\}$, thus obtaining

$$\Phi^{-1}(\mathbf{P}_*\{A + r\,\mathrm{Ell}\}) \geq \Phi^{-1}(\mathbf{P}\{A\}) + r = \Psi(\rho_0) + r.$$

Let us prove that $A + r\,\mathrm{Ell} \subset \{x \mid f(x) \leq \rho_0 + r\sigma\}$. Indeed,

$$\begin{aligned} f(y) &= \sup_{t\in T}[(\langle x, f_t\rangle + e_t) + r\langle l, f_t\rangle] \leq f(x) + \sup_{t\in T} r|\langle l, f_t\rangle| \\ &\leq \rho_0 + \sup_{t\in T} r|\langle Iz, f_t\rangle| \leq \rho_0 + r\sup_{t\in T}(z, I^* f_t) \\ &\leq \rho_0 + r|I^* f_t| \leq \rho_0 + r\sigma \end{aligned}$$

for $y = x + rl$, $x \in A$, and $l = Iz \in \mathrm{Ell}$. Therefore,

$$\Psi(\rho_0 + r\sigma) = \Phi^{-1}(\mathbf{P}\{x \mid f(x) \leq \rho_0 + r\sigma\}) \geq \Phi^{-1}(\mathbf{P}_*\{A + r\,\mathrm{Ell}\}) \geq \Psi(\rho_0) + r.$$

By setting $\rho = \rho_0 + r\sigma$, we rewrite this inequality in the form

$$\Psi(\rho) \geq \Psi(\rho_0) + (\rho - \rho_0)/\sigma,$$

or

$$\sigma\Psi(\rho) - \rho \geq \sigma\Psi(\rho_0) - \rho_0. \tag{11.15}$$

We have proved that the function $\sigma\Psi(\cdot) - \cdot$ increases. Therefore, $d > -\infty$.

Now let us obtain an upper bound for d. For any $t \in T$, we set $\sigma_t^2 = \mathsf{E}_{\mathbf{P}} f_t^2$. Then

$$F(r) = \mathbf{P}\{x \mid f(x) \leq r\} \leq \mathbf{P}\{x \mid f_t(x) + e_t \leq r\} = \Phi\big((r - e_t)/\sigma_t\big).$$

Therefore,

$$\Psi(r) \leq \inf_t (r - e_t)/\sigma_t \leq (r - e_-)/\sigma.$$

Hence,

$$\sigma\Psi(r) - r \leq -e_-,$$

and it follows that $d \leq -e_- < +\infty$. The proof is complete. □

COROLLARY (Borell's inequality). *Let m_F be the median of the distribution f; that is, let $F(m_F) = 0.5$. Then*

$$F(m_F + \tau) \geq \Phi(\tau/\sigma) \tag{11.16}$$

for any $\tau \geq 0$.

PROOF. It suffices to set $\rho_0 = m_F$ and $\rho = m_F + \tau$ in (11.15), and note that $\Psi(m_F) = 0$. □

PROBLEM 11.6. Suppose that $\mathbf{P}_n$, $n \in \overline{\mathbf{N}}$, are Gaussian measures with barycenter a_n in a Banach space $(\mathbf{X}, \|\cdot\|)$. Prove that if $\mathbf{P}_n \Rightarrow \mathbf{P}_\infty$, then $\|a_n - a_\infty\| \underset{n}{\to} 0$. (Inequality (11.16) allows us to cut off the parts of the measures $\mathbf{P}_n$ that are far away from zero.)

PROOF OF THEOREM 11.5. Taking into account the definition of the function Φ and the result of Lemma 11.2, we obtain

$$\begin{aligned}\mathbf{P}\{x \mid f(x) \geq r\} &= 1 - \Phi(\Psi(r)) = \exp\{-\Psi^2(r)/2 + o(\Psi(r))\} \\ &= \exp\{-(r+d)^2/2\sigma^2 + o(r)\}. \quad \square\end{aligned}$$

Theorem 11.5 covers some particularly important special situations, two of which will be considered shortly.

PROPOSITION 11.5. *Let A be a convex closed absorbing subset of $\mathbf{X}$. Let $\mathcal{I}\colon \mathbf{X} \to [0, \infty]$ be the action functional defined in* (7.17). *Then*

$$\mathbf{P}\{\mathbf{X} \setminus rA\} = \exp\{-\mathcal{I}(A) r^2 (1 + o(1))/2\}, \qquad \mathcal{I}(A) = \inf_{l \in \mathbf{X} \setminus A} \mathcal{I}(l) \tag{11.17}$$

as $r \to \infty$.

PROOF. Consider the Minkowski functional

$$f(x) = \inf_{r>0}\{r \mid x/r \in A\}.$$

Since A is convex, so is f. Since A is closed, f is lower semicontinuous.

Let $\pi_A \subset \mathbf{X}'$ be the polar of the set A,

$$\pi_A = \{f_\alpha \in \mathbf{X}' \mid \sup_{x \in A}\langle x, f_\alpha\rangle \leq 1\}.$$

Then the representation (11.2) of f takes the form

$$f(x) = \sup_{f_\alpha \in \pi_A} \langle x, f_\alpha\rangle.$$

By applying Theorem 11.5 to f and taking into account that $\{x \mid f(x) > r\} = \mathbf{X} \setminus rA$, we obtain

$$\mathbf{P}\{\mathbf{X} \setminus rA\} = \exp\{-(r+d)^2/2\sigma^2 + o(r)\},$$

where $\sigma^2 = \sup_{f_\alpha \in \pi_A} |I^* f_\alpha|^2$.

It remains to express σ in terms of the action functional.

Consider the set $A' = I^{-1}(A)$, which is a convex closed absorbing subset of $\mathbf{X}'_{\mathbf{P}}$, and the set $\pi'_A = \{u \in \mathbf{X}'_{\mathbf{P}} \mid \sup_{z \in A'}(u, z) \leq 1\}$, which is the polar of A'. Then, by the definition of the action functional and the polar, we have

$$\mathcal{I}(A) = \inf_{l \in \mathbf{X} \setminus A} \mathcal{I}(l) = \inf_{z \in \mathbf{X}'_{\mathbf{P}} \setminus A'} |z|^2 = \bigl(\sup_{u \in \pi'_A} |u|\bigr)^{-2}.$$

On the other hand, by the definition of σ and by the relation $I^*(\pi_A)^- = \pi'_A$, we obtain

$$\sigma^2 = \sup_{f_\alpha \in \pi_A} |I^* f_\alpha|^2 = \sup_{u \in \pi'_A} |u|^2.$$

Therefore, $\sigma^{-2} = \mathcal{I}(A)$, and the proposition is proved. $\square$

Note that if the set A is convex and closed but not absorbing (for example, a half-space) or closed and absorbing but not convex, then (11.17) need not hold.

PROPOSITION 11.6. *Suppose that* $(\mathbf{X}, \|\cdot\|)$ *is a Banach space,* $\mathbf{X}'$ *is the dual space of* $\mathbf{X}$, $\mathbf{P} \in \mathcal{G}_0(\mathbf{X})$, *and* K *is the correlation operator of the measure* $\mathbf{P}$. *Then*

$$\mathbf{P}\{x \mid \|x\| > r\} = \exp\{-r^2/2\sigma^2(1+o(1))\}$$

as $r \to \infty$, *where*

$$\sigma^2 = \sup_{\{f\in\mathbf{X}' \mid \|f\|=1\}} \mathsf{E}_{\mathbf{P}}\langle\cdot, f\rangle^2 = \|K\|.$$

PROOF. The functional in question (the norm) admits the following representation of the type of (11.2):

$$\|x\| = \sup_{\substack{\|f_t\|=1\\ f_t\in\mathbf{X}'}} \langle x, f_t\rangle.$$

Therefore, Theorem 11.5 readily implies the desired asymptotics. We only need to justify the relation between K and σ:

$$\|K\| = \sup_{\|f\|=1} \|Kf\| = \sup_{\substack{\|f\|=1\\ \|g\|=1}} \langle Kf, g\rangle = \sup_{\substack{\|f\|=1\\ \|g\|=1}} \mathsf{E}_{\mathbf{P}}\langle\cdot, f\rangle\langle\cdot, g\rangle = \sup_{\|f\|=1} \mathsf{E}_{\mathbf{P}}|\langle\cdot, f\rangle|^2. \qquad \square$$

Estimates of the distribution density. Let p be the density of the standard Gaussian distribution $\mathcal{N}(0,1)$. Elementary calculations show that for $r \geq 0$ one has

$$p(r) = r[1-\Phi(r)]/S(r), \tag{11.18}$$

where

$$S(r) = \int_0^\infty \exp\left\{-v - \frac{v^2 r^{-2}}{2}\right\} dv.$$

The function S is monotone increasing from 0 to 1.

In what follows, we derive a relation, similar to (11.18), between the density and the distribution function of a convex functional. By combining this relation with various estimates for the distribution function (for example, with Theorem 11.5), one can obtain lower and upper bounds for the density.

Let $\mathbf{P}$, f, σ, and d satisfy the assumptions of Theorem 11.5, and let F and m_F be, respectively, the distribution function and the median of the distribution $\mathbf{P}f^{-1}$. Just as in Lemma 11.2, we set $\Psi(r) = \Phi^{-1}(F(r))$.

THEOREM 11.6. *For any* $r \geq m_F$ *one has the estimates*

$$\frac{\Psi(r)[1-F(r)]}{\sigma S(\Psi(r))} \leq F'(r) \leq \frac{(r+d)^2[1-F(r)]}{\sigma^2(r-m_F)S(\Psi(r))}. \tag{11.19}$$

In particular,

$$\lim_{r\to\infty} F'(r)/r[1-F(r)] = \sigma^{-2}. \tag{11.20}$$

PROOF. We treat the distribution $\mathbf{P}f^{-1}$ as the image of the measure $\mathcal{N}(0,1)$ under the convex mapping $Q = \Psi^{-1} = F^{-1}\Phi$. Since Q is convex, it follows from Lemma 11.2 that

$$(r - m_F)/\Psi(r) = (Q(\Psi(r)) - Q(0))/\Psi(r) \le Q'(\Psi(r)) \le \lim_{\beta\to\infty} Q'(\beta) = \sigma. \tag{11.21}$$

We substitute (11.21) and (11.18) into the equation

$$F'(r) = p(\Psi(r))/Q'(\Psi(r))$$

and obtain

$$\frac{\Psi(r)[1 - \Phi(\Psi(r))]}{\sigma S(\Psi(r))} \le F'(r) \le \frac{\Psi^2(r)[1 - \Phi(\Psi(r))]}{(r - m_F)S(\Psi(r))}.$$

Recall that $\Phi(\Psi(r)) = F(r)$ and that Lemma 11.2 implies the inequality $\Psi(r) \le (r + d)/\sigma$. After the substitution, we obtain (11.19). The limit relation (11.20) follows from (11.19) and Lemma 11.2. □

For some r, the function F may have only one-sided derivatives. They also satisfy the estimates (11.19).

We point out that one can estimate the density only for sufficiently large values of the argument ($r \ge m_F$). It turns out that near the separation point r_0 the density may be either very small or very large. The corresponding examples are given in the following section.

By combining the right-hand side of (11.19) with Borell's inequality (11.16) and the representation (11.18), we obtain the following estimate.

COROLLARY. *For $r \ge m_F$, one has*

$$F'(r) \le \frac{(r+d)^2}{(r - m_F)^2\sqrt{2\pi}\sigma} \exp\left\{-\frac{(r - m_F)^2}{2\sigma^2}\right\}. \tag{11.22}$$

For example, if f is an affine functional, then $d = -m_F$ and (11.22) *becomes an equality.*

Appendix

1. Suppose that $\mathbf{P} \in \mathcal{G}_0(\mathbf{X})$, $f \in \mathcal{V}(\mathbf{X})$ is the functional defined by (11.2), and $\{g_t = \langle\cdot, f_t\rangle + e_t;\ t \in T\}$ is the family of affine functionals that defines f. The limit

$$\delta(g) = \lim_{\varepsilon\to 0} \sup_{s,s' \in T(g,\varepsilon)} (g_s - g_{s'})(x),$$

where $T(g,\varepsilon) = \{s \in T \mid \mathsf{E}_{\mathbf{P}}(g_s - g)^2 \le \varepsilon^2\}$, is called the *Itô–Nisio oscillation* [**140**] of the family $\{g_t\}$ and the affine functional g.

If the set $T(g,\varepsilon)$ is empty for small ε, then we set $\delta(g) = -\infty$.

It is well known that $\delta(g)$ coincides with some constant [**140**]; thus, the oscillation can be treated as a nonnegative number rather than a random variable.

Suppose that $\mathbf{P}\{f < \infty\} = 1$, $e_- = \inf_{t\in T} e_t > -\infty$, and the family $\{g_t\}$ forms a convex set in $\mathbf{X}'$. Then we have the representation

$$r_0 = \sup_{r\in\mathbf{R}^1} [r + \delta(r\mathbf{1})/2]$$

for the separation point of the distribution of f.

In particular, if $e_t \equiv 0$, then for $r \neq 0$ we have $\delta(r\mathbf{1}) = -\infty$, and hence,

$$r_0 = \delta(0)/2.$$

Example 11.3 shows that we cannot omit the condition $e_- > -\infty$, since in this case $\delta(r\mathbf{1}) = -\infty$ for all r, whereas $r_0 = 0$.

The convexity of $\{g_t\}$ is essentially not a restriction, since in (11.2) we can always replace this family by its convex hull. The formulas for recalculating the oscillations in terms of the convex hull can be found in [**5**]; see also [**54, 66**].

In terms of oscillations, one can sharpen the estimate for the tail of the distribution of f given in Theorem 11.5. Namely, under the assumptions of this theorem, we have

$$d = \sup\{\mathsf{E}_{\mathbf{P}} g - \delta(g)/2;\ g \in \mathbf{X}'_{\mathbf{P}}, \mathsf{D}_{\mathbf{P}} g = \sigma^2\}.$$

In particular, if $e_t \equiv 0$, then

$$d = -\frac{1}{2} \sup_{g \in \mathbf{X}'_{\mathbf{P}}} \{\delta(g),\ |g| = \sigma\}.$$

Thus, the parameter d is determined by the oscillation on the functionals of maximum variance (see [**48**]; previously, an estimate for d in terms of oscillations was proved in [**162**]).

2. Suppose that $\mathbf{P} \in \mathcal{G}_0(\mathbf{X})$, $f \in \mathcal{V}(\mathbf{X})$, and the assumptions of Theorem 11.4 are satisfied. We additionally assume that $e^0 \leq e^- = \inf_{t\in T} e_t$ (for example, $e_t \equiv 0$). Consider the set

$$U = \{f_t/|I^* f_t|,\ t \in T \setminus T_0\} \subset \mathbf{X}'.$$

The distribution $F = \mathbf{P}f^{-1}$ is absolutely continuous if and only if at least one of the following three conditions is satisfied:

(a) $\mathbf{P}\{\sup\limits_{x' \in U} \langle x, x'\rangle = \infty\} = 1$;

(b) the closed convex hull of the set $I^* U \subset \mathbf{X}'_{\mathbf{P}}$ contains zero;

(c) $\sup\limits_{t\in T}(e_t - e_-)/\sigma_t = +\infty$.

The following examples completely clarify the role of each of these conditions. Let $\mathbf{X} = \mathbf{C}[0,1]$.

A. Let $f(x) = \sup x(t)$ and let $\mathbf{P}$ be the distribution of a Wiener process. Only condition (a) is satisfied. The absolute continuity of F is due to the large amplitude of oscillations of the process near the singular point $T_0 = \{0\}$.

B. Let $f(x) = \sup x(t)$ and let $\mathbf{P}$ be the distribution of a continuously differentiable process ξ_t for which $\mathsf{E}\xi_t = 0$, $\xi_0 = \xi_1 = 0$, and $\xi'_0 = \xi'_1 \neq 0$. Only condition (b) is satisfied.

C. Let $f(x) = \sup(x(t) + \sqrt{t})$ and let $\mathbf{P}$ be the distribution of a continuously differentiable process ξ_t for which $\mathsf{E}\xi_t = 0$ and $\xi_0 = 0$ (see Lifshits [**59**], where some additional information about the absolute continuity of F without the assumption that $e^0 \leq e^-$ is given).

3. Let $\mathbf{P} \in \mathcal{G}(\mathbf{X})$ and let $f(x) = \sup_{t\in T}\langle x, f_t\rangle \in \mathcal{V}(\mathbf{X})$, $\mathbf{P}\{f < \infty\} = 1$. We introduce the following notation:

- $K(s,t) = \mathsf{E}_{\mathbf{P}}\langle \cdot, f_s\rangle\langle \cdot, f_t\rangle$ is the correlation function;

- $q(s) = \sup_{t\in T}\{|K(s,t)/K(s,s) - 1|\}$;
- $g_t = f_t - \dfrac{K(s,t)}{K(s,s)} f_s$ is the prediction error of f_t for a given f_s;
- $p_s(\cdot)$ is the distribution density of f_s, that is, the Gaussian density with parameters $(0, K(s,s))$.

If for some $s \in T$ we have $q(s) \le 0.5$, then the density F' of the distribution $F = \mathbf{P}f^{-1}$ satisfies the estimate

$$F'(r) \le (1-q(s))^{-1}\mathsf{E}_{\mathbf{P}}\big(p_s\big((r - \sup_T g_t)(1+q(s))^{-1}\big)\big).$$

The condition $q(s) \le 0.5$ implies that

$$\inf_{t\in T} K(t,t) \ge K(s,s)/4 > 0.$$

Therefore, this estimate is a sharper version of Proposition 11.4 (Weber [**164**]).

4. Suppose that $\mathbf{P} \in \mathcal{G}_0(\mathbf{X})$ and B is a subset of the Hilbert space $\mathbf{X}'_{\mathbf{P}}$ of measurable functionals. Let us define some entropy characteristics of B. By $N(B,\varepsilon)$ we denote the cardinality of the minimal ε-net for B. The function $H(B,\varepsilon) = \ln N(B,\varepsilon)$ is called the *metric entropy*. The expression

$$\Psi(B,\delta) = \int_0^\delta H^{1/2}(B,\varepsilon)\,d\varepsilon$$

is called the *Dudley integral*. The function

$$\Psi(B,\rho,\delta) = \sup_{z\in\mathbf{X}'_{\mathbf{P}}} \Psi(B\cap\{y \mid |y-z| \le \rho\},\delta)$$

can be used as a local entropy characteristic.

Consider a convex functional $f(\cdot) = \sup_{t\in T}\langle\cdot, f_t\rangle$. We set $B = \{I^*f_t,\, t\in T\} \subset \mathbf{X}'_{\mathbf{P}}$ and seek the entropy bounds for the tail of the distribution of f. If

$$\sigma^2 = \sup_{t\in T}|I^*f_t|^2 = \sup_{t\in T}\mathsf{E}_{\mathbf{P}} f_t^2 > 0$$

and $\Psi(B,\sigma) < \infty$, then for $r \ge 0$ we have

$$\mathbf{P}\{f \ge r\} \le 1 - \Phi\big((r^2\sigma^2 - 2\chi_1(r))^{1/2}\big),$$

where $\chi_1(r) = \inf_{\rho>0}\{4\sqrt{2}\sigma^{-2}\Psi(B,\rho,\rho/2)r + H(B,\rho)\}$.

If $\sigma \le 1$, then for $r \ge 4\sqrt{2}\Psi(B,1)+1$ we have the estimate

$$\mathbf{P}\{f \ge r\} \le 4.8\exp\Big\{-\frac{r^2}{2} + \chi_2(r)\Big\},$$

where $\chi_2(r) = \inf_{\rho>0}\{2\rho^2r^2 + 4\sqrt{2}\Psi(B,\rho,\rho)r + H(B,\rho) + \ln\rho\}$ [**46–48**].

A similar estimate can be written out for the density of the distribution $F = \mathbf{P}f^{-1}$. Specifically, if $r \ge 2 + 19\Psi(B,1)$ and $\Psi(B,r^{-1/2}) \le \sqrt{2}$, then

$$F'(r) \le Ar^{1/2}\Psi^{1/2}(B,r^{-1/2})[1 + |\ln\Psi(r^{-1/2})|]^{1/2} \times \exp\Big\{-\frac{r^2}{2} + 6\sqrt{2}\Psi(B,r^{-1/2})r + 9\Psi^2(B,1)\Big\},$$

where A is an absolute constant (Lifshits [**64**]).

§12. Distribution of the norm

Throughout this section, $(X, \|\cdot\|)$ is a Banach space, S and $\mathcal{D}$ are the unit sphere and the unit closed ball in $\mathbf{X}$, respectively, a is the barycenter of a Gaussian Radon measure $\mathbf{P} \in \mathcal{G}(X)$, $f(x) = \|x\|$, and $F = \mathbf{P}f^{-1}$ is the distribution of the norm with respect to $\mathbf{P}$. Since the norm is a convex functional, we can use the results obtained in §11 to study F. However, the properties of the distribution of the norm are very important in proving limit theorems [**8, 9, 11, 84**]; this is why we pay special attention to this case. In particular, estimates of the density F' and the higher derivatives of F, if they exist, are of interest.

The absolute continuity of F.

PROPOSITION 12.1. *Suppose that the measure* $\mathbf{P}$ *is not concentrated at a single point. Then each of the following three conditions is sufficient for the absolute continuity of* F *with respect to* $\lambda \in \mathbf{R}^1$:

(i) $0 \in \mathbf{N_P}$;

(ii) $a \in \mathbf{H}_\mathbf{P}^-$;

(iii) *the space* $\mathbf{X}$ *is strictly convex, that is, each point* $x \in S$ *is an extreme point of* $\mathcal{D}$.

PROOF. If $0 \in \mathbf{N_P}$ or $\mathbf{X}$ is strongly convex, then there exists at most one point $x \in \mathbf{N_P}$ such that $\|x\| = \inf_{y \in \mathbf{N_P}} \|y\|$. By Theorem 11.1 (ii) and Problem 11.3, we see that to verify that $F \ll \lambda$, it suffices to prove that $\mathbf{P}\{x\} = 0$. By assumption, we have $\mathbf{P}\{x\} < 1$. Since $\mathbf{P}\{\mathbf{N_P}\} = 1$, it follows that there exists a point $y \in \mathbf{N_P} \backslash \{\mathbf{X}\}$. Let $g \in \mathbf{X}'$ be a functional that separates the points x and y. Since $x, y \in \mathbf{N_P}$ and $\langle x, g \rangle \neq \langle y, g \rangle$, we see that the distribution of g is nondegenerate, that is, $\mathbf{P}g^{-1} = \mathcal{N}(\langle a, g \rangle, \sigma^2)$, $\sigma > 0$. In particular, $\mathbf{P}\{x\} \le \mathbf{P}\{g^{-1}(g(x))\} = 0$.

If $a \in \mathbf{H}_\mathbf{P}^-$, then $0 \in \mathbf{H}_\mathbf{P}^- = \mathbf{H}_\mathbf{P}^- + a = \mathbf{N_P}$; thus, the second condition implies the first. □

This statement need not hold if $\mathbf{P}$ is Gaussian but is not a Radon measure (see Example 11.2 and the comment on Problem 11.4).

If $0 \in \mathbf{N_P}$ and strong convexity is not assumed, then, generally speaking, Theorem 11.1 provides the strongest possible statement about the distribution of the norm. This is confirmed by the following result.

PROPOSITION 12.2. *Suppose that* $\mathbf{X} = \mathbf{R}^2$, $\mathbf{P} = \mathcal{N}(0,1) \times \mathcal{N}(1,0) \in \mathcal{G}(\mathbf{X})$, $\varphi\colon \mathbf{R}^1 \to (0, +\infty)$ *is a convex even function, and*

$$\lim_{t \to +\infty} \varphi'(t) \in (0, +\infty), \qquad \lim_{t \to +\infty} (\varphi(t) - t\varphi'(t)) \in [0, +\infty).$$

Then there exists a norm $U(\cdot)$ *on* $\mathbf{X}$ *such that* $\mathbf{P}U^{-1} = \mathcal{N}(0,1)\varphi^{-1}$.

PROOF. We define the unit ball of the norm U by

$$\mathcal{D}_U = \left\{ \left(\frac{\theta t}{\varphi(t)}; \frac{\theta}{\varphi(t)} \right),\ |\theta| \le 1,\ t \in \mathbf{R}^1 \right\} \subset \mathbf{R}^2.$$

Elementary calculations show that $\mathcal{D}_U$ is symmetric and convex. We set $U(x) = \inf\{r > 0, x \in r\mathcal{D}_U\}$. Then U is a norm, and $U(x) = \varphi(t)$ for $x = (t, 1)$. □

PROBLEM 12.1. Use Proposition 12.2 to construct a norm $U(\cdot)$ on $\mathbf{R}^2$ such that the distribution of U has an atom at the separation point and the density of U is unbounded at that point and has jumps at countably many points. Find a complete description of the class of distributions $\{\mathbf{P}U^{-1}, U \text{ is a norm on } \mathbf{R}^2\}$.

PROBLEM 12.2. Let $\mathbf{X}$ be a finite-dimensional normed space. Prove that a necessary and sufficient condition for $F \ll \lambda$ is that

$$\sup\{\|y\|, y \in \mathbf{N_P} \cap (x + \varepsilon\mathcal{D})\} > \inf\{\|y\|, y \in \mathbf{N_P}\}$$

for $x \in \mathbf{N_P}$ and $\varepsilon > 0$.

Construct a measure and a norm on $\mathbf{l}^2$ which prove that this condition is only necessary but not sufficient for absolute continuity in the infinite-dimensional case. (*Hint*: the unit ball of the new norm can be taken in the form $\{x \in \mathbf{l}^2 \mid x = \gamma y + (1-\gamma)z,\ 0 \le \gamma \le 1,\ y_1 = 0,\ \sum_2^\infty y_n^2 \le 1, |z_1| = 1,\ \sum_2^\infty n z_n^2 \le 1\}$; furthermore, one can set $\mathbf{N_P} = \{x \in \mathbf{l}^2 \mid x_1 = 1\}$.)

The boundedness of F'. By Theorem 11.1 (vi), F' is uniformly bounded on $\mathbf{R}^1 \setminus [r_0, r_0 + \varepsilon)$. However, to estimate the rate of convergence in the central limit theorem, we would like to have a uniform estimate of F' on $\mathbf{R}^1$.

PROPOSITION 12.3. *Suppose that $(\mathbf{X}, \|\cdot\|)$ is a finite-dimensional space and $\mathbf{P}\{a\} = 0$. If $0 \in \mathbf{N_P}$, then the density F' of the distribution F is uniformly bounded.*

PROOF. Without loss of generality, we can assume that $\mathbf{N_P} = \mathbf{X}$. Since 0 is the separation point of F, it suffices to verify that F' is bounded on the interval $[0,1]$. It follows from the condition $\mathbf{N_P} = \mathbf{X}$ that the density of the measure $\mathbf{P}$ with respect to the Lebesgue measure λ^m, $m = \dim \mathbf{X}$, is bounded. By passing to the radial system of coordinates, we obtain

$$F'(r) \le |S_{m-1}| \max_{\mathbf{X}} \frac{d\mathbf{P}}{d\lambda^m} \left(\max_{\mathbf{X}} \frac{|x|}{\|x\|} \right)^m r^{m-1},$$

where $|x|$ is the Euclidean norm on $\mathbf{X}$ and $|S_{m-1}|$ is the surface area of the unit Euclidean sphere in $\mathbf{X}$. This estimate shows that $\sup_{[0,1]} F' < \infty$, and moreover, $\lim_{r\to 0} F'(r) = 0$ for $m > 1$. □

The most typical cases covered by this result are centered ($a = 0$) and nondegenerate ($\mathbf{X} = \mathbf{N_P}$) measures. For the case of noncentered measures, see Problem 12.1.

In the infinite-dimensional case, the condition $0 \in \mathbf{N_P}$ is no longer sufficient for F' to be bounded. Here we must impose additional conditions either on the measure (essentially, these conditions imply closeness to the finite-dimensional case) or on the norm (by requiring the norm to be smooth and sufficiently "uniformly convex"). However, these additional conditions mostly serve to exclude pathological cases rather than prevent one from considering any specific measures arising in applications.

An example in which F' is unbounded. We shall construct an example of a Gaussian measure for which the distribution density F' is unbounded near zero (other forms of unboundedness are prohibited by Theorem 11.1 (vi). First, we prove the following elementary lemma, which contains the key point of our construction.

LEMMA 12.1. *Suppose that* ξ_k, $k \le n$, *are independent* $\mathcal{N}(0,1)$-*distributed random variables,* p_n *is the distribution density of the random variable* $M_n = \sup_{k\le n} |\xi_k|$, $\theta \in [0,1]$, *and* $r = (2\ln n - \ln\ln n + \theta)^{1/2}$. *Then*

$$p_n(r) \ge 0.13 \ln^{1/2} n \quad \text{and} \quad \mathbf{P}\{M_n \ge 2r\} \le 0.32 \ln^{3/2} n \cdot n^{-3}$$

for $n \ge 3$.

PROOF. We set $\Phi_+(r) = \mathbf{P}\{|\xi_1| \le r\} = 2\Phi(r) - 1$. The following formulas are obvious:

$$\mathbf{P}\{M_n \le r\} = \Phi_+^n(r), \qquad p_n(r) = n\Phi_+'(r)\Phi_+^{n-1}(r),$$
$$\Phi_+'(r) = 2\Phi'(r) = (2\ln n/\pi)^{1/2} n^{-1} \exp\{-\theta/2\},$$
$$\Phi_+^{n-1}(r) = [1-(1-\Phi_+(r))]^{n-1} \ge \exp\{-2(1-\Phi_+(r))(n-1)\}.$$

Moreover,

$$(1-\Phi_+(r))(n-1) \le 2\Phi'(\rho)n/\rho \le \max_n \left(2\ln n/\pi(2\ln n - \ln\ln n)^{1/2}\right) \le 0.625$$

for $\rho = (2\ln n - \ln\ln n)^{1/2}$. Therefore,

$$p_n(r) \ge (2\ln n/\pi)^{1/2} \exp\{-\theta/2 - 1{,}25\} \ge 0.13(\ln n)^{1/2}.$$

Furthermore,

$$\begin{aligned}\mathbf{P}\{M_n \ge 2r\} &\le \mathbf{P}\{M_n \ge 2\rho\} \le n\mathbf{P}\{|\xi_1| \ge 2\rho\} = n(1-\Phi_+(2\rho))\\ &\le n\Phi'(2\rho)/\rho = n(n^{-4}\ln^2 n)/(2\ln n - \ln\ln n)^{1/2}(2\pi)^{1/2}\\ &\le \ln^{3/2} n \cdot n^{-3} \max_n \left(\ln n/(2\pi(2\ln n - \ln\ln n))\right)^{1/2}\\ &\le 0.32\ln^{3/2} n \cdot n^{-3}. \quad \square\end{aligned}$$

REMARK. The sharper estimates in [**22**] show that

$$\mathbf{P}\{(M_n - a_n)/b_n \le r\} \to \exp(-\exp(-r))$$

as $n \to \infty$. Here

$$a_n = \left(2\ln(2n)\right)^{1/2} - \left(\ln\ln(2n) + \ln(4\pi)\right)/\left(2(2\ln(2n))^{1/2}\right),$$
$$b_n = (2\ln(2n))^{-1/2}.$$

Therefore,

$$\sup_{r>0} p_n(r) \sim b_n^{-1} \sup_{r>0} \left\{\exp(-\exp(-r))\exp(-r)\right\} = e^{-1}(2\ln(2n))^{1/2}$$

as $n \to \infty$.

Now let us construct the above-mentioned example. Consider a family of independent random variables $\xi_{k,m}$, $1 \le m < \infty$, $1 \le k \le n_m < \infty$. Suppose that $\xi_{k,m}$ is $\mathcal{N}(0,\sigma_m^2)$-distributed. We specify the lengths of the series n_m and the variances σ_m^2 later. We write $M_m = \max_k |\xi_{k,m}|$ and $M = \sup_{k,m} |\xi_{k,m}| = \sup_m M_m$. Let F be the distribution function and F' the distribution density of the random variable M (by Proposition 12.1, the density exists), and let F_m' be the distribution density M_m.

PROPOSITION 12.4. *There exist sequences σ_m, $y_m \to 0$, and $n_m \to \infty$ such that* $\mathbf{P}\{\lim_m M_m = 0\} = 1$ *and* $F'(y_m) \to \infty$.

PROOF. We proceed by induction. Set

$$\sigma_1 = 1, \qquad n_1 = 3, \qquad y_1 = \sigma_1(2\ln n_1 - \ln\ln n_1)^{1/2}.$$

Suppose that σ_j, n_j, and y_j are already defined for $j < m$. We set

$$y_m = 4^{1-m} y_1$$

and choose n_m so large that

$$n_m \geq \max\left\{n_{m-1} + 1, \exp\left(\mathbf{P}\{\sup_{q<m} M_q \leq y_m/2\}^{-1}\right)\right\}.$$

Finally, we define

$$\sigma_m = y_m(2\ln n_m - \ln\ln n_m)^{-1/2}.$$

The construction is complete. Now let us prove the desired properties. By Lemma 12.1, we have

$$\mathbf{P}\{M_m \geq 2y_m\} \leq 0.32 \ln^{3/2} n_m \, (n_m)^{-3}.$$

Therefore, $\sum_m \mathbf{P}\{M_m \geq 2y_m\} < \infty$. Since $y_m \to 0$, we have $M_m \to 0$ almost surely.

Now let us estimate F'. Since the M_m are independent, we have

$$F'(y_m) \geq \mathbf{P}\{\sup_{q<m} M_q \leq y_m/2\} F'_m(y_m) \mathbf{P}\{\sup_{q>m} M_q \leq y_m/2\}.$$

By construction,

$$\mathbf{P}\{\sup_{q<m} M_q \leq y_m/2\} \geq (\ln n_m)^{-1}.$$

By Lemma 12.1, one has

$$F'_m(y_m) \geq 0.13 \ln^{1/2} n_m \cdot \sigma_m^{-1} = 0.13 \ln^{1/2} n_m (2\ln n_m - \ln\ln n_m)^{1/2} y_m^{-1}.$$

Using the second assertion of the lemma, we obtain

$$\begin{aligned} \mathbf{P}\{\sup_{q>m} M_q > y_m/2\} &\leq \sum_{q>m} \mathbf{P}\{M_q > 2y_q\} \\ &\leq 0.32 \sum_{q>m} (\ln^{3/2} n_q) n_q^{-3} \leq 0.32 \sum_{n\geq 3} n^{-2} \leq 0.13. \end{aligned}$$

Therefore, $\mathbf{P}\{\sup_{q>m} M_q \leq y_m/2\} \geq 0.87$. By combining all these estimates, we obtain

$$F'(y_m) \geq 0.13 \cdot 0.87[(2\ln n_m - \ln\ln n_m)/\ln n_m]^{1/2} y_m^{-1} \geq 0.1 y_m^{-1} \to \infty. \quad \square$$

Let us arrange all the variables $\{\xi_{k,m}\}$ in a single sequence. Let $\mathbf{P}$ be the distribution of this sequence in the space $\mathbf{c}_0$. Proposition 12.4 shows that $\mathbf{P} \in \mathcal{G}_0(\mathbf{c}_0)$, and if $f(x) = \max_k |x_k|$ is the norm on $\mathbf{c}_0$, then the density of the distribution $\mathbf{P}f^{-1}$ is unbounded.

Sufficient conditions for the distribution density of the maximum of a Gaussian sequence to be bounded provide a natural addition to the last example.

Let ξ_k, $1 \leq k < \infty$, be a sequence of independent random variables with distribution $\mathcal{N}(0, \sigma_k^2)$, $\sigma_k > 0$. By $N(\varepsilon)$ we denote the number of indices k for which $\sigma_k \geq \varepsilon$. We introduce the *metric entropy* $H(\varepsilon) = \ln N(\varepsilon)$ and the *Dudley integral* $\Psi(\sigma) = \int_0^\sigma H^{1/2}(\varepsilon)\, d\varepsilon$.

PROPOSITION 12.5. *Suppose that for small σ the Dudley integral is finite and*

$$\lim_{\sigma\to 0} \sigma^{-\theta}\Psi(\sigma) = 0 \tag{12.1}$$

for some $\theta > 0$. Then the distribution density of the random variable $M = \sup_k |\xi_k|$ is uniformly bounded on $\mathbf{R}^1$.

Note that condition (12.1) is not too restrictive. Indeed, for this condition to hold, it suffices to have the estimate $H(\sigma) = o(\sigma^{2\theta-2})$, which, in turn, is close to a necessary condition for M to be finite.

PROOF. Suppose that F is the distribution function of M and r_0 is the separation point of F. Since the Dudley integral is finite, it readily follows that $r_0 = 0$. By Theorem 11.1 (vi), it suffices to examine the behavior of the density $F'(r)$ as $r \to 0$.

Let us choose small numbers r and Δ and estimate the probability of the event that M belongs to the interval $V = [r - \Delta, r]$. To this end, we divide the positive integers into three parts:

$T_1 = \{k \mid \sigma_k > r^{1/2}\}$;

$T_2 = \{k \mid \sigma_k \in [r^{A_1}, r^{1/2}]\}$, where $A_1 = \theta^{-1}$ (by the definition of the Dudley integral, we have $A_1 > 1$);

$T_3 = \{k \mid \sigma_k < r^{A_1}\}$.

Consider the corresponding maxima $M_i = \sup_{k\in T_i} |\xi_k|$. Then

$$\mathbf{P}\{M \in V\} \leq \mathbf{P}\{M_1 \in V\} + \mathbf{P}\{M_2 \in V\}\mathbf{P}\{M_1 \leq r\} + \mathbf{P}\{M_3 \in V\}. \tag{12.2}$$

Let us estimate each summand. We have

$$\begin{aligned}\mathbf{P}\{M_1 \in V\} &\leq \sum_{k\in T_1} \mathbf{P}\{|\xi_k| \in V;\ |\xi_t| \leq r,\ t \neq k\}\\ &\leq N(r^{1/2}) \max_{k\in T_1} \mathcal{N}(0,\sigma_k)\{V \cup -V\} \cdot \big(\mathcal{N}(0,\sigma_k)\{[-r,r]\}\big)^{N(r^{1/2})-1}\\ &\leq \frac{2}{\sqrt{2\pi}} N(r^{1/2}) r^{-1/2}\Delta\Big(\frac{2}{\sqrt{2\pi}} r^{1/2}\Big)^{N(r^{1/2})-1}.\end{aligned}$$

Therefore,

$$\mathbf{P}\{M_1 \in V\} \leq N(r^{1/2}) 2^{2-N(r^{1/2})}\Delta \leq \max_{n>0} n 2^{2-n}\Delta = 2\Delta$$

for $r \leq 0.25$. In a similar way, we obtain the inequality

$$\mathbf{P}\{M_1 \leq r\} \leq r^{N(r^{1/2})/2}. \tag{12.3}$$

We divide the most important region T_2 once more into very narrow regions so that each of them contains variables with almost equal variations. Namely, we have $T_2 = \bigcup T_{2,m}$, where

$$T_{2,m} = \Big\{k \mid \sigma_k \in \big[r^{1/2}(1 + r^{A_2})^{-m}, r^{1/2}(1 + r^{A_2})^{1-m}\big]\Big\},$$

$m = 1, 2, \ldots, m_{\max}(r)$, and $A_2 = 2A_1 + 1$.

One can readily obtain the following power-law estimate of the number of new regions:

$$m_{\max}(r) \le 2A_1 |\ln r| r^{-A_2}.$$

Therefore, in view of (12.3), we have

$$\mathbf{P}\{M_2 \in V\}\mathbf{P}\{M_1 \le r\} \le 2A_1 |\ln r| r^{N(r^{1/2})/2 - A_2} \max_m \mathbf{P}\Big\{M_{2,m} = \sup_{k \in T_{2,m}} |\xi_k| \in V\Big\}.$$

To estimate the distribution density of $M_{2,m}$, we need the converse of Lemma 12.1.

LEMMA 12.2. *Let ξ_k, $1 \le k \le n$, be independent random $\mathcal{N}(0, \sigma_k^2)$-distributed variables such that $\sigma \le \sigma_k \le (1 + \ln^{-1} n)\sigma$. Then the distribution density p_M of $M = \max_k |\xi_k|$ satisfies the estimate*

$$\sup_{v>0} p_M(v) \le 110\sigma^{-1} \ln^{1/2} n.$$

We omit the elementary proof and only note that for $r \le \min(1; \Psi(1)^{-2})$ this lemma can be applied to $M_{2,m}$, since the number of the ξ_k in $T_{2,m}$ has the bound

$$n \le N(r^{A_1}) \le \exp\{\Psi^2(1) r^{-2A_1}\},$$

and the difference between the maximum and minimum variations for the variables occurring in that region does not exceed

$$1 + r^{A_2} \le 1 + \ln^{-1} n.$$

By Lemma 12.2, we have

$$\mathbf{P}\{M_2 \in V\}\mathbf{P}\{M_1 \le r\} \le 220 A_1 |\ln r| r^{N(r^{1/2})/2 - 5A_2/2} \Delta.$$

Now let us examine the last summand in (12.2). Consider the functional $f = M_3$. Let F_3 be is its distribution function. To estimate the density, we apply the estimate (11.22). In our case, the parameters of the functional satisfy the estimates $d = 0$, $m_{F_3} \le 8\sqrt{2}\Psi(r^{A_1})$, and $\sigma \le r^{A_1}$. By condition (12.1), we have

$$m_{F_3} \le 8\sqrt{2}\Psi(r^{A_1}) \le 8\sqrt{2}(0.001)(r^{A_1})^{\theta} \le 0.1r$$

for small r. Substituting the bounds for d, m_{F_3}, and σ into (11.22) and additionally assuming that $\Delta \le 0.1r$, we obtain

$$F_3'(v) \le \frac{v^2}{(0.8)v^2} \cdot \frac{1}{\sqrt{2\pi}} \sup_{\sigma \le r^{A_1}} \sigma^{-1} \exp\left\{-\frac{(0.8v)^2}{2\sigma^2}\right\} \le r^{-A_1} \exp\{-0.32 r^{2-A_1}\}$$

for any $v \in V$. Now that all terms in (12.2) have been estimated, we obtain

$$\overline{\lim_{r \to 0}} \sup_{0 \le \Delta \le 0.1r} \mathbf{P}\{V\}/\Delta. \quad \square$$

Let us extend the above result to the case of a Gaussian process (we have proved the proposition for a sequence of independent random variables). Consider an arbitrary family of centered Gaussian variables $\{\xi_t, t \in T\}$. We set $d(t_1, t_2) = (\mathsf{E}(\xi_{t_1} - \xi_{t_2})^2)^{1/2}$.

The function d is called the *natural semimetric on* T. Without loss of generality, one can assume that (T, d) is a complete metric space, that is, to distinct t there correspond distinct ξ_t and the family $\{\xi_t\}$ is closed under passage to the limit. We introduce the following entropy characteristics:

$N(\varepsilon)$ is the minimum number of balls of radius ε in the space (T, d) which cover T;

$H(\varepsilon) = \ln N(\varepsilon)$ is the metric entropy;

$\Psi(\varepsilon) = \int_0^\varepsilon H^{1/2}(\sigma)\, d\sigma$ is the Dudley integral.

If $\Psi(\varepsilon) < \infty$, then (T, d) is a compact space. This is just the case that we shall consider. Along with the entropy, we need one more characteristic, which is specially adapted for estimating the Gaussian measure of a small ball. Let Ξ be a set in a Hilbert space. For $\varepsilon > 0$ we define the ε-dimension $\nu(\varepsilon)$ of Ξ as the maximum length of a chain $\xi_1, \dots, \xi_\nu$ of elements of Ξ such that $|\xi_1| \geq \varepsilon$ and the distance from ξ_i to $\mathcal{L}(\{\xi_1, \dots, \xi_{i-1}\})$ is at least ε.

Suppose that $\mathbf{P}$ is the distribution of the field ξ, that is, the Gaussian measure on the Banach space of continuous functions $\mathbf{X} = \mathbf{C}(T, d)$ with respect to which each coordinate functional $\pi_{t_1,t_2} : \mathbf{X} \to \mathbf{R}^2$ has the same distribution as the vector (ξ_{t_1}, ξ_{t_2}). Obviously, $\mathbf{P} \in \mathcal{G}_0(\mathbf{X})$. In the Hilbert space $\mathbf{L}^2(\mathbf{X}, \mathbf{P})$, we consider the set Ξ of one-dimensional coordinate functionals $\pi_t : x(\cdot) \to x(t)$. The ε-dimension $\nu(\varepsilon)$ of Ξ has a simple probability meaning. Namely, $\nu(\varepsilon)$ is the maximum length of a chain of time instants $t_1, \dots, t_\nu$ that satisfy the "renewal" condition

$$\mathsf{D}(\xi_{t_i} \mid \xi_{t_1}, \dots, \xi_{t_{i-1}}) \geq \varepsilon^2.$$

The left-hand side (the conditional variation) is nonrandom for a Gaussian field. It is the squared error of the linear prediction of ξ_{t_i} from known values of the preceding elements of the chain.

The following theorem deals with the distribution of the random variables $M = \sup_{t\in T} |\xi_t|$. Obviously, this distribution coincides (for a reasonable definition of M) with the distribution of the norm $\|x\| = \sup_{t\in T} |x(t)|$, $x \in \mathbf{X}$, with respect to the measure $\mathbf{P}$. As usual, let $F(r) = \mathbf{P}\{x \mid \|x\| \leq r\}$ be the distribution density of the norm.

THEOREM 12.1. *Suppose that either conditions* (i) *and* (ii), *or conditions* (iii) *and* (iv) *are satisfied*:

(i) $\overline{\lim\limits_{\sigma\to 0}}\, \Psi(\sigma)\sigma^{-\theta} < \infty$ *for some* $\theta > 0$;

(ii) $\lim\limits_{\sigma\to 0} \ln \nu(\sigma)/\nu(\sigma^\zeta) = 0$ *for some* $\zeta > 0$;

(iii) $\overline{\lim\limits_{\sigma\to 0}}\, N(\sigma)\sigma^{\theta} < \infty$ *for some* $\theta > 0$;

(iv) $\lim\limits_{\sigma\to 0} \nu(\sigma) > 0$.

Then the density F' is uniformly bounded on $\mathbf{R}^1$ except possibly a set of measure zero.

COROLLARY. *Let $T \subset \mathbf{R}^n$. Suppose that the field ξ_t is not identically zero and satisfies the Hölder condition*

$$d(t_1, t_2) \leq c|t_1 - t_2|^\alpha$$

for some c, $\alpha > 0$, and all $t_1, t_2 \in T$.

Then the distribution density of the random variable M is bounded.

PROOF. The assumptions of the corollary imply that conditions (iii) and (iv) are satisfied. □

PROOF OF THE THEOREM. We only outline this proof. We need to verify the assumptions of Theorem 9.1 for the situation in question. To this end, we construct the corresponding sets and directions. We set $\sigma_n = 2^{-n}$ and divide T into the "belts" $T_n = \{t \in T \mid \sigma_n \geq \mathsf{E}\xi_t^2 > \sigma_{n+1}\}$. Then we cover each belt by finitely many "caps" $T_{n,k}$.

We construct the caps as follows. Let $t_1, \ldots, t_{\nu(\sigma_n/3)}$ be the sequence occurring in the definition of $\nu = \nu(\sigma_n/3)$. We choose an orthonormal system $\{e_{n,k}, k \leq \nu\}$ in the space $\mathbf{L}_n \subset \mathbf{L}^2(\mathbf{X}, \mathbf{P})$ generated by coordinate functionals $\pi_{t_1}, \ldots, \pi_{t_\nu}$. For $k \in [\nu+1, 2\nu]$, we set $e_{n,k} = -e_{n,k-\nu}$. For $k \in [1, 2\nu]$, the cap $T_{n,k}$ is defined by the relation

$$T_{n,k} = \big\{t \in T_n \mid \mathsf{E}\xi_t e_{n,k} \geq \sigma_n / 3\sqrt{\nu}\big\}.$$

Let us construct the families of sets $\{E\}$ and $\{B\}$ and of admissible translations $\{l\}$ considered in Theorem 9.1. We define an admissible translation $l_{n,k} \in \mathbf{X}$ by setting $l_{n,k}(t) = \mathsf{E}\xi_t e_{n,k}$.

We set $B_{n,k} = \big\{x \in \mathbf{X} \mid \max_T |x(\cdot)| = \max_{T_{n,k}} |x(\cdot)|\big\}$. The set $B_{n,k}$ must be extended so that the sections of the extended set by lines parallel to the vector $l_{n,k}$ will be sufficiently long segments. Let z be a small number. We set

$$E_{n,k} = \big\{y \in \mathbf{X} \mid y = x + c\,\mathrm{sign}(x) l_{n,k},\ x \in E_{n,k},\ 0 \leq c \leq \sigma_n^z\big\},$$

where $\mathrm{sign}(x)$ is the sign of the function x at the point of maximum modulus. We leave it to the reader to verify that the sum (9.5) is finite. □

The boundedness of F' in a uniformly convex space. The function

$$\delta(\rho) = 1 - \sup\{\|x+y\|/2;\ \|x\| = \|y\| = 1,\ \|x-y\| \geq \rho\}, \qquad 0 < \rho < 2,$$

is called the *modulus* of uniform convexity of the norm $\|\cdot\|$.

If $\delta(\rho) > 0$ for all $\rho > 0$, then the space $(\mathbf{X}, \|\cdot\|)$ is said to be *δ-uniformly convex*. We need some geometric properties of a uniformly convex space which show that the norm varies sufficiently fast along any straight line in $\mathbf{X}$.

LEMMA 12.3. *Let x_0 and l be vectors in a δ-uniformly convex space $\mathbf{X}$. Let $\varphi(c) = \|x_0 + cl\|$. Suppose that $\inf_{c \in \mathbf{R}^1} \varphi(c) = \varphi(0)$. Then*

$$|c| \leq \varphi(c)\|l\|^{-1}\delta^*\left(1 - \frac{\varphi(0)}{\varphi(c)}\right), \tag{12.4}$$

$$|\varphi'(c)| \geq \|l\|\left(1 - \frac{\varphi(0)}{\varphi(c)}\right)\left(\delta^*\left(1 - \frac{\varphi(0)}{\varphi(c)}\right)\right)^{-1} \tag{12.5}$$

for all $c \in \mathbf{R}^1$.

Here $\delta^*(u) = \inf\{\rho \mid \delta(\rho) \geq u\}$ is the inverse function of δ. By φ' in (12.5) *one can understand any one-sided derivative of the convex function φ.*

PROOF. For any $c \neq 0$, there exists a c' such that $\varphi(c) = \varphi(c')$. By the choice of the initial point and the uniform convexity, we have

$$\varphi(0) \leq \varphi\left(\frac{c}{2} + \frac{c'}{2}\right) \leq \varphi(c)\left[1 - \delta\left(\frac{\|(c-c')l\|}{\varphi(c)}\right)\right] \leq \varphi(c)\left[1 - \delta\left(|c|\frac{\|l\|}{\varphi(c)}\right)\right].$$

Hence,

$$\delta\left(|c|\frac{\|l\|}{\varphi(c)}\right) \leq 1 - \frac{\varphi(0)}{\varphi(c)}.$$

This implies (12.4). To obtain (12.5), it suffices to substitute (12.4) into the estimate

$$|\varphi'(c)| \geq \big(\varphi(c) - \varphi(0)\big)|c|,$$

which holds since φ is convex. □

The following lemma characterizes the important class of "radial" Gaussian densities.

LEMMA 12.4. *Let $p(c)$ be a density of the form*

$$p(c) = A|c|^{m-1} \exp\left\{ -\frac{(c-\alpha)^2}{2\sigma^2} \right\} \tag{12.6}$$

in $\mathbf{R}^1$. Then

$$p(c) \leq \min\left\{ \frac{\sigma^{-m} 2^{1-m/2} |c|^{m-1}}{\Gamma(m/2)}; \frac{2}{\sigma \pi^{1/2}} \right\}. \tag{12.7}$$

Note that the uniform estimate for p depends on neither m nor α.

PROOF. The normalizing factor A has the obvious bound

$$A = \left[\int |c|^{m-1} \exp\left\{ -\frac{(c-\alpha)^2}{2\sigma^2} \right\} dc \right]^{-1} \leq \left[\int_0^\infty c^{m-1} \exp\left\{ -\frac{c^2}{2\sigma^2} \right\} dc \right]^{-1}$$
$$= \left[\int_0^\infty (2u)^{m/2-1} \sigma^m \exp(-u)\, du \right]^{-1} = \frac{2^{1-m/2} \sigma^{-m}}{\Gamma(m/2)}.$$

This yields the first part of (12.7).

Now let us obtain the uniform estimate. To be definite, we assume that $\alpha \geq 0$. If $c \leq 0$, then $p(c) \leq p(|c|)$. Therefore, it suffices to estimate $p(c)$ only for $c \geq 0$. By differentiating $p(\cdot)$, we find the point of maximum

$$c_* = \frac{\alpha}{2} + \left(\sigma^2(m-1) + \frac{\alpha^2}{4} \right)^{1/2}.$$

Let $c > c_*$. Then

$$0 \geq \frac{p'}{p}(c) = c^{-1}\left[(m-1) - \frac{c(c-\alpha)}{\sigma^2} \right] = (m-1)\left(\frac{1}{c} - \frac{1}{c_*} \right) - \frac{c - c_*}{\sigma^2}.$$

Taking into account the fact that $cc_* \geq c_*^2 \geq \sigma^2(m-1)$, we obtain

$$0 \geq \frac{p'}{p}(c) \geq -2\sigma^2(c - c_*).$$

By integrating this estimate, we arrive at the inequality

$$p(c) \geq \exp\left\{ -\frac{(c-c_*)^2}{\sigma^2} \right\} p(c_*).$$

Finally,

$$1 \geq \int_{c_*}^\infty p(c)\, dc \geq \int_0^\infty \exp\left\{ -\frac{u^2}{\sigma^2} \right\} du \cdot p(c_*),$$

$$p(c_*) \leq \left[\int_0^\infty \exp\left\{ -\frac{u^2}{\sigma^2} \right\} du \right]^{-1} = \frac{2}{\sigma \pi^{1/2}}. \quad □$$

THEOREM 12.2. *Let* $\mathbf{X}$ *be a uniformly convex space with*

$$\delta(\rho) \geq \Delta\rho^{\beta} \tag{12.8}$$

for some $\Delta > 0$, $\beta \geq 1$, *and* $\rho \in (0, 2]$.

Suppose that $\mathbf{P} \in \mathcal{G}(\mathbf{X})$ *and* $\dim \mathbf{N_P} \geq \beta$. *Then the distribution density of the norm with respect to* $\mathbf{P}$ *is bounded.*

PROOF. Let $m \geq \beta$, and let $z_1, \ldots, z_m$ be an orthogonal system in $\mathbf{X}'_{\mathbf{P},0}$. We set $l_k = Iz_k \in \mathbf{X}$ and construct a partition Γ of the space $\mathbf{X}$ into m-dimensional planes $\gamma = \{x_\gamma + \sum_{k=1}^m c_k l_k,\ c_k \in \mathbf{R}^1\}$. We choose the initial points x_γ so that

$$\|x_\gamma\| = \inf_{x \in \gamma} \|x\|.$$

Let us choose some γ. By virtue of the choice of z and l, the conditional measure $\mathbf{P}_\gamma$ is an m-dimensional Gaussian measure with unit correlation. Within γ, we construct one more partition, Θ, into lines passing through x_γ. Let $\theta = \{x_\gamma + cl,\ c \in \mathbf{R}^1\}$ be one of these lines. We normalize the direction $l \in \mathcal{L}(\{l_k\})$ by the condition $l = Iz$, $|z| = 1$. Then the conditional measure $\mathbf{P}_\theta$ on θ has the form

$$\mathbf{P}_\theta(dc) = p_\theta(c)\,dc = A_\theta |c|^{m-1} \exp\left\{ -\frac{(c - \alpha_\theta)^2}{2} \right\} dc.$$

Let us consider the conditional distribution of the norm with respect to $\mathbf{P}_\theta$. Since the function $f_\theta(c) = \|x_\gamma + cl\|$ is monotone both for $c \geq 0$ and for $c \leq 0$, it follows that the distribution $F_\theta = \mathbf{P}_\theta f_\theta^{-1}$ has the density

$$F'_\theta(r) = \sum_{\{c \mid f_\theta(c) = r\}} \frac{p_\theta(c)}{|f'_\theta(c)|}. \tag{12.9}$$

For $r > \|x_\gamma\|$, this sum contains exactly two terms. We apply Lemma 12.3 to the function f_θ. The substitution (12.5) yields

$$|f'_\theta(c)| \geq \|l\| \left(1 - \frac{\|x_\gamma\|}{r}\right) \left[\Delta^{-1}\left(1 - \frac{\|x_\gamma\|}{r}\right)\right]^{-1/\beta}.$$

We can apply (12.7) and (12.4) to the density p_θ:

$$\begin{aligned} p_\theta(c) &\leq \min\left\{ A_m |c|^{m-1}; \frac{2}{\pi^{1/2}} \right\} \\ &\leq \min\left\{ A_m r^{m-1} \|l\|^{1-m} \left[\Delta^{-1}\left(1 - \frac{\|x_\gamma\|}{r}\right)\right]^{(m-1)/\beta}; \frac{2}{\pi^{1/2}} \right\}, \end{aligned}$$

where $A_m = 2^{1-m/2}/\Gamma(m/2)$.

By combining these estimates with (12.9), we find that

$$\begin{aligned} F'_\theta(r) \leq 2\min\Bigg\{ & A_m r^{m-1} \|l\|^{-m} \Delta^{-m/\beta} \left(1 - \frac{\|x_\gamma\|}{r}\right)^{m/\beta - 1}; \\ & 2\|l\|^{-1} \Delta^{-1/\beta} \left(1 - \frac{\|x_\gamma\|}{r}\right)^{1/\beta - 1} \Big/ \pi^{1/2} \Bigg\}. \end{aligned}$$

Depending on the value of $1 - \|x_\gamma\|/r$, we use the first or the second expression for the minimum and obtain

$$F'_\theta(r) \le 4\pi^{-1/2}\Big(\frac{\pi^{1/2}A_m}{2}\Big)^{(1-\beta)/(1-m)}\|l\|^{-\beta}\Delta^{-1}r^{\beta-1} \equiv A(m,\beta)\|l\|^{-\beta}\Delta^{-1}r^{\beta-1}.$$

Let us average this estimate first with respect to θ and then with respect to γ. Using (3.6), we see that

$$F'(r) \le A(m,\beta)\Big\{\inf_{l\in\mathcal{L}(\{l_k\})}\|l\|\Big\}^{-\beta}\Delta^{-1}r^{\beta-1}; \tag{12.10}$$

that is, the density of the distribution $F = \mathbf{P}f^{-1}$ is bounded on any finite interval. By Theorem 11.1 (vi), the density is uniformly bounded on $\mathbf{R}^1$. □

The strong convexity of the norm alone without the power-law estimate (12.8) does not guarantee the boundedness of F' [**153**].

The spaces $\mathbf{L}^p$, $1 < p < \infty$, can serve as an example of strongly convex spaces in which Theorem 12.1 does not work. In this case, condition (12.8) holds with $\beta = \max\{p; 2\}$. For a Hilbert space, $\beta = 2$ and $\Delta = 0.25$.

Now we shall find out how the position of the barycenter of $\mathbf{P}$ affects uniform estimates of the density of the norm. We already have the estimate (12.10), which is independent of the barycenter. However, this estimate is not uniform.

Suppose that the measure $\mathbf{P}$ is centered, $\mathbf{P}_a\{\cdot\} = \mathbf{P}\{\cdot - a\}$, and $F_a = \mathbf{P}_a f^{-1}$ is the distribution of the norm with respect to the translated measure.

PROPOSITION 12.6. *If* $\mathbf{X}$ *satisfies the assumptions of Theorem* 12.1 *and* $\mathbf{P} \in \mathcal{G}_0(\mathbf{X})$, *then*

$$\sup_{r\in\mathbf{R}^1} F'_a(r) \le A(\mathbf{P},\mathbf{X})(\|a\|+1)^{\beta-1} \tag{12.11}$$

for all $a \in \mathbf{X}$. *Here* $A(\mathbf{P},\mathbf{X})$ *is a constant to be defined in the proof.*

PROOF. By Proposition 11.3, the function $\Psi_a(r) = \Phi^{-1}(F_a(r))$ is concave. Therefore, the derivative

$$\Psi'_a(r) = \frac{F'_a(r)}{\Phi'(\Phi^{-1}(F'_a(r)))} = (2\pi)^{1/2}F'_a(r)\exp\Big\{\frac{\Psi_a^2(r)}{2}\Big\}$$

is nonincreasing. Let $r \ge \|a\|+1$. Then it follows from the inequality $\Psi'_a(r) \le \Psi'_a(\|a\|+1)$ that

$$F'_a(r) \le F'_a(\|a\|+1)\exp\Big\{-\frac{\Psi_a^2(r)}{2} + \frac{\Psi_a^2(\|a\|+1)}{2}\Big\}. \tag{12.12}$$

Let us consider two cases.

1) $F_a(\|a\|+1) \ge 1/2$. Then $0 \le \Psi_a(\|a\|+1) \le \Psi_a(r)$ and $F'_a(r) \le F'_a(\|a\|+1)$.

2) $F_a(\|a\|+1) \le 1/2$. Then we use the estimate

$$F_a(\|a\|+1) = \mathbf{P}\{\|x-a\| \le \|a\|+1\} \ge \mathbf{P}\{\|x\| \le 1\}.$$

Now inequality (12.12) implies the estimate

$$F'_a(r) \le F'_a(\|a\|+1)\exp\Big\{\frac{[\Phi^{-1}\mathbf{P}\{\|x\|\le 1\}]^2}{2}\Big\}.$$

By substituting $r = \|a\| + 1$ into (12.10), we obtain

$$F_a'(r) \le A(\mathbf{P}, \mathbf{X})(\|a\| + 1)^{\beta - 1},$$

where

$$A(\mathbf{P}, \mathbf{X}) = A(m, \beta)\Big\{\inf_{l \in \mathcal{L}(\{l_k\})} \|l\|\Big\}^{-\beta} \Delta^{-1} \exp\left\{\frac{\min^2[\Phi^{-1}\mathbf{P}\{\|x\| \le 1\}, 0]}{2}\right\}. \qquad \square$$

The smoothness of F. We show that if the functional $f(x) = \|x\|$ is both uniformly convex and smooth, then its distribution function F is smooth. Specifically, we prove that the distribution of f belongs to one of the classes $\overline{\mathbf{W}}_1^M$. Recall that the inclusion $\mathbf{P}f^{-1} \in \overline{\mathbf{W}}_1^M$ means that the derivatives $F^{(1)}, \dots, F^{(M)}$ exist, the functions $F, F^{(1)}, \dots, F^{(M-1)}$ are absolutely continuous, and $F^{(M)}$ is a function of bounded variation.

We consider the derivatives of f as $\mathbf{R}^1$-valued polylinear forms. We denote the value of the ith-order derivative at a point $x \in \mathbf{X}$ by $f^{(i)}(x)[x_1, \dots, x_i]$. The norm of this derivative is denoted by $\|f^{(i)}(x)\|$ and is defined by the formula

$$\|f^{(i)}(x)\| = \sup_{\|x_1\| = \dots = \|x_i\| = 1} \big|f^{(i)}(x)[x_1, \dots, x_i]\big|.$$

We say that the norm f has bounded ith-order derivatives if

$$\big|f^{(i)}(x)[x_1, \dots, x_i]\big| \le c\|x\|^{1-i} \prod_{j=1}^{i} \|x_j\| \tag{12.13}$$

for some $c > 0$ and for any $x \ne 0$ and $x_j \in \mathbf{X}$.

This condition is equivalent to the boundedness of $\|f^{(i)}(\cdot)\|$ on the unit sphere in $\mathbf{X}$. It is well known that the spaces $\mathbf{L}^p$ satisfy this condition if p is even or $p > i$.

THEOREM 12.3. *Let the norm have bounded derivatives of orders* $1, 2, \dots, M+1$ *and satisfy condition* (12.8) *of power-law uniform convexity. Suppose that* $\mathbf{P} \in \mathcal{G}(\mathbf{X})$ *and* $\dim \mathbf{N_P} = \infty$. *Then* $\mathbf{P}f^{-1} \in \overline{\mathbf{W}}_1^M$.

PROOF. Theorem 9.3 provides sufficient conditions for the distribution of a functional to belong to $\overline{\mathbf{W}}_1^M$. To use these conditions, we need to verify the integrability of higher-order derivatives and to obtain a lower bound for $|f^{(1)}|$.

We start by estimating the higher-order derivatives. Under the assumptions of the theorem, we have an estimate for $\|f^{(i)}\|$, but we need to verify the integrability of the Hilbert–Schmidt norm $|f^{(i)}|$. Hence, we must establish the relation between the two norms. It is given in the following lemma, which is of interest in itself.

LEMMA 12.5. *Let* $\mathbf{L}\colon \mathbf{X}^i \to \mathbf{R}^1$ *be an i-linear form, and let* $\mathbf{P} \in \mathcal{G}_0(\mathbf{X})$. *Then*

$$|L|^2 = \int_{\mathbf{X}^i} \big|L[x_1, \dots, x_i]\big|^2 \mathbf{P}(dx_1) \dots \mathbf{P}(dx_i).$$

We omit the elementary proof, which includes treating the finite-dimensional case and passing to an infinite-dimensional $\mathbf{X}$ in the limit. We are interested in the following obvious consequence of this lemma:

$$|L|^2 \le \left(\int_{\mathbf{X}} \|x\|^2 \mathbf{P}(dx)\right)^i \|L\|^2.$$

Since, by Proposition 11.6, the norm is integrable with respect to the Gaussian measure, it suffices to verify the moment properties of the function $\|f^{(i)}(\cdot)\|$, $2 \le i \le M+1$. From the boundedness condition (12.13), we obtain the estimate

$$\|f^{(i)}(x)\| \le c\|x\|^{1-i}.$$

The variable $|f^{(i)}(\cdot)|$ has all moments, since

$$\int_{\mathbf{X}} \|x\|^{-d}\,\mathbf{P}(dx) < \infty \tag{12.14}$$

for all $d > 1$.

Let us prove this statement. First, we consider the case in which the measure $\mathbf{P}$ is centered. The support of $\mathbf{P}$ is infinite-dimensional; therefore, in $\mathbf{X}'$ we can choose continuous functionals $f_0, \dots, f_d$ that are orthonormal in $\mathbf{L}^2(\mathbf{X}, \mathbf{P})$. By continuity, we have

$$\|x\|^2 \ge c_1 \sum_{j=0}^{d} |\langle x, f_j\rangle|^2, \qquad c_1 = \min_j(\|f_j\|^{-2} d^{-1}).$$

By passing to the joint distribution of the functionals f_j (that is, the standard Gaussian distribution in $\mathbf{R}^{d+1}$) and then to the polar coordinates, we obtain

$$\begin{aligned}\int_{\mathbf{X}} \|x\|^{-d}\,\mathbf{P}(dx) &\le c_1^{-d/2} \int_{\mathbf{X}} \Big(\sum_{j=0}^{d} |\langle x, f_j\rangle|^2\Big)^{-d/2} \mathbf{P}(dx) \\ &= (2\pi)^{-(d+1)/2} c_1^{-d/2} \int_0^\infty r^{-d} r^d \exp\Big\{-\frac{r^2}{2}\Big\}\,dr < \infty.\end{aligned}$$

We have verified inequality (12.14) for a centered measure. To verify this inequality in the general case, it suffices to estimate the expression

$$\begin{aligned}\int_{\mathbf{X}} \|x+a\|^{-d}\,\mathbf{P}(dx) &= \int_0^\infty r^{-d}\,d\mathbf{P}\{\|x+a\| \le r\} \\ &= d\int_0^\infty \mathbf{P}\{\|x+a\| \le r\} r^{-d-1}\,dr\end{aligned} \tag{12.15}$$

for $\mathbf{P} \in \mathcal{G}_0(\mathbf{X})$. By the corollary of Theorem 10.6, we have

$$\mathbf{P}\{\|x+a\| \le r\} \le \mathbf{P}\{\|x\| \le r\}.$$

Returning to (12.15), we obtain

$$\int_{\mathbf{X}} \|x+a\|^{-d}\,\mathbf{P}(dx) \le \int_{\mathbf{X}} \|x\|^{-d}\,\mathbf{P}(dx) < \infty,$$

thus proving inequality (12.14) for a measure with an arbitrary barycenter.

It remains to obtain a lower bound for $f^{(1)}$. This bound implies the main assumption of Theorem 9.3, that is, the inequality

$$\int_{\mathbf{X}} |f^{(1)}(x)|^{-(2M+\varepsilon)}\,\mathbf{P}(dx) < \infty \tag{12.16}$$

for some $\varepsilon > 0$.

Let us take a positive integer $m > (\beta - 1)(2M + \varepsilon)$ and a system of vectors $\{l_j, 1 \le j \le m\} \subset \mathbf{H_P} \subset \mathbf{X}$ orthonormal in the sense of the Hilbert structure on $\mathbf{H_P}$ defined by the formula $|l|^2 = \mathcal{I}(l)$. Let us consider the truncated gradient

$V(x) = \sum_{j=1}^{m} f^{(1)}(x)[l_j]l_j$. This is the projection in $\mathbf{H_P}$ of the vector $f^{(1)}$ on the linear span of the system $\{l_j\}$. Therefore, $|V| \le |f^{(1)}|$, and for (12.16), it suffices to obtain a lower bound for $|V|$.

We shall use the notation introduced in the proof of Theorem 12.2. Suppose that $x = x_\gamma + cl \in \theta \in \gamma/\Theta$. Then

$$|V(x)|^2 = \sum_{j=1}^{m} f^{(1)}(x)[l_j]^2 \ge \Big|\sum_{j=1}^{m} f^{(1)}(x)[l_j](l,l_j)\Big|^2 = |f^{(1)}(x)[l]|^2 = |f'_\theta(c)|^2.$$

There are two lower bounds for $|f'_\theta(c)|$. If $f_\theta(c) \ge 2f_\theta(0)$, then the convexity of f and (12.4) imply

$$|f'_\theta(c)| \ge \frac{f_\theta(c) - f_\theta(0)}{|c|} \ge \frac{f_\theta(c)}{2|c|} \ge \frac{\|l\|}{2\delta^*(1)} \ge \frac{\|l\|}{2}. \tag{12.17}$$

But if $f_\theta(c) \le 2f_\theta(0)$, then by the definition of δ we have

$$f_\theta(0) \le f_\theta(c)\left[1 - \delta\left(\frac{\|cl\|}{f_\theta(c)}\right)\right] \le f_\theta(c) - f_\theta(0)\delta\left(\frac{\|cl\|}{2f_\theta(0)}\right).$$

Let us estimate $|f'_\theta(c)|$ with the help of inequality (12.8):

$$|f'_\theta(c)| \ge \frac{f_\theta(c) - f_\theta(0)}{|c|} \ge \frac{f_\theta(0)\delta(\|cl\|/2f_\theta(0))}{|c|} \ge \Delta\left(\frac{f_\theta(0)}{|c|}\right)^{1-\beta}\left(\frac{\|l\|}{2}\right)^{\beta}.$$

By integrating this estimate with respect to the radial conditional measure, we obtain

$$\int_{-\infty}^{\infty} |f'_\theta(c)|^{-(2M+\varepsilon)} p_\theta(c)\,dc \le \left(\frac{\|l\|}{2}\right)^{-(2M+\varepsilon)} + \left[\Delta f_\theta(0)^{1-\beta}\left(\frac{\|l\|}{2}\right)^{\beta}\right]^{-(2M+\varepsilon)} \int_{-\infty}^{\infty} |c|^{(1-\beta)(2M+\varepsilon)} p_\theta(c)\,dc.$$

Let us estimate the last integral. We write $A_m = 2^{1-m/2}/\Gamma(m/2)$ and apply (12.7) to p_θ:

$$\int_{-\infty}^{\infty} |c|^{(1-\beta)(2M+\varepsilon)} p_\theta(c)\,dc \le A_m \int_{-1}^{1} |c|^{(1-\beta)(2M+\varepsilon)+m-1}\,dc + 1.$$

We denote the right-hand side by $A(m,M,\varepsilon,\beta)$. It is finite by the choice of m. We integrate over the set of radial rays θ in the space γ and obtain

$$\int_\gamma |V(x)|^{-(2M+\varepsilon)}\,\mathbf{P}_\gamma(dx) \le \left(\inf_\theta \frac{\|l\|}{2}\right)^{-(2M+\varepsilon)} + \left[\Delta\left(\inf_\theta \frac{\|l\|}{2}\right)^{\beta}\right]^{-(2M+\varepsilon)} A(m,M,\varepsilon,\beta) f(x_\gamma)^{(\beta-1)(2M+\varepsilon)}.$$

This estimate is uniform with respect to $\gamma \subset \mathbf{X}$ except for the factor $f(x_\gamma)$. Let us obtain an integrable estimate for this factor.

If a_γ is the barycenter of the measure $\mathbf{P}_\gamma$ and $\beta_1 = (\beta - 1)(2M+\varepsilon)$, then by the definition of x_γ we have $f(x_\gamma) = \|x_\gamma\| \le \|a_\gamma\|$ and

$$f(x_\gamma)^{\beta_1} \le (1 + \|a_\gamma\|)^{1+\beta_1}.$$

Since the power-law function with exponent $1+\beta_1 \geq 1$ is convex, we can use the Jensen inequality to obtain

$$(1+\|a_\gamma\|)^{1+\beta_1} \leq \int_\gamma (1+\|x\|)^{1+\beta_1}\, \mathbf{P}_\gamma(dx).$$

The integration with respect to the quotient measure gives

$$\int_{\mathbf{X}/\Gamma} f(x_\gamma)^{\beta_1}\, \mathbf{P}_\Gamma(d\gamma) \leq \int_{\mathbf{X}} (1+\|x\|)^{1+\beta_1}\, \mathbf{P}_\gamma(dx).$$

By Proposition 11.6, the right-hand side is finite. By combining the above estimates, we see that

$$\int_{\mathbf{X}} |V(x)|^{-(2M+\varepsilon)}\, \mathbf{P}(dx) < \infty.$$

This implies (12.16). Now the application of Theorem 9.3 is justified, and therefore, we have $\mathbf{P}f^{-1} \in \overline{\mathbf{W}}_1^M$. □

Appendix

1. Suppose that $(\mathbf{X}, \|\cdot\|)$ is an infinite-dimensional separable Banach space. For each $\varepsilon > 0$, one can construct a norm $|\cdot|$ on $\mathbf{X}$ and a measure $\mathbf{P} \in \mathcal{G}_0(\mathbf{X})$ with the following properties:

a) $(1-\varepsilon)\|x\| \leq |x| \leq (1+\varepsilon)\|x\|$ for all $x \in \mathbf{X}$;

b) if $F(r) = \mathbf{P}\{x \mid |x| \leq r\}$, then the density F' is unbounded.

If $\mathbf{X}$ is a Hilbert space, then one can choose the norm $|\cdot|$ so that it is infinitely differentiable and its derivatives are bounded in the sense of (12.13) [**151, 152**].

By comparing this statement with Theorem 12.2, we see that the norm $|\cdot|$ cannot be uniformly convex of a power-law order.

2. There exist a sequence of independent Gaussian random variables $\{\xi_k\}$, $1 \leq k < \infty$, such that the distribution function F of the random variable $M = \sup_k |\xi_k|$ has a bounded derivative F', whereas F'' (the derivative of the distribution density) is unbounded [**77**].

For any function $\varphi\colon (0,\infty) \to [0,\infty)$ such that $\lim_{r\to 0}\varphi(r) = 0$, there exists a sequence $\{\xi_k\}$ such that [**75, 84**]

$$\sup_{r,\varepsilon>0} \frac{F(r+\varepsilon)-F(r)}{\varphi(\varepsilon)} = +\infty.$$

3. Suppose that $\mathbf{X}$ is a Hilbert space, $\mathbf{P} \in \mathcal{G}(\mathbf{X})$, a is the barycenter of $\mathbf{P}$, $\sigma_1^2 \geq \sigma_2^2 \geq \dots$ are the eigenvalues of the correlation operator, and F is the distribution function of the Hilbert norm with respect to $\mathbf{P}$. Then

$$\sup_{r\geq 0} F'(r) \leq c(\sigma_1,\sigma_2)^{-1}\left(\left(\sum_{k=1}^{\infty}\sigma_k^2\right)^{1/2} + \|a\|\right),$$

where c is an absolute constant. It is impossible to use only one eigenvalue in this estimate.

For the Hilbert space and the spaces $\mathbf{l}^p$, there are different, more complicated estimates of the dependence between F' and the correlation operator and the position of the barycenter [**80, 84**].

4. Suppose that $\{\xi_k\}$, $1 \leq k < \infty$, is a sequence of independent random $\mathcal{N}(0, \sigma_k^2)$-distributed variables and that the variances σ_k^2 decrease regularly:

$$\sigma_1 = b_1, \quad \sigma_k = (\ln k)^{-1/2} b_k, \quad k \geq 2,$$

where $\{b_k\}$ is a monotone sequence that tends to zero. Let $b(\cdot)\colon \mathbf{R}_+ \to \mathbf{R}_+$ be a monotone function with $b(k) = b_k$. We denote the inverse of b by b^*.

If for some $\alpha \in (0, 1/2)$ and any $\beta > 1$ we have

$$\lim_{t \to 0} \ln b^*(t) / \big(b^*(\beta t)\big)^{\alpha} = 0,$$

then the distribution density of the random variable $M = \max_k |\xi_k|$ is bounded [**82–84**].

Although this statement is close to Proposition 12.5, it neither follows from nor implies it. Indeed, the Paulauskas condition is based on the fact that the sequence σ_k decreases *regularly*, whereas (12.1) implies a restriction on the rate of decrease of σ_k.

5. Suppose that $\mathbf{X}$ is a Hilbert space and $\mathbf{P} \in \mathcal{G}_0(\mathbf{X})$ is the measure constructed in Example 7.3. Then the random variable $Q = \|x\|^2 = \sum_j \sigma_j^2 \xi_j^2$ is a quadratic form of independent $\mathcal{N}(0,1)$-distributed variables. Let $F(r) = \mathbf{P}\{x \mid Q(x) \leq r\}$. There exists extensive literature dealing with the properties of F; for example, see [**23, 50, 110, 117**]. It follows from Theorem 12.3 that F has derivatives of all orders. If $\sigma_j > 0$, then each derivative is uniformly bounded on $\mathbf{R}_+$.

Now we present an estimate for F' that can also be used for quadratic forms whose coefficients have alternating signs. Let

$$Q = \sum_j \alpha_j \xi_j^2, \qquad F = \mathbf{P}\{x \mid Q(x) \leq r\}.$$

We set

$$u(\delta) = \sup_{\alpha_j} \sup_{r \in \mathbf{R}^1} \Big\{ F'(r) \mid \sum \alpha_j^2 = 1, \sum |\alpha_j| < \infty, \sum \alpha_j^4 = \delta \Big\}$$

for $\delta > 0$. Then $\lim_{\delta \to 0} u(\delta) = (2\sqrt{\pi})^{-1}$. Moreover, if $1/3 < \delta < 1/2$, then

$$u(\delta) \leq c_0 - c_1 \ln(1 - 2\delta), \qquad c_0 = 0.77, \quad c_1 = 0.43.$$

For $\delta = 1/2$, the unpleasant singularity on the right-hand side is primarily due to the hyperbolic form $Q = (\xi_1^2 - \xi_2^2)/\sqrt{2}$, whose distribution density is unbounded [**89**].

6. Suppose that $(\mathbf{X}, \|\cdot\|)$ is a separable Banach space, $\mathbf{P} \in \mathcal{G}_0(\mathbf{X})$, and $F(r) = \mathbf{P}\{x \mid \|x\| \leq r\}$. Then there exists a $\mathbf{P}$-dependent constant c such that

$$1 - F(r) \geq c^{-1} \exp\{-c(r + \varepsilon)\varepsilon\}[1 - F(r - \varepsilon)]$$

for all $\varepsilon > 0$ and $r \geq \varepsilon$. If $r \geq \max\{1; \varepsilon\}$, then

$$F(r + \varepsilon) - F(r - \varepsilon) \leq c(r + 1)\varepsilon[1 - F(r - \varepsilon)].$$

If the norm has a bounded distribution density, then the last inequality holds for all $\varepsilon \geq 0$ and $r \geq \varepsilon$.

These statements can readily be derived from Proposition 11.3 and Theorem 11.6, respectively, (see [**2, 40**]; both inequalities occurred for the first time in [**2**]).

CHAPTER 4

Poisson Functionals

§13. The configuration space

Let $\mathbf{G}$ be a separable metric[1] space. Let us define some notions related to the space of configurations in $\mathbf{G}$. Informally speaking, a *configuration* is a locally finite subset of $\mathbf{G}$ with repetitions, that is, with multiple points. The main difference from the standard definition of configurations is that we allow repetitions.

Let $\mathfrak{B} = \mathfrak{B}_G$ be the Borel σ-algebra and $\mathfrak{B}_0 \subset \mathfrak{B}$ the ring of bounded Borel sets. A sequence $x_1, x_2, \dots$ (possibly finite) of points of $\mathbf{G}$ is said to be *locally finite* if any bounded set contains only finitely many terms of this sequence.

Two locally finite sequences $\{x_i\}$ and $\{y_i\}$ are said to be *equivalent* if there is a bijection σ between their indexing sets such that $y_i = x_{\sigma(i)}$ for all i.

An equivalence class of locally finite sequences is called a *configuration* in $\mathbf{G}$.

We denote the space of all configurations in $\mathbf{G}$ by $\mathfrak{X} = \mathfrak{X}(G)$. Elements of $\mathfrak{X}$ are usually denoted by X, Y, or Z. A sequence $\{x_i\}$ from an equivalence class X is called a *representing sequence* of X. In a natural way, we define the membership relation $x \in X$ (a point belongs to a configuration), the multiplicity of a point in a configuration, the intersection $X \cap V$ ($V \subset \mathbf{G}$), and the number $|X \cap V|$ of points in this intersection. In the last two cases, we count the points with their multiplicities.

Furthermore, let Σ be the minimal possible σ-algebra on $\mathfrak{X}(\mathbf{G})$ with respect to which all mappings of the form $X \mapsto |X \cap V|$, $X \in \mathfrak{X}$, where $V \in \mathfrak{B}_0$, are measurable.

If $\mathbf{G}$ is a bounded set, then we have the simplest configuration space in which all configurations are finite and $\mathfrak{X}(\mathbf{G})$ is naturally identified with the set $\{\varnothing\} \cup \bigcup_{n=1}^{\infty} \mathbf{G}^n/\pi_n$. Here $\mathbf{G}^n/\pi_n$ is the quotient of $\mathbf{G}^n$ with respect to the permutation group π_n. Note that in this case the σ-algebra Σ is the union of the σ-algebras $\overline{\pi}_n \mathfrak{B}(\mathbf{G}^n)$, where $\overline{\pi}_n \colon \mathbf{G}^n \to \mathbf{G}^n/\pi_n$ is the natural projection. In the general situation, when the space $\mathbf{G}$ is unbounded, for any bounded $\mathbf{G}_1 \subset \mathbf{G}$ we have the natural bijection $X \leftrightarrow Y \cup Z$ between $\mathfrak{X}(\mathbf{G})$ and the set

$$\mathfrak{X}(\mathbf{G} \setminus \mathbf{G}_1) \cup \Big(\{\varnothing\} \cup \bigcup_{n=1}^{\infty} \mathbf{G}_1^n/\pi_n\Big). \tag{13.1}$$

Let us define the *canonical bijection* $\sigma_{s,t}$ that relates any two representing sequences $s = \{x_i\}$ and $t = \{y_i\}$, $s, t \in X$. This bijection is uniquely determined by the following two conditions:

1) $\sigma_{s,t}$ takes s to t; that is, $y_i = x_{\sigma_{s,t}(i)}$ for each i;

[1] The metric structure on $\mathbf{G}$ is inessential to our considerations. What really matters is the assumption that singletons are measurable and a class of "bounded" subsets of G is reasonably defined.

2) $\sigma_{s,t}$ preserves the order at each point of the configuration; that is, if $i < j$ and $x_i = x_j$, then $\sigma_{s,t}(i) < \sigma_{s,t}(j)$.

Now let us give the definition of a function $f\colon X \to M$, where X is a configuration and M is a set. First, we define the *representation* of f on each representing sequence of X, and then impose the consistency condition on these representations. A sequence $i \mapsto f(s,i) \in M$ is called the *representation* of f on a sequence $s = \{x_i\}$. Furthermore, $f(s,i)$ is called the *value* of f at the element x_i of the representing sequence s. The representations of f on representing sequences s and t is said to be *consistent* if $f(t,i) = f(s, \sigma_{s,t}(i))$ for any i. We say that a function f is defined on the configuration X if a system of consistent representations of f is given on all representing sequences of X. The meaning of this definition is obvious: if a point $x \in G$ belongs (with multiplicity k) to the configuration X, then f assigns to x an ordered sequence of k elements of M. If X does not have multiple points, then we have the ordinary definition of a function. In the general case, we can assume that we take several copies of each multiple point and assign a value of the function to each of these copies. Obviously, each function f is uniquely determined by its representation on any representing sequence.

Poisson random measures. Let Π be a nonnegative measure on $\mathfrak{B}$ such that Π is finite on $\mathfrak{B}_0$. A family of random variables $\{\nu(A),\ A \in \mathfrak{B}_0\}$ is called a *Poisson random measure of intensity* Π if for any $A \in \mathfrak{B}_0$ the random value $\nu(A)$ has the Poisson distribution with parameter $\Pi(A)$, that is,

$$\mathbf{P}\{\nu(A) = n\} = e^{-\Pi(A)} \frac{(\Pi(A))^n}{n!}$$

for $n = 0, 1, \dots$, and for any finite family $\{A_i\}$ of disjoint sets $A_i \in \mathfrak{B}_0$, the variables $\nu(A_i)$ are independent.

A Poisson random measure can be realized on the configuration space $(\mathfrak{X}, \Sigma)$. First, by setting

$$\mathbf{P}\big(\{X \mid |X \cap V| = k\}\big) = e^{-\Pi(V)} \frac{(\Pi(V))^k}{k!}, \tag{13.2}$$

we define a set function $\mathbf{P}$ on the sets of the form $\{X \mid |X \cap V| = k\}$, where $V \in \mathfrak{B}_0$ and $k \geq 0$.

One can prove that this function can be extended to a probability measure $\mathbf{P}$ on Σ (e.g., see [**53**]).

We define a family of random variables $\{\nu(V), V \in \mathfrak{B}_0\}$ on the probability space $(\mathfrak{X}, \Sigma, \mathbf{P})$ by setting

$$\nu(V, X) = |X \cap V|.$$

One can readily show that the family $\{\nu(V), V \in \mathfrak{B}_0\}$ is a Poisson random measure. We also note that ν is indeed a random measure: for a fixed $X \in \mathfrak{X}$, $\nu(\cdot, X)$ is the measure on $(\mathbf{G}, \mathfrak{B})$ equal to $\sum_{x \in X} m(x)\delta_x$, where $m(x)$ is the multiplicity of the point x. The measure $\mathbf{P}$ is also called a *Poisson measure of intensity* Π.

Along with ν, we often consider the centered measure $\widetilde{\nu} = \nu - \Pi$, which satisfies $\mathsf{E}\widetilde{\nu}(V) = 0$ for any $V \in \mathfrak{B}_0$.

As was already mentioned, for bounded $\mathbf{G}$ it is natural to identify the space $\mathfrak{X}(\mathbf{G})$ with the set $\{\varnothing\} \cup \{\bigcup_{n=1}^{\infty} \mathbf{G}^n/\pi_n\}$. Let us show that in this case we have

$$\mathbf{P} = e^{-\Pi(\mathbf{G})}\Big(\mathbf{1}_{\{\varnothing\}} + \sum_{n=1}^{\infty} \frac{1}{n!}(\Pi^n)\pi_n^{-1}\Big). \tag{13.3}$$

To this end, it suffices to verify that (13.2) is satisfied for the measure $\mathbf{P}$ defined by (13.3). Indeed, let $V \subset \mathbf{G}$. Then

$$\begin{aligned} \mathbf{P}\big(\{X \mid |X \cap V| = k\}\big) &= e^{-\Pi(\mathbf{G})} \sum_{n=k}^{\infty} C_n^k (\Pi(V))^k (\Pi(\mathbf{G} \setminus V))^{n-k} \frac{1}{n!} \\ &= e^{-\Pi(\mathbf{G})} \frac{(\Pi(\mathbf{G}))^k}{k!} \sum_{n=k}^{\infty} (\Pi(\mathbf{G} \setminus V))^{n-k} \frac{1}{(n-k)!} \\ &= e^{-\Pi(V)} \frac{(\Pi(V))^k}{k!} \end{aligned}$$

for $k > 0$ (for $k = 0$, (13.2) is obvious).

Stochastic integrals. Stochastic integrals provide the most important examples of functionals on the configuration space. We need the following definitions.

Let $(\mathbf{B}, \|\cdot\|)$ be a separable Banach space, $\varphi\colon \mathbf{G} \to \mathbf{B}$ a measurable function vanishing outside some bounded set (we denote the class of such functions by $\mathbf{I}_0(\mathbf{G}, \mathbf{B})$), and ν a Poisson random measure of intensity Π defined on the probability space $(\mathfrak{X}, \Sigma, \mathbf{P})$. The random variable

$$X \mapsto \sum_{x \in X} \varphi(x) = \int_{\mathbf{G}} \varphi(x) \nu(dx, X) \tag{13.4}$$

is called the *stochastic integral* of φ with respect to ν and is denoted by $\int \varphi\, d\nu$. If $\varphi \in \mathbf{L}^1(\mathbf{G}, \Pi)$, then, by definition, we set

$$\int \varphi\, d\tilde{\nu} = \int \varphi\, d\nu - \int \varphi\, d\Pi.$$

Now let us calculate the mean value of the stochastic integral.

PROPOSITION 13.1. *For any* $\varphi \in \mathbf{I}_0(\mathbf{G}, \mathbf{B}) \cap \mathbf{L}^1(\mathbf{G}, \Pi)$, *one has*

$$\mathsf{E} \int \varphi\, d\nu = \int \varphi\, d\Pi.$$

PROOF. Let $\mathbf{G}_1 = \{x \in \mathbf{G} \mid \varphi(x) \neq 0\}$. Since the values of a random measure on disjoint sets are independent, it follows from (13.3) that

$$\begin{aligned} \mathsf{E} \int \varphi\, d\nu &= e^{-\Pi(\mathbf{G}_1)} \sum_{n=1}^{\infty} \frac{1}{n!} \int_{\mathbf{G}_1^n} \sum_{i=1}^{n} \varphi(x_i) \Pi(dx_1) \dots \Pi(dx_n) \\ &= e^{-\Pi(\mathbf{G}_1)} \sum_{n=1}^{\infty} \frac{(\Pi(\mathbf{G}_1))^{n-1}}{(n-1)!} \int_{\mathbf{G}_1} \varphi\, d\Pi = \int \varphi\, d\Pi. \quad \square \end{aligned}$$

Note that (13.4) gives a function defined on the entire configuration space $\mathfrak{X}$. Obviously, this is possible only for $\varphi \in \mathbf{I}_0(\mathbf{G}, \mathbf{B})$. In what follows, we extend the class of integrable functions. The price we have to pay is that the integral will be defined on a subset of $\mathbf{P}$-full measure in $\mathfrak{X}$ rather than on the entire $\mathfrak{X}$.

A function $\varphi\colon \mathbf{G} \to \mathbf{B}$ is said to *belong to the class* $\mathbf{I}(\mathbf{G}, \Pi)$ if for any $\varepsilon > 0$ there exists a bounded set $V \in \mathfrak{B}_0$ such that

$$\mathbf{P}\Big\{X \in \mathfrak{X} \mid \Big\| \int \varphi \mathbf{1}_{V'}\, d\nu \Big\| > \varepsilon \Big\} < \varepsilon \tag{13.5}$$

for any bounded measurable subset $V' \subset \mathbf{G} \setminus V$. Likewise, φ is said to *belong to* $\widetilde{\mathbf{I}}(\mathbf{G}, \Pi)$ if φ is locally integrable (that is, integrable on each bounded set) and for each $\varepsilon > 0$ there exists $V \in \mathfrak{B}_0$ such that

$$\mathbf{P}\Big\{X \in \mathfrak{X} \mid \Big\| \int \varphi \mathbf{1}_{V'}\, d\widetilde{\nu} \Big\| > \varepsilon \Big\} < \varepsilon$$

for any $V' \in \mathfrak{B}_0$ with $V' \subset \mathbf{G} \setminus V$. Now let us choose an increasing sequence $\{V_n\}$ such that $V_n \in \mathfrak{B}_0$, $V_n \uparrow \mathbf{G}$, and (13.5) is satisfied for $V = V_n$ and $\varepsilon = 1/n$. In this case, the sequence of random vectors $\int \varphi \mathbf{1}_{V_n}\, d\nu$ is a Cauchy sequence in probability. The corresponding limit is called the *stochastic integral* and is denoted by $\int \varphi\, d\nu$. In a similar way we define $\int \varphi\, d\widetilde{\nu}$ for $\varphi \in \widetilde{\mathbf{I}}(\mathbf{G}, \Pi)$. One can readily show that these integrals are independent of the choice of the sequence V_n. We also note that in this case convergence in probability, as well as the weak convergence of the corresponding distributions, automatically follows from convergence almost everywhere (e.g., see [**16,** Ch. V, §2]).

PROPOSITION 13.2. *Let* $\mathbf{B}$ *be a Hilbert space, and let* φ *be a function such that* $\|\varphi\| \in \mathbf{L}^2(\mathbf{G}, \Pi)$. *Then* $\varphi \in \widetilde{\mathbf{I}}(\mathbf{G}, \Pi)$.

PROOF. Let $V \in \mathfrak{B}_0$. Then

$$\begin{aligned}
\mathsf{E}\Big\| \int \mathbf{1}_V \varphi\, d\widetilde{\nu} \Big\|^2 &= \int \Big(\sum_{x \in X \cap V} \varphi(x), \sum_{x \in X \cap V} \varphi(x) \Big)\, d\mathbf{P} - \Big\| \int_V \varphi\, d\Pi \Big\|^2 \\
&= e^{-\Pi(V)} \sum_{n=1}^{\infty} \frac{1}{n!} \Big(\int_{V^n} \sum_{i,j=1}^{n} (\varphi(x_i), \varphi(x_j)) \Pi(dx_1) \dots \Pi(dx_n) \Big) - \Big\| \int_V \varphi\, d\Pi \Big\|^2 \\
&= e^{-\Pi(V)} \sum_{n=1}^{\infty} \frac{1}{n!} \Big(n \int_V \|\varphi\|^2\, d\Pi (\Pi(V))^{n-1} \\
&\qquad + n(n-1) \int_V (\varphi(x), \varphi(y))\, \Pi(dx)\Pi(dy) \Big) - \Big\| \int_V \varphi\, d\Pi \Big\|^2 \\
&= \int_V \|\varphi\|^2\, d\Pi.
\end{aligned}$$

This identity together with Chebyshev's inequality readily implies (13.5). □

The notion of a stochastic integral proves useful for the description of infinitely divisible distributions. Specifically, suppose that $\mathbf{B}$ is a separable Banach space and $u \in \mathbf{B}$ is an infinitely divisible random vector. (Just as in the one-dimensional case, the infinite divisibility means that for each n the random vector u can be represented in the form $u = u_1 + \dots + u_n$, where the u_i are independent identically distributed random vectors; here and in what follows the equality of vectors is understood in the sense that their distributions coincide.)

Infinitely divisible random vectors admit the *Lévy–Khinchin representation*; namely, any such vector u can be uniquely represented as the sum of three independent terms,

$$u = u_0 + u_1 + u_2, \tag{13.6}$$

where $u_0 \in \mathbf{B}$ is a constant vector, u_1 is a centered Gaussian vector, and

$$u_2 = \int_{\{\|x\|\le 1\}} x\, d\widetilde{\nu} + \int_{\{\|x\|>1\}} x\, d\nu. \tag{13.7}$$

Here ν is a Poisson random measure[2] of intensity Π on $\mathbf{B}$. (The measure Π is called the *Lévy measure* of the vector u). Note that Π satisfies the conditions $\Pi(\{0\}) = 0$ and $\int \min(\|x\|^2, 1)\, d\Pi(x) < \infty$.

Let us consider two important classes of infinitely divisible random vectors.

Stable vectors. A random vector u is said to be *stable* if the following assertion holds for independent random vectors u_1 and u_2 whose distributions coincide with the distribution of u. For any $a, b > 0$, there exists a number c and a nonrandom vector x such that the distributions of the vectors $au_1 + bu_2$ and $cu + x$ coincide. If $x = 0$, then u is said to be *strongly stable*. The distribution of a stable random vector is also called *stable*.

A vector u is said to be *stable with exponent* α (α*-stable*) if the numbers a, b, and c satisfy the relation $a^\alpha + b^\alpha = c^\alpha$.

PROPOSITION 13.3. *Each stable random vector is α-stable for some $\alpha \in [0, 2]$. The case $\alpha = 2$ corresponds to Gaussian vectors; in this case, the strong stability means that the mean value is zero.*

The proof can be found in [**145**].

Note that if a vector u is α-stable, then

$$u = \frac{1}{n^{1/\alpha}}(u_1 + \cdots + u_n) + x = \frac{1}{n^{1/\alpha}}\Big((u_1 + n^{1/\alpha-1}x) + \cdots + (u_n + n^{1/\alpha-1}x)\Big)$$

(here the u_i are independent copies of u and $x = x(n) \in \mathbf{B}$), and hence, each stable vector is infinitely divisible.

We denote the unit sphere by $S = \{x \in \mathbf{B} \mid \|x\| = 1\}$ and introduce the coordinates (r, θ) in $\mathbf{B} \setminus \{0\}$, where $r = r(x) = \|x\| \in (0, \infty)$ and $\theta = \theta(x) = x\|x\|^{-1} \in S$.

THEOREM 13.1. *Let u be an α-stable vector with $\alpha \ne 2$. Then $u_1 = 0$ in the representation* (13.6), *and the Lévy measure Π has the following form in the coordinates (r, θ):*

$$\Pi(dr, d\theta) = \frac{dr}{r^{1+\alpha}}\pi\,(d\theta), \tag{13.8}$$

where π is some finite measure on S, called the spectral measure.

The proof of this theorem can also be found in [**145**].

---

[2] When this measure is defined, the metric on $\mathbf{B}$ must be chosen so that the sets bounded away from zero are bounded.

Processes with independent increments. A process $\xi(t)$, $t \in \mathbf{R}_+$, is called a *process with independent increments* if the random variables $\xi(t_0), \xi(t_1) - \xi(t_0), \dots, \xi(t_k) - \xi(t_{k-1})$ are independent for all $t_0 < t_1 < \cdots < t_k$ in $\mathbf{R}_+$.

If the distribution of $\xi(t+h) - \xi(h)$ depends only on h, then the process is said to be *homogeneous*. In what follows, we assume that the process takes values in $\mathbf{R}^d$, $d \geq 1$.

The following theorem describes the structure of an arbitrary stochastically continuous process with independent increments.

THEOREM 13.2. *If $\xi(t)$ is a separable stochastically continuous process with independent increments, then there exist a Gaussian process $\xi_1(t)$ with trajectories continuous with probability* 1 *and a Poisson random measure ν of intensity* Π *independent of ξ_1 on $\mathbf{R}_+ \times \mathbf{R}^d$ such that*

$$\xi(t) = \xi_1(t) + \iint_{\substack{0 \leq s < t \\ |x| \leq 1}} x\widetilde{\nu}\,(ds\,dx) + \iint_{\substack{0 \leq s < t \\ |x| > 1}} x\nu\,(ds\,dx). \tag{13.9}$$

The measure Π is called the *Lévy measure* of the process ξ. It satisfies the condition

$$\iint_{[0,t) \times \mathbf{R}^d} \min(|x|^2, 1)\Pi(ds\,dx) < \infty$$

for any $t \in \mathbf{R}_+$. We also note that for homogeneous processes the measure $\Pi(ds\,dx)$ has the form $ds\,\Pi_1(dx)$.

PROBLEM 13.1. Let $\mathfrak{X}_1 \subset \mathfrak{X}$ be the set of configurations that have no multiple points, and let the measure Π vanish on singletons in $\mathbf{G}$. Prove that $\mathfrak{X}_1 \in \Sigma$ and $\mathbf{P}\{\mathfrak{X}_1\} = 1$.

PROBLEM 13.2. Let Π be some finite measure in a Banach space $\mathbf{B}$. Consider the probability measure $\exp \Pi = e^{-\Pi(\mathbf{B})} \sum_{k=0}^{\infty} \Pi^k/k!$ (the power is understood in the sense of convolution). Prove that $\exp \Pi$ coincides with the distribution of the random variable $\int x\,d\nu$, where ν is a Poisson random measure of intensity Π.

PROBLEM 13.3 (change of variables in the stochastic integral). Let $g\colon \mathbf{G} \to \mathbf{G}_1$ be a mapping of $\mathbf{G}$ into some set $\mathbf{G}_1$. A subset $A \subset \mathbf{G}_1$ is said to be *bounded* if $g^{-1}(A)$ is a bounded set. Let $\mathfrak{X}_1$ be the set of configurations in $\mathbf{G}_1$ (with respect to the class of bounded sets). We set $\mathcal{G}(X) = \{g(x), x \in X\}$ and thus define a mapping $\mathcal{G}\colon \mathfrak{X} \to \mathfrak{X}_1$. Furthermore, suppose that $\mathbf{P}$ is a Poisson measure of intensity Π on Σ and ν is the corresponding Poisson random measure. Prove that $\mathbf{P}_1 = \mathbf{P}\mathcal{G}^{-1}$ is a Poisson measure of intensity Πg^{-1} on $\mathcal{G}(\Sigma)$ and

$$\int \varphi(g)\,d\nu = \int \varphi\,d\nu_1$$

(where ν_1 is a Poisson random measure of intensity Πg^{-1}).

PROBLEM 13.4. Choose a metric on $\mathbf{B}$ so that on the space $\mathfrak{X}(\mathbf{B})$ corresponding to this metric one can define a Poisson measure $\mathbf{P}$ whose intensity has the form (13.8).

§14. Differential calculus on the configuration space

The local structure of the configuration space. We make some additional assumptions about the structure of the space $\mathbf{G}$. Namely, suppose that $\mathbf{S}$ is a complete separable metric space and $I \subset \mathbf{R}^m$ ($m \geq 1$) is an open set (also treated as a metric space with respect to the Euclidean metric). We set $\mathbf{G} = \mathbf{S} \times I$ and equip this space with some metric[3] consistent with the direct product structure and the Borel σ-algebra $\mathfrak{B}_G$.

On the configuration space $\mathfrak{X}(\mathbf{G})$, we introduce the structure of a "plane Hilbert manifold along I"; namely, for each configuration $X \in \mathfrak{X}(\mathbf{G})$ we consider the Hilbert space $\mathbf{H}(X)$ of square integrable mappings $X \to \mathbf{R}^m$,

$$\mathbf{H}(X) = \Big\{h\colon X \to \mathbf{R}^m \mid |h|^2_{\mathbf{H}(X)} = \sum_{x\in X} |h(x)|^2_{\mathbf{R}^m} < \infty\Big\}. \tag{14.1}$$

The space $\mathbf{H}(X)$ has the meaning of the *tangent space along* I at a point $X \in \mathfrak{X}$. Indeed, for $h \in \mathbf{H}(X)$, we can define the configuration $X + h \in \mathfrak{X}$ that consists of the points $x+h(x)$, $x \in X$. In this formula, the summation "along I" is understood as follows: we represent each point $x \in \mathbf{G}$ in the form $x = (\theta, r)$, $\theta \in \mathbf{S}$, $r \in I$, and, by definition, set $x + a = (\theta, r + a)$ for $a \in \mathbf{R}^m$ such that $r + a \in I$.

The family

$$Q = Q(V) = \{Q(X)\colon \mathbf{H}(X) \to \mathbf{H}(X) \mid X \in \mathfrak{X}\}$$

of mappings $Q(X)$ of the form $h \mapsto \mathbf{1}_{X\cap V}h$ is called the *orthogonal projection* defined by a Borel set V. The orthogonal projections corresponding to the bounded sets V are said to be *finite-dimensional* (in this case, all the spaces $\mathbf{H}_V(X) = Q\mathbf{H}(X)$ are finite-dimensional).

Furthermore, a mapping $K\colon X \mapsto K(X) \in \mathbf{H}(X)$ is called a *vector field* if it is measurable in the following sense. For any bounded open set $V \subset \mathbf{G}$, the mapping $X \mapsto X + Q(V)K(X)$ is Σ-measurable (as a mapping $\mathfrak{X}(\mathbf{G}) \to \mathfrak{X}(\mathbf{S} \times \mathbf{R}^m)$). Note that if K_1 and K_2 are two measurable vector fields, then $X \mapsto \big(K_1(X), K_2(X)\big)_{\mathbf{H}(X)}$ is a measurable function. We also use the *fields of multilinear functionals* $L\colon X \mapsto L(X) \in \mathcal{L}^j(\mathbf{H}(X))$, *fields of multilinear Hilbert–Schmidt functionals* $L\colon X \mapsto L(X) \in \mathcal{L}^j_{(2)}(\mathbf{H}(X))$, and *fields of trace class operators* $L\colon X \mapsto L(X) \in \mathcal{L}_{(1)}(\mathbf{H}(X))$. A j-linear functional L is said to be *measurable* if the function $X \mapsto L(X)\big(K_1(X), \dots, K_j(X)\big)$ is measurable for any j-tuple $K_1, \dots, K_j$ of measurable vector fields.

The topology of the configuration space. First, suppose that $\mathbf{G}$ is a *bounded* space. Then all configurations on $\mathbf{G}$ are finite and, as was already noted, $\mathfrak{X}(G)$ is naturally identified with $\{\varnothing\} \cup \bigcup_{n=1}^{\infty} \mathbf{G}^n/\pi_n$, which defines the *natural topology* on $\mathfrak{X}(\mathbf{G})$.

If G is *unbounded*, then $\mathfrak{X}(G)$ bears no "canonical" topology. Let us consider some topologies which will be used in this case.

[3]Even the case in which $\mathbf{S}$ is a singleton is meaningful, but this is insufficient for studying the distributions of functionals of stable random vectors (§§16, 17).

1. The coarse topology. The *coarse topology* w on $\mathfrak{X}$ is the *weakest topology* in which the mapping

$$X \mapsto \sum_{x \in X} \varphi(x)$$

is continuous for any continuous nonnegative function φ with bounded support.

It is well known [**53**] that $(\mathfrak{X}, w)$ is a complete separable metrizable space. A sequence $\{X_n\}$ converges to X in w if the sequence of measures $\sum_{x \in X_n \cap V} \delta_x$ weakly converges to $\sum_{x \in X \cap V} \delta_x$ for any bounded V. If $\mathbf{G}$ is bounded, then the coarse topology coincides with the natural topology. For our goals, the coarse topology is too poor, that is, the supply of w-continuous functions is too small.

2. The $\mathcal{O}$-topology. This topology describes the properties of the space $\mathfrak{X}$ "along I." A set $\mathfrak{U} \subset \mathfrak{X}$ is said to be *$\mathcal{O}$-open* ($\mathfrak{U} \in \mathcal{O}$) if for any configuration $X \in \mathfrak{U}$ and any bounded open subset $V \subset \mathbf{G}$ there exists an $\varepsilon > 0$ such that $X + h \in \mathfrak{U}$ for any $h \in \mathbf{H}_V(X)$ with $|h| < \varepsilon$. In other words, a set $\mathfrak{U}$ is $\mathcal{O}$-open if for any configuration in $\mathfrak{U}$ we can slightly "perturb" (along I) the points of this configuration that lie in a bounded open set V so that the perturbed configuration still remains in $\mathfrak{U}$.

Let us consider a useful property of the class of open sets.

PROPOSITION 14.1. *Suppose that $\mathfrak{U} \in \mathcal{O}$, $X \in \mathfrak{U}$, and $V \subset \mathbf{G}$ is a bounded open set. Then the sets*

$$\{h \in \mathbf{H}_V(X) \mid X + h \in \mathfrak{U}\},$$
$$\{h \in \mathbf{H}_V(X) \mid (X \cap V) + h \subset V, X + h \in \mathfrak{U}\}$$

are open in $\mathbf{H}_V(X)$.

We leave the proof to the reader.

We shall mainly use the $\mathcal{O}$-topology. For $\mathfrak{U} \subset \mathfrak{X}(\mathbf{G})$, we denote by $\mathfrak{U}^-$ its $\mathcal{O}$-closure and by $\partial\mathfrak{U}$ its boundary in the space $\mathfrak{X}(\mathbf{S} \times \mathbf{R}^m) \supset \mathfrak{X}(\mathbf{G})$. Obviously, $\mathfrak{U}^- \subset \mathfrak{X}(\mathbf{S} \times I^-)$. If $\mathfrak{U}$ is an open set and $\mathfrak{U}^- \subset \mathfrak{X}(\mathbf{G})$, then $\mathfrak{U}$ is called an *inner open set.*

Let us make some remarks about the $\mathcal{O}$-topology. A sequence $\{X_n\}$ converges to X in $\mathcal{O}$ if and only if there exist a finite-dimensional projection $Q = Q(V)$ and a number n_0 such that for any $n \geq n_0$ the terms of this sequence have the form $X_n = X + h_n$, where $h_n \in \mathbf{H}_V(X)$ and $|h_n| \to 0$.

The $\mathcal{O}$-topology is so rich that it fails to have a countable basis (even locally). Moreover, this topology does not take into account any properties of the space $\mathbf{S}$. For example, an $\mathcal{O}$-open set need not be Σ-measurable. Nevertheless, the $\mathcal{O}$-topology is well adapted for studying objects that do not have good properties "along $\mathbf{S}$."

3. Poisson topologies. The topologies w and $\mathcal{O}$ are, in a sense, extreme topologies: w is the weakest and $\mathcal{O}$ is the strongest "reasonable" topology. To get rid of their disadvantages, we define "intermediate" topologies, which will be called *Poisson topologies.* Although they are defined not on the entire configuration space but on some subset of $\mathbf{P}$-full measure, this is sufficient for our goals.

First, we give some auxiliary definitions.

By setting

$$v(X) = X \cap (\mathbf{G} \setminus V), \tag{14.2}$$

we assign a mapping $v\colon \mathfrak{X}(\mathbf{G}) \to \mathfrak{X}(\mathbf{G} \setminus V)$ to each measurable set $V \subset \mathbf{G}$. Thus, each configuration X is represented in the form $X = (X \cap V) \cup v(X)$.

Two configurations $X, Y \in \mathfrak{X}$ are said to be *close to each other* if there exists a bounded set $V \subset \mathbf{G}$ such that X and Y coincide outside V. A subset $\mathfrak{X}_0$ of the configuration space is said to be *saturated* if $\mathfrak{X}_0$ contains the empty set and if for each configuration $X \in \mathfrak{X}_0$, all configurations close to X belong to $\mathfrak{X}_0$. (In particular, all finite configurations belong to a saturated set.)

If $\mathfrak{X}_0$ is saturated and $V \subset \mathbf{G}$ is bounded, then we set $v(\mathfrak{X}_0) = \{X \in \mathfrak{X}_0 \mid X \cap V = \varnothing\} \subset \mathfrak{X}_0$. Then the mapping $i_V\colon (Y, Z) \mapsto Y \cup Z$ is a bijection of $\mathfrak{X}(V) \times v(\mathfrak{X}_0)$ onto $\mathfrak{X}_0$. Now if τ is a topology on $\mathfrak{X}_0$, then by $v(\tau)$ we denote the induced topology on the subspace $v(\mathfrak{X}_0)$. Furthermore, by $\tau(V)$ we denote the topology induced by τ on $\mathfrak{X}(V)$ (note that $\mathfrak{X}(V)$ is naturally identified with the set $\{X \in \mathfrak{X}_0 \mid X \cap (\mathbf{G} \setminus V) = \varnothing\}$).

Now we can give the definition of a Poisson space.

Let $\mathfrak{X}_0$ be a Σ-measurable saturated set. A topological space $(\mathfrak{X}_0, \tau)$ (and its topology τ) is called a *Poisson* space with respect to a measure $\mathbf{P}$ if

(1) $\mathfrak{X}_0$ is a set of full measure, that is, $\mathbf{P}\{\mathfrak{X}_0\} = 1$;
(2) the measure $\mathbf{P}$ is dense on $(\mathfrak{X}_0, \tau)$;
(3) τ is a Lindelöf topology (that is, each open cover contains a countable subcover) and is measurable ($\tau \subset \Sigma$);
(4) for any open bounded set $V \subset \mathbf{G}$, the topology $\tau(V)$ is not stronger than the natural topology $\rho(V)$ on $\mathfrak{X}(V)$;
(5) τ is not stronger than $\mathcal{O}$ ($\tau \subset \mathcal{O}$);
(6) for any open bounded set $V \subset \mathbf{G}$, the mapping i_V is continuous with respect to the topologies $(\tau(V) \times v(\tau), \tau)$.

There is a rather general method for constructing Poisson topologies. One considers an additive mapping of $\mathfrak{X}_0$ into an appropriate linear topological space $\mathbf{B}$ and defines τ as the preimage of the topology of $\mathbf{B}$. Let us describe this construction in more detail.

Suppose that $\mathbf{B}$ is a separable Banach space (for our goals this is sufficient) and $\mathfrak{X}_0$ is a measurable saturated set of full measure. A mapping $\Lambda\colon \mathfrak{X}_0 \to \mathbf{B}$ is said to be *additive* if for any measurable bounded set $V \subset \mathbf{G}$ we have

$$\Lambda(X) = \Lambda(X \cap V) + \Lambda(v(X)), \qquad X \in \mathfrak{X}_0. \tag{14.3}$$

Let us additionally assume that the mapping Λ is surjective, that is, $\Lambda(\mathfrak{X}_0) = \mathbf{B}$. Furthermore, let $\varkappa$ be the strong topology in $\mathbf{B}$, and let $\tau = \Lambda^{-1}(\varkappa)$. Obviously, τ is a Lindelöf topology and the measure $\mathbf{P}$ is dense on $\mathfrak{X}_0$. For τ not to be stronger than $\mathcal{O}$, it suffices to require that Λ be $\mathcal{O}$-continuous.

Furthermore, it is obvious that the continuity of i_V is equivalent to the continuity of the composite mapping $\Lambda \circ i_V$. Since $(\Lambda \circ i_V)(Y, Z) = \Lambda(Y) + \Lambda(Z)$, we see that the latter continuity follows from the continuity of addition in $\mathbf{B}$ and from the definition of τ.

As an example of a measurable additive mapping, we take the stochastic integral $\Lambda = \int g\, d\nu$ (or $\int g\, d\widetilde{\nu}$). This integral can be viewed as a function defined $\mathbf{P}$-almost everywhere on $\mathfrak{X}$. Let us show that by varying Λ on a set of zero probability we can ensure that the domain $\mathfrak{X}_0$ of this function is saturated and Λ is additive. Let $g \in \mathbf{I}(\mathbf{G}, \Pi)$. Let $\mathfrak{X}_0$ be the set of $X \in \mathfrak{X}$ for which $\int \mathbf{1}_{V_n} g\, d\nu$ is convergent (for some sequence $\{V_n\}$ chosen in advance). Obviously, if $X \in \mathfrak{X}_0$, then configurations close to X also belong to $\mathfrak{X}_0$. By passing to the limit, we obtain the additivity relation (14.3). Of course, the reasoning remains valid for integrals of the form $\int g\,d\widetilde{\nu}$.

Differentiation in the configuration space. Suppose that $\mathfrak{U} \in \mathcal{O} \cap \Sigma$ and $V \subset \mathbf{G}$ is a bounded open set. A measurable function $f\colon \mathfrak{U} \to \mathbf{R}^1$ is said to be l times $\mathbf{H}_V$-*differentiable* ($f \in \mathbf{C}_V^{(l)}(\mathfrak{U})$) if for any $X \in \mathfrak{U}$ the mapping $h \mapsto f(X+h)$ is l times continuously differentiable in a neighborhood of zero in $\mathbf{H}_V(X)$ (obviously, this mapping is then differentiable in the same sense on the entire set $\{h \in \mathbf{H}_V(X) \mid X + h \in \mathfrak{U},\ (X \cap V) + h \subset V\}$). We denote the jth-order derivative at a point h by $f_V^{(j)}(X, h)$ and omit the argument h if $h = 0$. By definition, the derivative is a multilinear functional on $\mathbf{H}_V(X)$.

The notion of differentiability means that the function varies smoothly if the configuration points belonging to V are slightly varied along the real coordinate.

Let us consider the simplest situation, in which one deals with the configuration space on an open bounded set $\mathbf{G} \subset \mathbf{R}^m$. Then $\mathfrak{X}(\mathbf{G})$ can be represented in the form $\{\varnothing\} \cup \bigcup_{n=1}^{\infty} G^n/\pi_n$, and one can readily show that a function f is $\mathbf{H}_{\mathbf{G}}$-differentiable if and only if the restriction of this function to each set G^n/π_n is differentiable, that is, the function $f \circ \overline{\pi}_n$ is differentiable in the usual sense on the set $G^n \subset (\mathbf{R}^m)^n$ (here $\overline{\pi}_n$ is the natural projection). The notion of the derivative in this case is also transparent. Namely,

$$(f^{(1)}(X))(x_i) = [\partial_i(f \circ \overline{\pi}_n)](x_1, \dots, x_n)$$

for the configuration $X = \{x_1, \dots, x_n\}$.

The classes $\mathbf{C}_V^{(l)}(\mathfrak{U})$ of differentiable vector fields $K\colon X \mapsto K(X) \in \mathbf{H}_V(X)$ are defined almost similarly. The difference is that the mapping $h \mapsto f(X+h)$ is replaced by the mapping $h \mapsto T(X,h)K(X+h)$, where the *transport operator* $T(X,h)\colon \mathbf{H}(X+h) \to \mathbf{H}(X)$ is defined by the natural formula

$$[T(X,h)g](x) = g(x + h(x)), \qquad x \in X, \quad g \in \mathbf{H}(X+h).$$

Suppose that $\mathfrak{U} \in \mathcal{O} \cap \Sigma$ and $f\colon \mathfrak{U} \to \mathbf{R}^1$ is a measurable function. We say that f is l times differentiable ($f \in \mathbf{C}^{(l)}(\mathfrak{U})$) if $f \in \mathbf{C}_V^{(l)}(\mathfrak{U})$ for any bounded open subset $V \subset \mathbf{G}$ and there exist "globalizing" measurable vector fields $f^{(j)}\colon X \mapsto f^{(j)}(X) \in \mathcal{L}^j(H(X))$ ($j = 1, \dots, l$, $X \in \mathfrak{U}$) such that for any V and any $X \in \mathfrak{U}$ we have

$$f_V^{(j)}(X)(Q\cdot, \dots, Q\cdot) = f^{(j)}(X)(Q\cdot, \dots, Q\cdot),$$

where $Q = Q(V)$, $j = 1, \dots, l$.

We say that f is *l times differentiable up to the boundary* ($f \in \mathbf{C}^{(l)}(\mathfrak{U}^-)$) if there exist an open set $\mathfrak{U}_1 \supset \mathfrak{U}^-$ in $\mathfrak{X}(\mathbf{S} \times \mathbf{R}^m)$ and a function $g \in \mathbf{C}^{(l)}(\mathfrak{U}_1)$ such that $f = g$ on $\mathfrak{U}$.

The differentiability of a vector field $K\colon X \mapsto K(X) \in \mathbf{H}(X)$ is defined in a similar way. Moreover, we can readily generalize the definition of differentiability to functions ranging in some Banach space $\mathbf{B}$. In this case, the derivatives are multilinear $\mathbf{B}$-valued functionals.

Let us introduce the notation for the normalized derivative. We set

$$n(X; f) = f^{(1)}(X)/|f^{(1)}(X)|_{\mathbf{H}(X)} \in \mathbf{H}(X) \tag{14.4}$$

for $f \in \mathbf{C}^{(1)}(\mathfrak{U})$.

Suppose that $l \geq 1$, V is a bounded open set, and $Q = Q(V)$. We say that a vector field $K \in \mathbf{C}_V^{(l)}(\mathfrak{U})$ belongs to the class $\mathbf{C}_V^{(l,0)}(\mathfrak{U})$ if there exists a bounded open set V_1 such that $\overline{V}_1 \subset V$ and $Q(V_1)K = K$.

Let us give an example of a differentiable function. Suppose that $\varphi \in \mathbf{I}_0(\mathbf{G}, \mathbf{R}^1)$ and for any $\theta \in \mathbf{S}$ the function $\varphi(\theta, \cdot\,)$ belongs to $\mathbf{C}^{(l)}(I)$. We set

$$f(X) = \sum_{x \in X} \varphi(x).$$

One can readily show that $f \in \mathbf{C}^{(l)}(\mathfrak{X})$ and

$$(f^{(1)}(X))(x) = \varphi^{(1)}(x)$$

for all $x \in X$ (here we naturally identify a linear functional on $\mathbf{H}(X)$ with an element of $\mathbf{H}(X)$ and write

$$(f^{(j)}(X))[h_1, \dots, h_j] = \sum_{x \in X} \varphi^{(j)}(x)[h_1(x), \dots, h_j(x)] \tag{14.5}$$

for all $j = 1, \dots, l$).

Formulas (14.5) remain valid if φ ranges in a Banach space $\mathbf{B}$; moreover, the composite function differentiation formula

$$(g \circ f(X))^{(1)}(x) = g^{(1)}(f(X))\varphi^{(1)}(x) \tag{14.6}$$

holds for any Fréchet differentiable function $g\colon \mathbf{B} \to \mathbf{R}^1$. By passing to the limit, one can readily prove formulas (14.5) and (14.6) for the case in which the functional φ belongs to one of the classes $\mathbf{I}(\mathbf{G}, \Pi)$ or $\widetilde{\mathbf{I}}(\mathbf{G}, \Pi)$.

Let us introduce one more useful notion of differentiability of functions.

Suppose that $q = (q(0), q(1), \dots, q(l))$, $q(j) \geq 1$, $j = 0, \dots, l$, is a multi-index of length $l + 1$, $V \in \mathbf{G}$ is an open set (not necessarily bounded), and $Q = Q(V)$. For $Y \in v(\mathfrak{X})$ (where v is defined by (14.2)), by $\mathbf{P}_Y$ (respectively, E_Y) we denote the conditional probability (respectively, the conditional mean value) for $\mathbf{P}$ corresponding to the condition $v = Y$.

We say that a function $f\colon \mathfrak{X} \to \mathbf{R}^1$ belongs to the class $\mathcal{W}_V^q(\mathfrak{X}, \mathbf{P})$ if for any increasing sequence of bounded open sets $\{V_n\}$, $\bigcup_n V_n = V$, there exists a sequence of functions $\{f_n\}$, $f_n \in \mathbf{C}_{V_n}^{(l)}$, with the following properties:

1) $f_n \to f$ in $\mathbf{L}^{q(0)}(\mathfrak{X}, \mathbf{P}_Y)$ for $\mathbf{P}v^{-1}$-almost all Y;

2) the sequence of mappings

$$X \mapsto f_n^{(j)}(X)[Q_n(\cdot), \dots, Q_n(\cdot)] \in \mathcal{L}_{(2)}^j(\mathbf{H}(X)), \qquad Q_n = Q_n(V_n),$$

converges in $\mathbf{L}^{q(j)}(\mathfrak{X}, \mathbf{P}_Y, \mathcal{L}_{(2)}^j(\mathbf{H}))$ for each $j = 1, \dots, l$ and for $\mathbf{P}v^{-1}$-almost all Y.

The limits of these sequences will be called the *derivatives of f along* $\mathbf{H}_V$ and denoted by $f^{(j)}$. They are measurable fields of Hilbert–Schmidt forms, $f^{(j)}(X) \in \mathcal{L}_{(2)}^j(\mathbf{H}(X))$. One can show that they are independent of the choice of V_n and f_n. In particular, the first derivative $f^{(1)}$ can be treated as a measurable vector field.

In a similar way, we can define the classes $\mathcal{W}_V^q(\mathfrak{X}, \mathbf{P}, \mathbf{H})$ and $\mathcal{W}_V^{q,0}(\mathfrak{X}, \mathbf{P}, \mathbf{H})$ for vector fields (in the latter case, the approximating sequences of vector fields are taken from $\mathbf{C}_{V_n}^{l,0}$).

Next, we say that a function f belongs to the class $\mathcal{W}^q_{V,\mathrm{loc}}(\mathfrak{X},\mathbf{P})$ (the subscript "loc" means that the function and its derivatives are integrable only locally) if there exist an increasing sequence $\{A_n\}$ of sets $A_n \in \mathcal{O}\cap\Sigma$ and a sequence of functions $f_n \in \mathcal{W}^q_V(\mathfrak{X},\mathbf{P})$ such that

1) $\mathbf{P}\{\bigcup_n A_n\} = 1$;

2) the restrictions of the functions f and f_n to A_n coincide for each n.

In this case, the limits $f^{(j)} = \lim_n f_n^{(j)}$ exist ($\mathbf{P}$-almost surely) for $j = 1,\dots,l$ and are independent of the choice of the sequences $\{A_n\}$ and $\{f_n\}$.

We conclude this section with a very important remark. The norm in the tangent space $\mathbf{H}(X)$ was chosen rather arbitrarily; we have never used the explicit form of this norm. All our statements remain valid if we assume that for each $X \in \mathfrak{X}$ we have a Hilbert space $\mathbf{H}(X)$ (a subset of $\mathbf{R}^X$) equipped with some norm $|\cdot|_{\mathbf{H}(X)}$. For example, we can set

$$|h|_{\mathbf{H}(X)} = \Big(\sum_{x\in X} \varkappa(x)|h(x)|^2_{\mathbf{R}^m}\Big)^{1/2}, \tag{14.7}$$

where $\varkappa\colon \mathbf{G} \to (0,\infty)$ is some measurable function.

In what follows, we need norms of the form (14.7), but, unless otherwise specified, we consider the $\mathbf{l}^2$-norm (that is, $\varkappa = 1$). The same is true of the choice of the metric on $I \subset \mathbf{R}^m$: we have nowhere used the explicit form of the metric.

In all applications, some Poisson random measure is given beforehand and for this measure we must choose an appropriate metric in I and a norm on $\mathbf{H}(X)$ (see Problem 13.4). If the Poisson measure has a Lebesgue intensity (this case is just the case that will be considered in most detail), then it is most convenient to use the Euclidean metric in I and the $\mathbf{l}^2$-norm in $\mathbf{H}(X)$. If the intensity Π is not Lebesgue, then we can either choose an appropriate norm in $\mathbf{H}(X)$ (so that the functions in question will be differentiable) and a metric on I, or, vice versa, choose a transformation $\mathbf{R}^m \to I$ that takes the Lebesgue measure to the measure Π (see Problem 13.3). We shall use both methods.

§15. The Gauss–Ostrogradskii formula

This section deals with the "geometry" of the configuration space $\mathfrak{X}$. In the first part, we define the notion of a smooth surface in $\mathfrak{X}$ (with respect to the differential structure introduced in §14). Then for a Poisson measure $\mathbf{P}$ whose intensity has the special form $\Pi = \pi \times \lambda$, where λ is the Lebesgue measure on I, we introduce the notion of the surface measure corresponding to $\mathbf{P}$ and prove the "factorization theorem," which states that the integral with respect to the measure $\mathbf{P}$ can be expressed in terms of integrals with respect to the surface measures on the level surfaces of a smooth function. This theorem readily implies existence conditions for the distribution density of some class of stochastic functionals. Moreover, the density can be expressed in terms of the integral with respect to the surface measure on the corresponding level surface. Finally, we prove a formula that can naturally be called a Poisson analog of the classical Gauss–Ostrogradskii formula.

Surfaces in the configuration space. A set $\Gamma \subset \mathfrak{X}$ is called an *elementary surface of codimension* k (for $k = 1$, we have an *elementary hypersurface*) if there exist an open measurable set $\mathfrak{U}$ containing Γ ($\mathfrak{U} \in \mathcal{O}\cap\Sigma$, $\mathfrak{U} \supset \Gamma$) and a function $f =$

$(f_1, \ldots, f_k) \in \mathbf{C}^{(1)}(\mathfrak{U}, \mathbf{R}^k)$ such that $\Gamma = f^{-1}(0)$ and for some finite-dimensional projection $Q = Q(V)$ (where V is an open set) we have

$$\inf_{X \in \mathfrak{U}} \det \Big((Qn(X; f_i), Qn(X; f_j))_{\mathbf{H}(X)} \Big) > 0,$$

where $n(X; f_i) = f_i^{(1)}(X)/|f_i^{(1)}(X)| \in \mathbf{H}(X)$, $i = 1, \ldots, k$.

More generally, a set $\Gamma \subset \mathfrak{X}$ is called a *surface of codimension* k if there exists an at most countable family $\{\Gamma_j\}$ of elementary surfaces of codimension k such that $\bigcup_j \Gamma_j = \Gamma$ and each finite intersection of Γ_j is an elementary surface of the same codimension.

Subsets of surfaces, as well as at most countable disjoint unions of subsets of surfaces of the same codimension, are naturally called *surface sets.*

LEMMA 15.1. *Suppose that $\mathcal{U} \in \mathcal{O} \cap \Sigma$, $f \in \mathbf{C}^{(1)}(\mathfrak{U})$, and $t \in f(\mathfrak{U})$. Then $f^{-1}(t) \cap \{X \mid f^{(1)}(X) \neq 0\}$ is a hypersurface.*

PROOF. Let $\{S_j\}$ be an increasing sequence of bounded subsets of $\mathbf{S}$, $\bigcup_j S_j = \mathbf{S}$. Let $\{\psi_j\}$ be a sequence of smooth functions $\mathbf{R}^m \to \mathbf{R}^1$ such that $\psi_j([-j, j]^m) = 1$, $\psi_j(\mathbf{R}^m \setminus [-2j, 2j]^m) = 0$, and $\psi_j \leq 1$ for all $j = 1, 2, \ldots$. By setting

$$g_j(\theta, r) = \mathbf{1}_{S_j}(\theta)\psi_j(r),$$

we define a sequence $\{g_j\}$ of functions $g_j \colon G \to \mathbf{R}^1$. Next, we set

$$D_j(X) = \sum_{x \in X} g_j(x)|f^{(1)}(X)(x)|^2, \qquad X \in \mathfrak{U},$$

$$W_j = \Big\{ X \in \mathfrak{U} \mid D_j(X) > \frac{1}{2}|f^{(1)}(X)|^2 > 0 \Big\}.$$

The desired statement follows from the representation

$$f^{-1}(t) \cap \{X \mid f^{(1)}(X) \neq 0\} = \bigcup_{j=0}^{\infty} (f^{-1}(t) \cap W_j). \quad \square$$

Surface measures on the configuration space related to a Poisson random measure. On $(G, \mathfrak{B})$, let us consider a Poisson random measure ν of intensity $\Pi = \pi \times \lambda$, where π is some locally finite measure on $\mathbf{S}$ and λ is the Lebesgue measure on I. Let $\mathbf{P}$ be the corresponding Poisson measure on $(\mathfrak{X}, \Sigma)$.

Now let us construct surface measures corresponding to $\mathbf{P}$ on elementary surfaces in $\mathfrak{X}$. It will be shown that this construction can naturally be generalized to arbitrary surfaces (the measure on a surface can be obtained by the standard "sewing" of measures on elementary surfaces) and to measurable surface sets.

Thus, suppose that $\Gamma = f^{-1}(0)$ ($f \in \mathbf{C}^{(1)}(\mathfrak{U}, \mathbf{R}^k)$) is an elementary surface of codimension k and $Q = Q(V)$ is the corresponding finite-dimensional orthogonal projection, whose existence is required by the definition of an elementary surface. Since an arbitrary orthogonal projection $Q(V_1)$ with $V_1 \supset V$ can be taken instead of $Q(V)$, we assume that the open set V has the form $V = B \times \Delta$. We use formula (14.1) to define a mapping $v \colon \mathfrak{X}(\mathbf{G}) \to \mathfrak{X}(\mathbf{G} \setminus V)$.

Let us choose a configuration $X \in \Gamma$ and set $l = |X \cap V|$ (obviously, $lm \geq k$). In what follows, let $(\theta_1, r_1), \ldots, (\theta_l, r_l)$ ($\theta_i \in B$, $r_i \in \Delta$) be a given representing sequence of the configuration $X \cap V$. For each function $h \in \mathbf{H}_V(X)$, by $h_1, \ldots, h_l$

we denote the corresponding representation of the restriction $h\big|_{X\cap V}$. If we consider the m-dimensional vectors $h_1, \ldots, h_l$ as coordinates in $(\mathbf{R}^m)^l$, then we can identify the spaces $\mathbf{H}_V(X)$ and $(\mathbf{R}^m)^l$. This identification, the definition of an elementary surface, and the implicit function theorem readily imply that the level sets of the function $h \mapsto f(X+h)$ defined on $\{h \in \mathbf{H}_V(X) \mid X+h \in \mathfrak{U},\ (X\cap V)+h \subset V\}$ are $C^{(1)}$-manifolds of codimension k in $\mathbf{H}_V(X)$. By $\widetilde{\sigma}^l_{v(X),\theta_1,\ldots,\theta_l}$ we denote the standard Lebesgue surface measure defined on these manifolds. The mapping $h \mapsto (X\cap V)+h$ takes this measure to some measure $\sigma^l_{v(X),\theta_1,\ldots,\theta_l}$ on the space $\mathfrak{X}(V)$.

Choose $Z \in \mathfrak{X}(\mathbf{G}\setminus V)$ and define a measure σ_Z on the space $\mathfrak{X}(V)$ by setting

$$\sigma_Z(dY) = e^{-\Pi(V)} \sum_{l\geq k/m} \frac{1}{l!}\mathbf{1}_{\{|Y|=l\}} \int \pi(d\theta_1)\ldots\pi(d\theta_l)\mathbf{1}_{\mathfrak{U}}(Z\cup Y)\sigma^l_{Z,\theta_1,\ldots,\theta_l}(dY).$$

Now we can define the surface measure on Γ.

Let Γ be an elementary hypersurface. The measure

$$\sigma_\Gamma(dX) = \mathbf{1}_\Gamma(X)|Qn|^{-1}\sigma_{v(X)}(d(X\cap V))\mathbf{P}v^{-1}(dv(X)),$$

where $\mathbf{P}v^{-1}$ is the image of $\mathbf{P}$ under the mapping v $\big(v(X) = X\cap(\mathbf{G}\setminus V)\big)$ and $n(X) = n(X;f) = f^{(1)}(X)/|f^{(1)}(X)|$, is called the *surface measure* on Γ.

More generally, if Γ is an elementary surface of codimension k, then the surface measure σ_Γ is defined by the formula

$$\sigma_\Gamma(dX) = \mathbf{1}_\Gamma(X)\left[\frac{\det\big(n_i(X), n_j(X)\big)^k_{i,j=1}}{\det\big(Qn_i(X), Qn_j(X)\big)^k_{i,j=1}}\right]^{1/2} \sigma_{v(X)}(d(X\cap V))\mathbf{P}v^{-1}(dv(X)).$$

Here $n_i(X) = n_i(X; f_i)$.

Now we shall try to explain these formulas informally. The measure σ_Z, $Z \in \mathfrak{X}(\mathbf{G}/V)$, is a surface measure on the section $\Gamma_Z = \{Y \in \mathfrak{X}(V) \mid Y\cup Z \in \Gamma\}$ of the surface Γ. Recall that the standard measure on a surface in $\mathbf{R}^l = \mathbf{R}^{l_1+l_2}$ can be expressed in terms of surface measures on the sections of this surface by affine subspaces parallel to $\mathbf{R}^{l_1}$. A simple geometric argument shows that it suffices to multiply the surface measure of the section by $|Qn|^{-1}$, where n is the normal to the surface and Q is the projection on $\mathbf{R}^{l_1}$, and integrate with respect to the remaining l_2 variables. In our case, we need to integrate over the measure $\mathbf{P}v^{-1}(dZ)$.

The reader can readily show that the surface measure is well defined (that is, independent of the choice of V and f).

Suppose that $\mathfrak{U}$ is an open measurable set, $f\colon \mathfrak{U} \to \mathbf{R}^1$ is a function from the class $\mathbf{C}^{(1)}(\mathfrak{U})$, and $f^{(1)}$ does not vanish on $\mathfrak{U}$. By Lemma 15.1, the level sets $f^{-1}(t) = \{X \in \mathfrak{U} \mid f(X) = t\}$ are hypersurfaces. By σ_t we denote the surface measure on $f^{-1}(t)$.

THEOREM 15.1. *The measure* $\mathbf{P}$ *admits the representation*

$$d\mathbf{P} = |f^{(1)}|^{-1}\,dt\,d\sigma_t; \tag{15.1}$$

in other words, for any $\mathbf{P}$*-integrable function* $\Psi\colon \mathfrak{U} \to \mathbf{R}^1$ *the following conditions are satisfied:*

(1) *the function* $X \mapsto \Psi(X)|f^{(1)}(X)|^{-1}$ *is* σ_t*-integrable for Lebesgue almost all* $t \in \mathbf{R}^1$;

(2) *the function* $t \mapsto \int \Psi(X)|f^{(1)}(X)|^{-1}\sigma_t(dX)$ *is integrable, and*

$$\int_{\mathbf{R}^1} dt \int_{f^{-1}(t)} \Psi|f^{(1)}|^{-1}\, d\sigma_t = \int_{\mathfrak{U}} \Psi\, d\mathbf{P}.$$

REMARK. The vector case $f \in \mathbf{C}^{(1)}(\mathfrak{U}, \mathbf{R}^k)$ is a natural generalization of this theorem. The corresponding representation has the form

$$d\mathbf{P} = \left[\det(f_i^{(1)}, f_j^{(1)})_{i,j=1}^k\right]^{-1/2} dt\, d\sigma_t.$$

PROOF OF THEOREM 15.1. Let us choose an approximating sequence $Q_j = Q_j(V_j)$, $V_j = B_j \times \Delta_j$, of finite-dimensional orthogonal projections and define the functions $D_j\colon \mathfrak{U} \to \mathbf{R}^1$ just as in the proof of Lemma 15.1. We also choose an auxiliary continuous increasing function $\rho\colon \mathbf{R}^1 \to [0,1]$ such that $\rho(1/3) = 0$ and $\rho(2/3) = 1$. We set $\widetilde{\varkappa}_j(X) = \rho\big(|f^{(1)}(X)|^{-2}D_j(X)\big)$ $(X \in \mathfrak{U})$ and define a partition of unity $\{\varkappa_j\}_{j=1}^\infty$ by setting

$$\varkappa_1 = \widetilde{\varkappa}_1, \quad \varkappa_2 = (1-\widetilde{\varkappa}_1)\widetilde{\varkappa}_2, \quad \varkappa_3 = (1-\widetilde{\varkappa}_1)(1-\widetilde{\varkappa}_2)\widetilde{\varkappa}_3, \quad \dots.$$

Note that $f^{-1}(t) \cap \{\varkappa_j \neq 0\}$ are elementary hypersurfaces, and we can take Q_j to be the finite-dimensional projection occurring in the definition of an elementary hypersurface. Obviously, it suffices to prove the theorem with Ψ replaced by $\Psi_j = \Psi\varkappa_j$. For brevity, we omit the subscript j everywhere in what follows. First, we have the obvious relation $\int_{\mathfrak{U}} \Psi\, d\mathbf{P} = \sum_{l=1}^\infty \int_{\mathfrak{U}_l} \Psi\, d\mathbf{P}$, where $\mathfrak{U}_l = \{X \in U \mid |X \cap V| = l\}$ (if $X \cap V = \varnothing$, then $\Psi(X) = 0$; therefore, the term with $l = 0$ is lacking). By the properties of Poisson measures, we can factor the measure in each term on the right-hand side,

$$\mathbf{P}(dX) = \mathbf{P}v^{-1}(dv(X)) \cdot \frac{(\Pi(V))^l}{l!} e^{-\Pi(V)} \cdot \frac{\pi(d\theta_1)\dots\pi(d\theta_l)\, dr_1 \dots dr_l}{(\Pi(V))^l}.$$

As before, $(\theta_1, r_1), \dots, (\theta_l, r_l)$ is the representing sequence of the configuration $X \cap V$.

Thus,

$$\int_{\mathfrak{U}_l} \Psi(X)\, \mathbf{P}(dX) = \frac{1}{l!} e^{-\Pi(V)} \int_{\mathfrak{X}(G\setminus V)} \mathbf{P}v^{-1}(dZ) \\ \times \int_{B^l} \pi(d\theta_1)\dots\pi(d\theta_l) \int_{\Delta^l} \mathbf{1}_{\mathfrak{U}_l}(X)\Psi(X)\, dr_1 \dots dr_l.$$

Here $X = Z \cup \{(\theta_1, r_1), \dots, (\theta_l, r_l)\}$ and $Z = v(X)$.

By a well-known formula, the inner (finite-dimensional!) integral can be expressed in terms of the surface measure $\widetilde{\sigma}^l_{Z,\theta_1,\dots,\theta_l}$ as follows:

$$\int_{\Delta^l} \mathbf{1}_{\mathfrak{U}_l}(X)\Psi(X)\, dr_1 \dots dr_l = \int_{\mathbf{R}^1} dt \int \Psi(X)|Qf^{(1)}(X)|^{-1}\widetilde{\sigma}^l_{Z,\theta_1,\dots,\theta_l}(dr_1, \dots, dr_l)$$

(the inner integral is taken over the level set $\{(r_1, \dots, r_l) \in \Delta^l \mid X \in \mathfrak{U}_l \cap f^{-1}(t)\}$).

By substituting the last integral into the preceding formula, replacing the measure $\widetilde{\sigma}^l_{Z,\theta_1,\dots,\theta_l}$ by the corresponding measure $\sigma^l_{Z,\theta_1,\dots,\theta_l}$ in the configuration space $\mathfrak{X}(V)$, and summing over l, we readily obtain

$$\begin{aligned}\int_{\mathfrak{U}} \Psi\, d\mathbf{P} \int_{\mathbf{R}^1} dt \int_{\mathfrak{X}(G\setminus V)} &\mathbf{P}v^{-1}(dZ)\\ &\times \int_{\mathfrak{X}(V)} \mathbf{1}_{f^{-1}(t)}(Z\cup Y)\Psi(Z\cup Y)|Qf^{(1)}(Z\cup Y)|^{-1}\sigma_Z(dY)\\ &= \int_{\mathbf{R}^1} dt \int_{f^{-1}(t)} \Psi(X)|Qf^{(1)}(X)|^{-1}|Q_n(X;f)|\sigma_t(dX)\\ &= \int_{\mathbf{R}^1} dt \int_{f^{-1}(t)} \Psi(X)|f^{(1)}(X)|^{-1}\sigma_t(dX). \quad \square\end{aligned}$$

Surface measures in the configuration space which correspond to other measures. For some applications, the above notions are not sufficiently general. For example, in what follows we shall consider functionals of a process with independent increments and obtain some conditions under which the distributions of such functionals are absolutely continuous. To this end, we need to define surface measures that correspond to Poisson random measures and whose intensities are different from that of the Lebesgue measure.

Here we consider a very useful generalization of the surface measure. In this case the surface measure is defined by "locally" the same measure. In particular, this class contains Poisson measures whose intensities are absolutely continuous with respect to the Lebesgue measure.

Just as before, we suppose that $\mathbf{G} = \mathbf{S}\times I$, but the metric in I is not necessary Euclidean (see Problem 13.4), $\mathfrak{X} = \mathfrak{X}(\mathbf{G})$, and for each $X \in \mathfrak{X}(\mathbf{G})$ there is a Hilbert space $\mathbf{H}(X)$ (a subset of $\mathbf{R}^X$) equipped with the norm $|\cdot|^\sim = |\cdot|^\sim_{\mathbf{H}(X)} = ((\cdot,\cdot)^\sim)^{1/2}$. We only require that the norm $|\cdot|^\sim_{\mathbf{H}(X)}$ is in some sense "consistent" with the σ-algebra Σ. Namely, if K_1 and K_2 are two measurable vector fields, then the function $X \mapsto (K_1(X), K_2(X))^\sim$ must also be measurable.

Furthermore, suppose that $\widetilde{\mathbf{P}}$ is some probability measure on $(\mathfrak{X},\Sigma)$, satisfying the following condition. For some σ-finite measure π on $\mathbf{S}$ and each bounded measurable set $V \subset \mathbf{G}$, there exists a measurable function $q_V(Y,Z)$, $Y \in v(\mathfrak{X})$, $Z \in \mathfrak{X}(V)$, such that for $\widetilde{\mathbf{P}}v^{-1}$-almost all Y we have the relations

$$\widetilde{\mathbf{P}}^Y_V \ll \mathbf{P}_V, \qquad \frac{d\mathbf{P}^Y_V}{d\mathbf{P}_V}(Z) = q_V(Y,Z), \tag{15.2}$$

where $\widetilde{\mathbf{P}}^Y_V$ is the conditional measure (for $\widetilde{\mathbf{P}}$) on $\mathfrak{X}(V)$ corresponding to the condition $v = Y$ and $\mathbf{P}_V$ is a Poisson measure of intensity $\pi\times\lambda$ on $\mathfrak{X}(V)$.

For each measure $\widetilde{\mathbf{P}}$ of this type and for each surface Γ, we define a surface measure $\widetilde{\sigma}_\Gamma$ related to $\widetilde{\mathbf{P}}$. As was already noted, it suffices to define $\widetilde{\sigma}_\Gamma$ for elementary surfaces Γ.

Thus, suppose that Γ is an elementary hypersurface and that $Q = Q(V)$ and f are, respectively, finite-dimensional projection and the function occurring in the definition of an elementary surface. We set

$$n(X) = n(X; f) = \frac{f^{(1)}(X)}{|f^{(1)}(X)|^{\sim}_{\mathbf{H}(X)}}.$$

Next, for $Y \in v(\mathfrak{X})$, by Γ^Y we denote the corresponding section of Γ, that is, the set $\{Z \in \mathfrak{X}(V) \mid Y \cup Z \in \Gamma\}$, and by σ_{Γ^Y} we denote the surface measure on Γ^Y corresponding to the measure $\mathbf{P}_V$. Moreover, by $|\cdot|_V$ we denote the norm on $\mathbf{H}_V(X)$ defined by (14.1) and by $(\cdot,\cdot)_V$ the inner product corresponding to this norm. Note that $\mathbf{P}_V$ and $|\cdot|_V$ exist only locally and need not have a continuation to all $\mathfrak{X}$ and $\mathbf{H}(X)$ (that is, the corresponding limit as $V \uparrow G$ need not exist).

Now we define the surface measure $\widetilde{\sigma}_\Gamma$ by

(15.3)
$$\begin{aligned}\widetilde{\sigma}_\Gamma(dY, dZ) &= \widetilde{\mathbf{P}}v^{-1}(dY)(|Qn|^{\sim})^{-1}|Qf^{(1)}|^{\sim}|Qf^{(1)}|_V^{-1}q_V(Y,Z)\sigma_{\Gamma^Y}(dZ)\\ &= \widetilde{\mathbf{P}}v^{-1}|Qn|_V^{-1}q_V(Y,Z)\sigma_{\Gamma^Y}(dZ).\end{aligned}$$

Straightforward calculations show that this measure is well defined, that is, the definition is independent of the choice of the set V. We omit these calculations.

Similarly, for a surface of codimension k we define the surface measure $\widetilde{\sigma}_\Gamma$ as follows:

$$\widetilde{\sigma}_\Gamma(dY, dZ) = \widetilde{\mathbf{P}}v^{-1}(dY)\frac{(\det\{(n_i, n_j)^{\sim}\})^{1/2}}{(\det\{(Qn_i, Qn_j)\})^{1/2}}q_V(Y,Z)\sigma_{\Gamma^Y}(dZ).$$

Here $n_i(X) = n_i(X; f) = f_i^{(1)}(X)/|f_i^{(1)}(X)|^{\sim}$.

Now let us show that the measure $\widetilde{\mathbf{P}}$ satisfies the assumptions of Theorem 15.1, that is,

$$\int_{\mathbf{R}^1} dt \int_{f^{-1}(t)} \Psi(|f^{(1)}|^{\sim})^{-1}\, d\widetilde{\sigma}_t = \int_{\mathfrak{U}} \Psi\, d\widetilde{\mathbf{P}} \tag{15.4}$$

for any $\widetilde{\mathbf{P}}$-integrable function Ψ. Just as in the proof of Theorem 15.1, it suffices to prove (15.4) with Ψ replaced by $\Psi_j = \Psi\varkappa_j$. We have (the subscript j is omitted)

$$\begin{aligned}\int_{\mathfrak{U}} \Psi\, d\widetilde{\mathbf{P}} &= \int \widetilde{\mathbf{P}}v^{-1}(dY)\int \Psi(Y\cup Z)\mathbf{1}_{\mathfrak{U}}(Y\cup Z)q_V(Y\cup Z)\mathbf{P}_V(dZ)\\ &= \int \widetilde{\mathbf{P}}v^{-1}(dY)\int_{\mathbf{R}^1} dt\\ &\quad\times \int_{\Gamma_t^Y} \Psi(Y\cup Z)\mathbf{1}_{\mathfrak{U}}(Y\cup Z)|Qf^{(1)}|_V^{-1}q_V(Y,Z)\sigma_{\Gamma_t^Y}(dY)\\ &= \int \widetilde{\mathbf{P}}v^{-1}(dY)\int_{\mathbf{R}^1} dt\\ &\quad\times \int_{\Gamma_t^Y} \Psi(Y\cup Z)\mathbf{1}_{\mathfrak{U}}(Y\cup Z)(|f^{(1)}|^{\sim})^{-1}|Qn|_V^{-1}q_V(Y,Z)\sigma_{\Gamma_t^Y}(dY)\\ &= \int_{\mathbf{R}^1} dt\int_{\Gamma_t^Y}(|f^{(1)}|^{\sim})^{-1}\, d\widetilde{\sigma}_t.\end{aligned}$$

In a similar way, we can prove the following analog of (15.1):

$$\int_{\mathfrak{U}} \Psi \, d\widetilde{\mathbf{P}} = \int_{\mathbf{R}^k} dt \int_{f^{-1}(t)} \Psi \Big[\det \big[(f_i^{(1)}, f_j^{(1)})^{\sim} \big]_{i,j=1}^k \Big]^{-1/2} d\widetilde{\sigma}_t. \tag{15.4$'$}$$

Absolute continuity of distributions of functionals.

PROPOSITION 15.1. *Suppose that* $f = (f_1, \dots, f_k) \in \mathbf{C}^{(1)}(\mathfrak{U}, \mathbf{R}^k)$,

$$\dim \mathcal{L}\{f_1^{(1)}(X), \dots, f_k^{(1)}(X)\} = k$$

for all $X \in \mathfrak{U}$, *and the probability measure* $\widetilde{\mathbf{P}}$ *satisfies the assumptions stated in the previous subsection. Then the image of the measure* $\widetilde{\mathbf{P}}_{\mathfrak{U}} = \widetilde{\mathbf{P}}(\cdot \cap \mathfrak{U})$ *under the mapping* f *is absolutely continuous with respect to the Lebesgue measure on* $\mathbf{R}^k$.

PROOF. In (15.4′), we set $\Psi = \mathbf{1}_{\{f \le t\}}$ (the inequality $f \le t$ is understood coordinatewise) and obtain the following representation for the distribution function F of the measure $\widetilde{\mathbf{P}}_{\mathfrak{U}} f^{-1}$:

$$F(t) = \int_{(-\infty, t]} ds \int_{f^{-1}(s)} \Big(\det \big((f_i^{(1)}, f_j^{(1)})_{i,j=1}^k \big) \Big)^{-1/2} d\widetilde{\sigma}_s.$$

Proposition 15.1 is thereby proved. □

For example, let us consider a functional of a process with independent increments. Let $\xi(t)$, $t \in [0, T]$, be a separable stochastically continuous $\mathbf{R}^d$-valued process with independent increments, and let $g\colon \mathbf{D}^d[0,T] \to \mathbf{R}^k$ be a functional defined on some subset of $\mathbf{D}^d[0,T]$ containing almost all trajectories of ξ. By $\mathcal{P}^\xi$ we denote the measure generated by ξ on $\mathbf{D}^d[0,T]$.

The representation (13.9) of the process ξ allows us to choose a convenient probability space of the form $\mathbf{L} \times \mathfrak{X}([0,T] \times \mathbf{R}^d)$, equipped with the measure $\mathbf{P_L} \times \mathbf{P}$, on which we define the process ξ and the stochastic functional $g(\xi(\cdot))$. Here $(\mathbf{L}, \mathbf{P_L})$ is some probability space for the Gaussian process ξ_1 (usually, $\mathbf{L}$ is a vector space) and $(\mathfrak{X}([0,T] \times \mathbf{R}^d), \mathbf{P})$ is a configuration space with Poisson measure of intensity Π.

In the sequel, for the mapping $g = (g_1, \dots, g_k)$ we assume that the functional $g_i\colon \mathbf{D}^d[0,T] \to \mathbf{R}^1$ satisfies the following differentiability condition for all $i = 1, \dots, k$. There exists an open set $\mathbf{D}_0^d$, $\mathcal{P}^\xi(\mathbf{D}_0^d) = 1$, such that the limit

$$\lim_{\varepsilon \to 0} \frac{g_i(y + \varepsilon l) - g_i(y)}{\varepsilon} = \frac{dg_i}{dl}(y)$$

exists for all $y \in \mathbf{D}_0^d$ and any $l \in \mathbf{D}$. In this case, the functional $l \mapsto dg_i(y)/dl$ (we denote it by $\nabla g_i(y)$) is linear for all $y \in \mathbf{D}_0^d$, and the functional $(y, l) \mapsto (\nabla g_i(y), l)$ is continuous on $\mathbf{D}_0^d \times \mathbf{D}^d$.

One can readily see that these assumptions are sufficient for the functions of the form $X \mapsto g_i(\xi_\cdot(X))$ to belong to the class $\mathbf{C}_V^{(1)}(\mathfrak{X}(\mathbf{S} \times I))$, $\mathbf{S} = [0,T]$, $I = \mathbf{R}^d \setminus \{0\}$, for any fixed value of the Gaussian coordinate and for any open set $V \subset [0,T] \times \mathbf{R}^d$ such that the projection of its closure on $\mathbf{R}^d$ is bounded away from zero.

Now let us write out explicit expressions for the derivatives of the functions of the form $X \mapsto g_i(\xi_\cdot(X))$. For simplicity, we restrict our consideration to the case

$d = 1$. For $h \in \mathbf{H}_V(X)$, we have

$$\left[\frac{d}{d\varepsilon} g_i(\xi_\cdot(X+\varepsilon h))\right]_{\varepsilon=0} = \Big(\nabla g_i(\xi_\cdot(X)), \sum_{\substack{(s,y)\in X\cap V\\ s\le \cdot}} h(s,y)\Big) \tag{15.5}$$
$$= \sum_{(s,y)\in X\cap V} \big(\nabla g_i(\xi_\cdot(X)), \mathbf{1}_{[s,T]}(\cdot)\big) h(s,y).$$

The expression on the right-hand side in (15.5) is finite, since $|X \cap V| < \infty$.

Now let us construct the "tangent space" $\mathbf{H}(X)$ at each point $X \in \mathfrak{X}$ by introducing the norm $|\cdot|^{\sim}$ on this space as follows:

$$|h|^{\sim}_{\mathbf{H}(X)} = \Big(\sum_{(s,y)\in X} h^2(s,y)(\min(y^2,1))^{-1}\Big)^{1/2}.$$

In what follows, we show that for $\mathbf{P}$-almost all $X \in \mathfrak{X}$ and any $h \in \mathbf{H}(X)$, the sum on the right-hand side in (15.5) is finite for $V = G = [0,T] \times (\mathbf{R}^1 \setminus \{0\})$ as well. Indeed,

$$\sum_{(s,y)\in X} \big(\nabla g_i(\xi_\cdot(X)), \mathbf{1}_{[s,T]}(\cdot)\big) h(s,y)$$
$$\le \Big(\sum_{(s,y)\in X} \big(\nabla g_i(\xi_\cdot(X)), \mathbf{1}_{[s,T]}(\cdot)\big)^2 \min(y^2,1)\Big)^{1/2} \Big(\sum_{(s,y)\in X} \frac{h^2(s,y)}{\min(y^2,1)}\Big)^{1/2}.$$

Note that

$$\sup_{s\in[0,T]} \big|\big(\nabla g_i(\xi_\cdot(X)), \mathbf{1}_{[s,T]}(\cdot)\big)\big| < \infty$$

for any fixed X. On the other hand, $\sum_{(s,y)\in X} \min(y^2,1) < \infty$ for almost all X. Thus, we have shown that the functions $X \mapsto g_i(\xi_\cdot(X))$ belong to the class $\mathbf{C}^{(1)}(\mathfrak{X}(\mathbf{G})))$, so that

$$\big(g_i(\xi(X))\big)^{(1)}(s,y) = \big(\nabla g_i(\xi_\cdot(X)), \mathbf{1}_{[s,T]}(\cdot)\big) \min(|y|^2,1). \tag{15.6}$$

Now we can use Proposition 15.1.

THEOREM 15.2. *Suppose that the measure* Π *is absolutely continuous with respect to the Lebesgue measure on* $[0,T] \times \mathbf{R}^d$,

$$\int_{t_1}^{t_2} \int_{\mathbf{R}^d} \Pi(ds\, dx) = \infty \tag{15.7}$$

for any distinct $t_1, t_2 \in [0,T]$, *and the functional* $g\colon \mathbf{D}^d[0,T] \to \mathbf{R}^k$ *satisfies the condition*

$$\dim \mathcal{L}\{\nabla g_1(y), \dots, \nabla g_k(y)\} = k \tag{15.8}$$

for $\mathcal{P}^{\xi}$*-almost all* y.

Then the distribution of the stochastic functional $g(\xi_\cdot)$ *is absolutely continuous with respect to the Lebesgue measure on* $\mathbf{R}^k$.

PROOF. For simplicity, we consider only the case $d = 1$.

First, let us show that the assumptions of the theorem imply

$$\dim \mathcal{L}\{(g_1 \circ \xi)^1(X), \dots, (g_k \circ \xi)^{(1)}(X)\} = k \tag{15.9}$$

for $\mathbf{P}$-almost all X.

Indeed, it readily follows from (15.6) that (15.9) is equivalent to linear independence in $\mathbf{H}(X)$ of the vectors $u_i = \{u_i(x),\ x = (s, y) \in X\}$, $i = 1, \dots, k$, where

$$u_i(x) = \big(\nabla g_i(\xi.), \mathbf{1}_{[s,T]}(\cdot)\big) \min(|y|^2, 1).$$

Since (15.8) holds and the functions $s \mapsto \big(\nabla g_i(\xi.(X)), \mathbf{1}_{[s,T]}\big)$ are continuous, it follows that there exist points $s_1^0, \dots, s_k^0$ and a number $\varepsilon > 0$ (obviously, depending on X) such that for any $s_1, \dots, s_k$, $s_j \in (s_j^0 - \varepsilon,\ s_j^0 + \varepsilon)$, the vectors

$$b_i = \Big(\big(\nabla g_i(\xi.(X)), \mathbf{1}_{[s_1,T]}(\cdot)\big), \dots, \big(\nabla g_i(\xi.(X)), \mathbf{1}_{[s_k,T]}(\cdot)\big)\Big) \in \mathbf{R}^k, \tag{15.10}$$

$i = 1, \dots, k$, are linearly independent. Moreover, by (15.7), $\mathbf{P}$-almost all configurations X satisfy the condition $X \cap [(s_j^0 - \varepsilon,\ s_j^0 + \varepsilon) \times \mathbf{R}^1] \neq \varnothing$ for all $j = 1, \dots, k$.

This fact together with the linear independence of the vectors b_i implies the linear independence of the vectors u_i and (15.9) (the details are left to the reader).

Now it follows from Proposition 15.1 that the conditional distribution of the stochastic functional $g(\xi.)$ is absolutely continuous (provided that the trajectory of the Gaussian process is fixed). Hence, the unconditional distribution is also absolutely continuous. □

The Gauss–Ostrogradskii formula. In what follows, we return to the consideration of the "usual" situation. Thus, $\mathfrak{X}$ is the configuration space on $\mathbf{S} \times I$, the metric in I is Euclidean, the norm on $\mathbf{H}(X)$ is defined by (14.1), and $\mathbf{P}$ is a Poisson measure of intensity $\pi \times \lambda$.

Suppose that $(\mathfrak{X}_0, \tau)$ is a Poisson space and $\mathfrak{U}$ is an inner τ-open set. The set $\mathfrak{U}$ is called a *set with smooth boundary* if for each configuration $X \in \partial\mathfrak{U}$ there exist a τ-neighborhood of $\mathfrak{U}_X$, a function $f \in \mathbf{C}^{(1)}(\mathfrak{U}_X)$, and a finite-dimensional projection Q such that

1) $\mathfrak{U} \cap \mathfrak{U}_X = \{Y \in \mathfrak{U}_X \mid f(Y) < 0\}$;

2) $\inf\limits_{Y \in \mathfrak{U}_X} |Qn(Y; f)| > 0$.

Since the topology τ is Lindelöf, one can readily see that $\partial\mathfrak{U}$ is a hypersurface. We also note that for each set with smooth boundary, the field of the outward normal n is well defined by the formulas $n(Y) = n(Y; f)$, $Y \in \mathfrak{U}_X$.

If $\mathfrak{U}$ is a set with smooth boundary and $K \in \mathbf{C}^{(1)}(\mathfrak{U}^-)$ is a vector field, then the relation

$$\int_{\mathfrak{U}} \operatorname{tr} K^{(1)}\, d\mathbf{P} = \int_{\partial\mathfrak{U}} (K, n)\, d\sigma_{\partial\mathfrak{U}}, \tag{15.11}$$

where $\sigma_{\partial\mathfrak{U}}$ is a surface measure on $\partial\mathfrak{U}$, is called the *Gauss–Ostrogradskii formula*.

The conditions under which the Gauss–Ostrogradskii formula holds are stated in the following two theorems.

THEOREM 15.3. *Suppose that K is a finite-dimensional vector field, that is, $Q_0K = K$ for some finite-dimensional projection Q_0. Then the inclusions* $\operatorname{tr} K^{(1)} \in \mathbf{L}^1(\mathfrak{U}, \mathbf{P})$ *and* $(K, n) \in \mathbf{L}^1(\partial\mathfrak{U}, \sigma_{\partial\mathfrak{U}})$ *are sufficient for* (15.11) *to hold.*

THEOREM 15.4. *Suppose that*

$$K^{(1)}(X) \in \mathcal{L}_{(1)}(\mathbf{H}(X)), \qquad \int_{\mathfrak{U}} \|K^{(1)}\|_1 \, d\mathbf{P} < \infty$$

for any configuration $X \in \mathfrak{U}$.

Then the inclusions $|K| \in \bigcup_{\delta>0} \mathbf{L}^{1+\delta}(\mathfrak{U}, \mathbf{P})$ *and* $|K| \in \mathbf{L}^1(\partial\mathfrak{U}, \sigma_{\partial\mathfrak{U}})$ *are sufficient for* (15.11) *to hold.*

PROOF OF THEOREM 15.3. First, we assume that the union of $\mathfrak{U}$ with finitely many τ-open sets $\mathfrak{U}^1, \dots, \mathfrak{U}^n$ from the definition of a set with smooth boundary contains the set $\{X \in \mathfrak{U}^- \mid K(X) \neq 0\}$. Let f_i and Q_i $(i = 1, \dots, n)$ be the corresponding functions and orthogonal projections from this definition. We choose some finite-dimensional orthogonal projection $Q = Q(V) \geq Q_0 \vee Q_1 \vee \dots \vee Q_n$ for which $V = B \times \Delta$ is open. We can write

$$\int_{\mathfrak{U}} \operatorname{tr} K^{(1)} \, d\mathbf{P} = \sum_{l=1}^{\infty} \int_{\mathfrak{U}_l} \operatorname{tr} K^{(1)} \, d\mathbf{P}, \qquad \mathfrak{U}_l = \{X \in \mathfrak{U} \mid |X \cap V| = l\}$$

and factor the measure $\mathbf{P}$ in each term:

$$\int_{\mathfrak{U}_l} \operatorname{tr} K^{(1)} \, d\mathbf{P} = \frac{1}{l!} e^{-\Pi(V)} \int_{\mathfrak{X}(\mathbf{G}\setminus V)} \mathbf{P}v^{-1}\,(dZ) \int_{\mathbf{B}^l} \pi(d\theta_1) \dots \pi(d\theta_l)$$
$$\times \int_{\Delta^l} \mathbf{1}_{\mathfrak{U}_l}(X) \operatorname{tr} K^{(1)}(X) \, dr_1 \dots dr_l$$

(we again use the notation $X = Z \cup \{(\theta_1, r_1), \dots, (\theta_l, r_l)\}$ and $Z = v(X)$). Let us perform some transformations in the inner integral. We set

$$\mathfrak{U}_l^0 = \{(r_1, \dots r_k) \in \Delta^l \mid X \in \mathfrak{U}_l\}$$

(here and in what follows, we omit the dependence on the outer variables Z, $\theta_1, \dots, \theta_l$) and define a vector field $K^l \colon \overline{\mathfrak{U}}_l^0 \to (\mathbf{R}^m)^l$ by

$$\left(K^l(r_1, \dots, r_l)\right)_j = K(X)((\theta_j, r_j)), \qquad j = 1, \dots, l,$$

where $\overline{\mathfrak{U}}_l^0$ is the Euclidean closure of $\mathfrak{U}_l^0$. We also set

$$\mathfrak{U}_l^i = \{(r_1, \dots, r_l) \in \Delta^l \mid X \in \mathfrak{U}_i, \ |X \cap V| = l\}, \qquad i = 1, \dots, n,$$

and consider a (smooth) partition of unity $\varphi_0, \varphi_1, \dots, \varphi_n$ subordinate to the open cover $\{\mathfrak{U}_l^i\}_{i=0}^n$ of the set $\{K^l \neq 0\}$. Then the desired integral is equal to

$$\int_{\mathfrak{U}_l^0} \operatorname{tr} K^{(1)}(X) \, dr_1 \dots dr_l = \sum_{i=0}^{n} \int_{\mathfrak{U}_l^0 \cap \mathfrak{U}_l^i} \operatorname{tr}(K^l \varphi_i)^{(1)} \, dr_1 \dots dr_l.$$

We apply the Gauss–Ostrogradskii formula to each term on the right-hand side. Then the term with $i = 0$ disappears. Indeed, the boundary of the set $\mathfrak{U}_l^0$ contains two parts, the boundary $\partial\mathfrak{U}_l$ and the boundary $\partial\Delta^l$ of the cube. On the first part, φ_0 and hence $K^l\varphi_0$ vanish. Since the vector field K is finite-dimensional $(Q_0K = K)$, it follows from the inequality $Q \geq Q_0$ that the normal component of the field K^l is zero on the second part of the boundary of the set $\mathfrak{U}_l^0$. Thus, the integral over the boundary of $\mathfrak{U}_l^0$ is zero.

For $i \geq 1$, similar arguments show that

$$\int_{\mathfrak{U}_l^0 \cap \mathfrak{U}_l^i} \operatorname{tr}(K^l \varphi_i)^{(1)}\, dr_1 \dots dr_m = \int \left(K\varphi_i, \frac{Qn}{|Qn|} \right) d\widetilde{\sigma} = \int_{\partial \mathfrak{U}_l^0} (K\varphi_i, n)|Qn|^{-1}\, d\widetilde{\sigma}.$$

Here $\widetilde{\sigma}$ is the Euclidean surface measure. In the passage to the last integral, we have applied the relations $QK = K$ and $\operatorname{supp} \varphi_i \subset \mathfrak{U}_l^i$.

By combining all these formulas and using the definition of the surface measure $\sigma_{\partial \mathfrak{U}}$, we arrive at the Gauss–Ostrogradskii formula (15.11).

Before we consider the general case, note that the previous arguments remain valid if we assume that $(K^l)_j$ is continuously differentiable only with respect to r_j ($j = 1, \dots, l$). In this case, we can again refer to the classical Gauss–Ostrogradskii formula.

Let us consider the general case. We choose some finite-dimensional orthogonal projection $Q = Q(V) \geq Q_0$ for which $V = B \times \Delta$ is open and an increasing sequence $\{F_n'\}$ of measurable compact sets in $(v(\mathfrak{X}_0), v(\tau))$ such that $\mathbf{P}v^{-1}\{F_n'\} \to 1$. We also note that S contains an increasing sequence $\{B_n\}$ of compact sets such that $B_n \subset B$ and $\pi(B_n) \to \pi(B)$ (this is a consequence of the fact that all finite measures on $\mathfrak{B}(S)$ are regular). We set

$$F_n'' = \{Y \in \mathfrak{X}(B_n \times \bar{\Delta}) \mid |Y| \leq n\}, \qquad F_n = i_V(F_n'' \times F_n')$$

(recall that the mapping i_V is defined by the relation $i_V(Y, Z) = Y \cup Z$). Obviously, the sets F_n are compact in the topology $i_V\big(\tau(V) \times v(\tau)\big)$ and, therefore, in the topology τ. Moreover, $\mathbf{P}_n\{F_n\} \to 1$.

Consider the vector fields $K_n = K\mathbf{1}_{F_n}$. By the above-treated special case and the subsequent remark for the fields K_n, we have (15.11). By passing to the limit with respect to n, we prove (15.11) for the vector field K. The proof of the theorem is complete. □

PROOF OF THEOREM 15.4. Let g_n be the functions introduced in the proof of Lemma 15.1. We set $K_n(X)(x) = g_n(x)K(X)(x)$. Obviously, the vector fields K_n satisfy the assumptions of Theorem 12.3, and hence, (15.11) holds for K_n. By passing to the limit, we prove (15.11) for the field K. □

PROBLEM 15.1. Suppose that $g\colon \mathbf{D}[0, T] \to \mathbf{R}^k$ is an integral functional, that is,

$$g(x(\cdot)) = \int_0^T u(x(s))\, ds = \left(\int_0^T u_1(x(s))\, ds, \dots, \int_0^T u_k(x(s))\, ds \right),$$

where the functions $u_j \in \mathbf{C}^{(1)}(\mathbf{R})$ satisfy the following condition:

$$\det \big(u_j'(x_i)\big)_{i,j=1}^k \neq 0$$

for Lebesgue almost all $x_1, \dots, x_k$. Prove that g satisfies assumption (15.8) in Theorem 15.2.

§16. Smooth functionals

Suppose that $f\colon \mathfrak{U} \to \mathbf{R}^1$ is a smooth function (in the sense of §14) defined on a subset of the configuration space $\mathfrak{X}(\mathbf{S} \times I)$ and $\mathbf{P}$ is a Poisson measure of intensity $\pi \times \lambda$. To obtain some information about the smoothness of the density of the measure $\mathbf{P}f^{-1}$ (Theorem 15.2 gives existence conditions for the density), it is convenient to use the method of differential operators (see §6).

Here the differential operators used in Theorem 6.1 will be the operators of differentiation along vector fields. Specifically, to a measurable vector field $K\colon X \mapsto K(X) \in \mathbf{H}(X)$ we assign the differential operator D defined by

$$(16.1)\qquad (Dg)(X) = \big(g^{(1)}(X), K(X)\big)_{\mathbf{H}(X)}.$$

To obtain conditions under which the measure $\mathbf{P}$ belongs to the classes $\overline{\mathbf{W}}{}_1^l(D)$ introduced in §6, we need integration by parts formulas.

PROPOSITION 16.1. *Suppose that a function f and a vector field K satisfy the following conditions*:

(1) $f \in \mathbf{C}^{(1)}(\mathfrak{X})$, $K \in \mathbf{C}^{(1)}(\mathfrak{X})$;

(2) $f \cdot K \in \mathbf{C}^{(1)}(\mathfrak{X}^-)$;

(3) *the field K is finite-dimensional, that is, $QK = K$ for some finite-dimensional projection Q*;

(4) $\int \big(|(f^{(1)}, K)| + |f \operatorname{tr} K^{(1)}|\big)\, d\mathbf{P} < \infty$.

Then

$$(16.2)\qquad \int (f', K)\, d\mathbf{P} = -\int f \operatorname{tr} K^{(1)}\, d\mathbf{P}.$$

PROOF. Equation (16.2) can be rewritten in the form

$$\int \operatorname{tr}(fK)^{(1)}\, d\mathbf{P} = 0,$$

which is a special case of the Gauss–Ostrogradskii formula. □

Now we assume that the vector field K defining the differential operator D belongs to the class $\mathbf{C}^{(l)}(\mathfrak{X})$.

We define functions $a_0, a_1, \dots, a_l\colon \mathfrak{X} \to \mathbf{R}^1$ as follows:

$$(16.3)\qquad a_0 \equiv 1, \qquad a_{i+1} = (a_i^{(1)}, K) + a_i \operatorname{tr} K^{(1)} = Da_i + \operatorname{tr} K^{(1)}.$$

One can readily show that a_i $(1 \le i \le l)$ is a linear combination of the expressions

$$(16.4)\qquad (\operatorname{tr} K^{(1)})^{\beta_1} (D \operatorname{tr} K^{(1)})^{\beta_2} \dots (D^{i-1} \operatorname{tr} K^{(1)})^{\beta_i},$$

where $\beta_i \ge 0$ and $\sum_{j=1}^i j\beta_j = i$.

Set

$$(16.5)\qquad \begin{aligned} A_i = \Big\{ g \in \mathbf{C}^{(i)}(\mathfrak{X}) \mid \forall j = 1, \dots, i\ (a_{j-1} D^{i-j} g) K \in \mathbf{C}^{(1)}(\mathfrak{X}^-),\\ \int \Big[|D^{i-j} g| \big(|a_{j-1} \operatorname{tr} K^{(1)}| + |(a'_{j-1}, K)| \big) \Big]\, d\mathbf{P} < \infty,\\ \int |D^{i-j+1} g|\, |a_{j-1}|\, d\mathbf{P} < \infty \Big\}. \end{aligned}$$

PROPOSITION 16.2. *Suppose that a finite-dimensional vector field K belongs to $\mathbf{C}^{(l)}(\mathfrak{X})$ and D is the differential operator* (16.1). *Then the measure $\mathbf{P}$ belongs to $\mathbf{W}_1^{(l)}(D \mid L^1(\mathbf{P}), A_1, \dots, A_l)$, where $A_1, \dots, A_l$ are defined in* (16.5). *Moreover, $d\mu_i = a_i\, d\mathbf{P}$ are the representing measures of the functionals $D^i\mathbf{P}$ on A_i, $i = 1, \dots, l$.*

To prove this theorem, it suffices to note that for $g \in A_i$ relations (16.2) and (16.3) yield

$$\int D^i g\, d\mathbf{P} = \dots = (-1)^j \int a_j D^{i-j} g\, d\mathbf{P} = \dots = (-1)^i \int a_i g\, d\mathbf{P}. \tag{16.6}$$

To state the following proposition, we use the multi-indices introduced in §8. Suppose that $l \geq 1$ and $d > l$ is an even number. Let $p_i = \big(d/(d-i), \dots, d/(d-i)\big)$ be a multi-index of length $i+1$ ($i = 1, \dots, l$), and $p = (d, \dots, d)$ a multi-index of length $l+1$.

PROPOSITION 16.3. *Suppose that $V \subset \mathbf{G}$ is an open set, $K \in \mathbf{H}_V^{p,0}$, and D is the corresponding differential operator* (16.1). *Then for $\mathbf{P}v^{-1}$-almost all Y (the mapping v is defined by* (14.1)*), the conditional measure $\mathbf{P}_Y$ belongs to the class $\overline{\mathbf{W}}^l(D \mid \mathbf{L}^1(\mathbf{P}_Y), \mathbf{H}_V^{p_1}, \dots, \mathbf{H}_V^{p_l})$. For $i = 1, \dots, l$, the representing measure μ_i of the functional $D^i\mathbf{P}_Y$ is absolutely continuous with respect to $\mathbf{P}_Y$ and satisfies*

$$d\mu_i = a_i\, d\mathbf{P}_Y, \qquad a_i \in \mathbf{L}^{l/i}(\mathbf{P}_Y),$$

where the functions a_i are defined in (16.3).

Before we prove this proposition, let us make two remarks. First, the definition of a_i contains the traces of some operators. The proof implies that these traces are defined almost surely (recall that by our definitions these operators are only Hilbert–Schmidt operators). Second, a similar statement for the measure $\mathbf{P}$ readily follows from Proposition 16.3.

COROLLARY. *Under the assumptions of Proposition* 16.3, *the measure $\mathbf{P}$ belongs to the class*

$$\overline{\mathbf{W}}_1^l(D \mid \mathbf{L}^1(\mathbf{P}),\ A_1, \dots, A_l),$$

where $A_i = \{g \in \mathbf{H}_V^{p_i} \mid \int (a_{i-j} D^j g)\, d\mathbf{P} < \infty,\ j = 0, 1, \dots, i\}$, $i = 1, \dots, l$.

To prove this corollary, it suffices to integrate the relation

$$\int D^i g\, d\mathbf{P} = \dots = (-1)^j \int a_j D^{i-j} g\, d\mathbf{P}_Y = \dots = (-1)^i \int a_i g\, d\mathbf{P}_Y$$

with respect to the quotient measure $\mathbf{P}v^{-1}$.

PROOF OF PROPOSITION 16.3. First, we prove two auxiliary lemmas. The first lemma can be derived from the Stokes theorem and hence can be applied to a wider class of integration domains. To state this lemma, we need some auxiliary notation. By the letter τ we denote a permutation of the numbers $1, \dots, d$, and $\sum_\tau$ denotes the sum over all $d!$ permutations. By $\Gamma(\tau)$ we denote the set of cycles comprising the permutation τ. For a cycle $\gamma = (i_1, \dots, i_s) \in \Gamma(\tau)$, we write $\prod_{i\in\gamma} A_i$ instead of $A_{i_1} \cdot \ldots \cdot A_{i_s}$. (Since we consider matrices, the order of the factors is important). By $\mathfrak{U}$ we denote a bounded open parallelepiped in $\mathbf{R}^r$. A set of mappings $v_1, \dots, v_d \colon \mathfrak{U} \to \mathbf{R}^r$ is said to be *admissible for* $\mathfrak{U}$ if all these mappings belong to the class $\mathbf{C}^{(1)}$, can be continuously extended to $\partial\mathfrak{U}$ together with their derivatives, and possess the following properties: if $x \in \partial\mathfrak{U}$ is not a corner point (this means that the normal n to $\partial\mathfrak{U}$ at this point is determined uniquely), then

1) $(v_i, n)(x) = 0$, $i = 1, \dots, d$;
2) $(v_i^{(1)}(v_j), n)(x) = 0$, $i, j = 1, \dots, d$, $i \neq j$.

LEMMA 16.1. *Any set $v_1, \dots, v_d$ of $\mathbf{C}^{(2)}$ mappings admissible for $\mathfrak{U}$ satisfies the relation*

$$\sum_{\tau}(-1)^{|\Gamma(\tau)|}\int_{\mathfrak{U}}\prod_{\gamma\in\Gamma(\tau)}\operatorname{tr}\Big(\prod_{i\in\gamma}v_i^{(1)}\Big) = 0. \tag{16.7}$$

From now on, integration over $\mathfrak{U}$ is with respect to Lebesgue measure.

PROOF OF LEMMA 16.1. For $d = 1$, (16.7) becomes

$$\int_{\mathfrak{U}}\operatorname{tr} v^{(1)} = 0. \tag{16.8}$$

Indeed, we have the more general relation

$$\int_{\mathfrak{U}} f\operatorname{tr} v^{(1)} = -\int_{\mathfrak{U}}(f^{(1)}, v), \tag{16.9}$$

which can be obtained by applying the Stokes theorem to the vector field fv. The conditions on v imply that the surface integral is equal to zero.

For an arbitrary d, the lemma is proved by induction. We restrict our consideration to the passage from $d = 2$ to $d = 3$ (the passage from $d = 1$ to $d = 2$ is obvious and hence not instructive).

It follows from (16.9) that

$$\begin{aligned}\int_{\mathfrak{U}}\operatorname{tr} v_1^{(1)}\operatorname{tr} v_2^{(1)}\operatorname{tr} v_3^{(1)} &= -\int_{\mathfrak{U}}((\operatorname{tr} v_1^{(1)}\operatorname{tr} v_2^{(1)})^{(1)}, v_3)\\ &= -\int_{\mathfrak{U}}[\operatorname{tr} v_1^{(2)}(v_3)\operatorname{tr} v_2^{(1)} + \operatorname{tr} v_2^{(2)}(v_3)\operatorname{tr} v_1^{(1)}]\\ &= -\int_{\mathfrak{U}}[\operatorname{tr}(v_1^{(1)}(v_3))^{(1)}\operatorname{tr} v_2^{(1)} + \operatorname{tr}(v_2^{(1)}(v_3))^{(1)}\operatorname{tr} v_1^{(1)}]\\ &\quad + \int_{\mathfrak{U}}[\operatorname{tr}(v_1^{(1)}v_3^{(1)})\operatorname{tr} v_2^{(1)} + \operatorname{tr}(v_2^{(1)}v_3^{(1)})\operatorname{tr} v_1^{(1)}].\end{aligned}$$

By applying the induction assumption (for $d = 2$) to the first two terms and by using (16.9), we obtain

$$\begin{aligned}&-\int_{\mathfrak{U}}\{\operatorname{tr}[(v_1^{(1)}(v_2))^{(1)}v_2^{(1)}] + \operatorname{tr}[(v_2^{(1)}(v_3))^{(1)}v_1^{(1)}]\}\\ &\qquad + \int_{\mathfrak{U}}[\operatorname{tr}(v_1^{(1)}v_3^{(1)})\operatorname{tr} v_2^{(1)} + \operatorname{tr}(v_2^{(1)}v_3^{(1)})\operatorname{tr} v_1^{(1)}]\\ &= -\int_{\mathfrak{U}}\{\operatorname{tr}[v_1^{(2)}(v_3)v_2^{(1)} + v_1^{(1)}v_3^{(1)}v_2^{(1)}] + \operatorname{tr}[v_2^{(2)}(v_3)v_1^{(1)} + v_2^{(1)}v_3^{(1)}v_1^{(1)}]\}\\ &\qquad + \int_{\mathfrak{U}}[\operatorname{tr}(v_1^{(1)}v_3^{(1)})\operatorname{tr} v_2^{(1)} + \operatorname{tr}(v_2^{(1)}v_3^{(1)})\operatorname{tr} v_1^{(1)}]\end{aligned}$$

$$= -\int_{\mathfrak{U}} ((\operatorname{tr}(v_1^{(1)} v_2^{(1)}))^{(1)}, v_3) - \int_{\mathfrak{U}} [\operatorname{tr}(v_1^{(1)} v_3^{(1)} v_2^{(1)}) + \operatorname{tr}(v_2^{(1)} v_3^{(1)} v_1^{(1)})]$$
$$+ \int_{\mathfrak{U}} [\operatorname{tr}(v_1^{(1)} v_3^{(1)}) \operatorname{tr} v_2^{(1)} + \operatorname{tr}(v_2^{(1)} v_3^{(1)}) \operatorname{tr} v_1^{(1)}]$$
$$= -\int_{\mathfrak{U}} [\operatorname{tr}(v_1^{(1)} v_3^{(1)} v_2^{(1)}) + \operatorname{tr}(v_1^{(1)} v_2^{(1)} v_3^{(1)})]$$
$$+ \int_{\mathfrak{U}} [\operatorname{tr}(v_1^{(1)} v_2^{(1)}) \operatorname{tr} v_3^{(1)} + \operatorname{tr}(v_1^{(1)} v_3^{(1)}) \operatorname{tr} v_2^{(1)} + \operatorname{tr}(v_2^{(1)} v_3^{(1)}) \operatorname{tr} v_1^{(1)}]. \quad \square$$

LEMMA 16.2. *Suppose that, under the assumptions of Lemma 16.1, d is even and $v_i \equiv v$, $i = 1, \ldots, d$. Then*

$$\int_{\mathfrak{U}} (\operatorname{tr} v^{(1)})^d \leq c_d \int_{\mathfrak{U}} \|v^{(1)}\|_2^d \tag{16.10}$$

with some constant c_d independent of $\mathfrak{U}$ and r (here $\|\cdot\|_2$ is the Hilbert–Schmidt norm).

PROOF OF LEMMA 16.2. We consider only the case $d = 4$; the general case can be considered in a similar way. We set

$$\alpha = \int_{\mathfrak{U}} (\operatorname{tr} v^{(1)})^4, \qquad \beta = \int_{\mathfrak{U}} \|v^{(1)}\|_2^4.$$

By Lemma 16.1, we have

$$\alpha = 6\int_{\mathfrak{U}} \operatorname{tr}((v^{(1)})^4) - 8\int_{\mathfrak{U}} \operatorname{tr}((v^{(1)})^3) \operatorname{tr} v^{(1)} + 6\int_{\mathfrak{U}} \operatorname{tr}(v^{(1)})^2 (\operatorname{tr} v^{(1)})^2 - 3\int_{\mathfrak{U}} (\operatorname{tr}(v^{(1)})^2)^2.$$

Since the absolute value of the trace of any operator does not exceed the trace norm of this operator, we have the estimate $|\operatorname{tr}(v^{(1)})^n| \leq \|v_1\|_1^n$ for any $n \geq 1$. Now we apply the Hölder inequality and obtain

$$\alpha \leq 8\alpha^{1/4}\beta^{3/4} + 6\alpha^{1/2}\beta^{1/2} + (6+3)\beta \leq 9(\alpha^{1/4}\beta^{3/4} + \alpha^{1/2}\beta^{1/2} + \beta).$$

Thus, $\alpha \leq 729\beta$. $\square$

An obvious approximation shows that to ensure (16.10), one need not require the second derivatives to exist.

Now let us return to the proof of Proposition 16.3. First, we consider the case $l = 1$.

Suppose that $f \in \mathbf{H}_V^{d/(d-1), d/(d-1)}$. We choose approximating sequences $\{V_n\}$, $\{f_n\}$, and $\{K_n\}$ for V, f, and K, respectively; let Q_n be the orthogonal projection corresponding to V_n. By Proposition 16.1, we have

$$\mathsf{E}_Y(f_n^{(1)}, K_n) = -\mathsf{E}_Y(f_n \operatorname{tr} K_n^{(1)}) \tag{16.11}$$

for $\mathbf{P}v^{-1}$-almost all Y. Since $f_n^{(1)} \to f$ in $\mathbf{L}^{d/(d-1)}$ and $K_n \to K$ in $\mathbf{L}^d$, it follows that the left-hand side of (16.11) converges to $\mathsf{E}_Y(f, K)$. To pass to the limit in the expression on the right-hand side in (16.11), we need to show that the sequence $\{\operatorname{tr} K_n^{(1)}\}$ is fundamental in $\mathbf{L}^d$. Let us choose n and n_1 ($n < n_1$) and estimate $\mathsf{E}_Y(\operatorname{tr} K_{n_1}^{(1)} - \operatorname{tr} K_n^{(1)})^d$. Let us additionally assume that all sets V_n have the form $V_n = B \times J$, with $B \in S$ and $J \subset I$. We shall see that there is no loss of generality due to this assumption. We have

$$
\begin{aligned}
\mathsf{E}_Y\big(\operatorname{tr} K_{n_1}^{(1)} - \operatorname{tr} K_n^{(1)}\big)^d &= \mathsf{E}_Y\big(\operatorname{tr}[K_{n_1}^{(1)} \circ Q_{n_1} - \operatorname{tr} K_n^{(1)} \circ Q_n]\big)^d \\
&= \int (\mathbf{P}_Y v_{n_1}^{-1})(dX) \sum_{r=0}^{\infty} \frac{[\Pi(V_{n_1})]^r}{r!} e^{-\Pi(V_{n_1})} \\
&\quad \times \int_{V_{n_1}^r} \Big(\operatorname{tr}\big[K_{n_1}^{(1)} \circ Q_{n_1}(X \cup \{x_1, \dots, x_r\}) - K_n^{(1)} \circ Q_n(X \cup \{x_1, \dots, x_r\})\big]\Big)^d \\
&\quad \times \frac{\Pi(dx_1) \dots \Pi(dx_r)}{[\Pi(V_{n_1})]^r} \\
&= \int (\mathbf{P}_Y v_{n_1}^{-1})(dX) \sum_{r=0}^{\infty} \frac{[\Pi(V_{n_1})]^r}{r!} e^{-\Pi(V_{n_1})} \int_{\mathbf{B}^r} \frac{\pi(d\theta_1) \dots \pi(d\theta_r)}{[\Pi(V_N)]^r} \\
&\quad \times \int_{J^r} \Big(\operatorname{tr}\big[K_{n_1}^{(1)} \circ Q_{n_1}(X \cup \{(\theta_1, r_1), \dots, (\theta_r, r_r)\}) \\
&\qquad - K_n^{(1)} \circ Q_n(X \cup \{(\theta_1, r_1), \dots, (\theta_r, r_r)\})\big]\Big)^d \, dr_1 \dots dr_r.
\end{aligned}
$$

By applying the estimate (16.10) to the inner integral $\int_{J^r}$ and transforming the resulting expression backward, we obtain

$$
\mathsf{E}_Y(\operatorname{tr} K_{n_1}^{(1)} - \operatorname{tr} K_n^{(1)})^d \le c_d \mathsf{E}_Y \|K_{n_1}^{(1)} \circ Q_{n_1} - K_n^{(1)} \circ Q_n\|_2^d.
$$

By the definition of an approximating sequence, the right-hand side of the last estimate tends to zero as $n, n_1 \to \infty$. Thus, we have proved that in (16.10) we can pass to the limit as $n \to \infty$. This proves our proposition in the case $l = 1$.

For an arbitrary l, we need to pass to the limit in the chain of equations

$$
\mathsf{E}_Y(D_n^l f_n) = \dots = (-1)^i \mathsf{E}_Y(a_{i,n} D_n^{l-i} f_n) = \dots = (-1)^l \mathsf{E}_Y(a_{l,n} f_n).
$$

Here D_n is the differential operator defined by the field K_n and $a_{1,n}, \dots, a_{l,n}$ is the sequence of the form (16.3) that corresponds to the field K_n. This passage to the limit can be justified just as in the case $l = 1$. Proposition 16.3 is proved. □

By using Theorem 6.1 and Propositions 16.2 and 16.3, we can obtain some results about the smoothness of the image of a Poisson measure $\mathbf{P}$ under the action of smooth (in the sense of §14) functionals.

There are differential operators of two types. The operators of the first type are defined by "finite-dimensional" vector fields (see Proposition 16.2) and the operators of the second type are defined by "infinite-dimensional" fields that satisfy some restrictions on the moments (see Proposition 16.3).

Sometimes it is necessary to use differential operators that are sums of first- and second-type operators. We present the corresponding statement.

THEOREM 16.1. *Suppose that $l \ge 1$ and $f\colon \mathfrak{X} \to \mathbf{R}^1$ is a measurable mapping. The following conditions are sufficient for the distribution $\mathbf{P}f^{-1}$ to belong to the class $\overline{\mathbf{W}}_1^l$.*

1. $f \in \mathbf{C}^{l+1}(\mathfrak{X})$.

2. *There exist open sets $V, \widetilde{V} \in \mathbf{G}$ (V is bounded) and a vector field $K = K_1 + K_2$ such that $K_1, K_2 \in \mathbf{C}^{(l)}(\mathfrak{X}^-)$, $QK_1 = K_1$, $\widetilde{Q}K_2 = K_2$ (here Q and $\widetilde{Q}$*

are the orthogonal projections corresponding to V and $\widetilde{V}$, respectively), and the following properties hold[4]:

(a) $\mathbf{P}\{Df \neq 0\} = 1$;

(b) $f \in \bigcap_{p<\infty} \mathbf{H}_V^p$, $p = (p_0, p_1, \dots, p_{l+1})$; $K \in \bigcap_{p<\infty} \mathbf{H}_V^p$, $p = (p_0, p_1, \dots, p_l)$;

(c) *we have* $(ga_i(Df)^{-1})K \in \mathbf{C}(\mathfrak{X}^-)$ *(the sets $S_{t,i}$ were defined in §6) for all* $t = 0, \dots, l-1$, $i = 0, \dots, t$, *and* $g \in S_{t,i}(D, f)$.

3. *All mathematical expectations of the form*

$$\mathsf{E}\big|(Df)^{-l-|k|}(D^2 f)^{k_1} \dots (D^{l+1} f)^{k_l} (\operatorname{tr} K^{(1)})^{\beta_1} (D \operatorname{tr} K^{(1)})^{\beta_2} \dots (D^{i-1} \operatorname{tr} K^{(1)})^{\beta_i}\big|,$$

$$i = 0, \dots, l, \qquad \sum_{j=1}^{l} j k_j = l - i, \qquad \sum_{j=1}^{i} j \beta_j = i,$$

are finite.

PROOF. This theorem readily follows from Theorem 6.1 and Propositions 16.2 and 16.3. □

Now we use this result for studying the distribution of quadratic functionals of a stable vector.

Suppose that $\mathbf{B}$ is a separable Banach space, μ a stable measure on $\mathbf{B}$ with the exponent α $(0 < \alpha < 2)$, $f\colon \mathbf{B} \to \mathbf{R}^1$ a quadratic functional of the form

$$\widetilde{f}(u) = \langle Au, u \rangle + \langle b^*, u \rangle,$$

where $b^* \in \mathbf{B}^*$, and $A\colon \mathbf{B} \to \mathbf{B}^*$ a continuous self-adjoint linear operator.

Let B_0 be the affine support of the measure μ. There exists a unique operator $A_0\colon B_0 \to B_0^*$ such that $\langle Ab_1, b_2 \rangle = \langle A_0 b_1, b_2 \rangle$, $b_1, b_2 \in B_0$.

THEOREM 16.2. *If* $\dim A_0(B_0) > 2l$, *then* $\mu\widetilde{f}^{-1} \in \overline{\mathbf{W}}_1^l$.

PROOF. By S we denote the unit sphere in $\mathbf{B}$; that is, $S = \{b \in \mathbf{B} \mid \|b\| = 1\}$. Furthermore, consider the configuration space $\mathfrak{X}$ on $G = S \times (0, \infty)$ and a Poisson random measure ν of intensity $\Pi = \pi \times \lambda$, where λ is the Lebesgue measure on $(0, \infty)$ and π is the spectral measure of a stable vector (see §13).

Just as in §13, we assume that the random measure ν is determined on the probability space $(\mathfrak{X}, \Sigma, \mathbf{P})$ by the identity mapping $\mathfrak{X} \to \mathfrak{X}$. The Poisson random measure ν allows us to construct a stable random vector u with distribution μ by using stochastic integration:

$$u = b_0 + \alpha^{-1}\Big\{ \int_{\{x \mid r(x) \geq 1\}} \frac{\theta(x)}{[r(x)]^{1/\alpha}}\, d\widetilde{\nu}(x) + \int_{\{x \mid 0 < r(x) < 1\}} \frac{\theta(x)}{[r(x)]^{1/\alpha}}\, d\nu(x) \Big\}. \tag{16.12}$$

Here $\theta(x) \in S$ and $r(x) \in \mathbf{R}^1$ are the coordinates of a point $x \in G$. By the change of variables $r \mapsto r^{-\alpha}$, we can obtain the stochastic integral (16.12) from the usual representation (13.6) with $u_1 = 0$ and the Lévy measure (see Problem 13.4). We slightly simplify the notation in (16.12) and write $|x|$ instead of $r(x)$; thus, $x = (\theta(x), |x|)$. Now we define a mapping $f\colon \mathfrak{X} \to \mathbf{R}^1$ by the formula $f(X) = \widetilde{f}(u(X))$. Then we have the obvious equation $\mu\widetilde{f}^{-1} = \mathbf{P}f^{-1}$, which allows us to use Theorem 16.1 for studying the smoothness properties of the measure $\mu\widetilde{f}^{-1}$ (as

[4] In this theorem, D is the differential operator (16.1) corresponding to the field K.

was already shown in §14, for any $n = 1, 2, \dots$ we have $f \in \mathbf{C}^n(\mathfrak{X}_0)$, where $\mathfrak{X}_0$ is some measurable saturated set of full measure.) First, let us obtain an explicit expression for the first derivative (gradient) of f. By (14.5), we have

$$(f^{(1)}(X))(x) = -\frac{1}{\alpha^2|x|^{1+1/\alpha}}\langle 2Au + b^*, \theta(x)\rangle, \qquad u = u(X). \tag{16.13}$$

In the statement of Theorem 16.1, we set $V = S \times (0, 2)$ and $\widetilde{V} = S \times (1, \infty)$. Obviously, $f \in \bigcap_{p<\infty} \mathbf{H}^p_V$. Now, to verify assumption 2 of the theorem, we need to choose an appropriate vector field. To this end, we choose and fix some infinitely differentiable function ρ such that $\rho(z) = z^{-1-1/\alpha}$ for $z \geq 2$, $\rho(z) = z^{1+1/\alpha}$ for $z \leq 1$, and ρ is bounded away from zero on $[1, 2]$.

We construct a vector field K by setting

$$(K(X))(x) = -\alpha^2(2Au + b^*, \theta(x))\rho(|x|). \tag{16.14}$$

It is easy to see that the field K can be represented as the sum $K_1 + K_2$ of vector fields that satisfy assumption 2 of Theorem 16.1. It suffices to take some partition of unity $\{\varphi_1, \varphi_2\}$ corresponding to the cover $((-\infty, 2), (1, \infty))$ of the real axis and set

$$(K_1(X))(x) = (K(X))(x)\varphi_1(|x|), \qquad (K_2(X))(x) = (K(X))(x)\varphi_2(|x|).$$

The motivation of this choice of K is quite obvious. We take an "almost" gradient of f (it is well known that the gradient of a function is the direction of its steepest descent) rather than the gradient itself, because the gradient has singularities at the points where $|x| = 0$. The function ρ is used just to get rid of these singularities.

Thus, to prove the theorem, we need to verify assumptions 2 (a), 2 (c), and 3 of Theorem 16.1. Assumption 2 (c) is the simplest since ρ rapidly decays at zero (the details are left to the reader).

Let us verify assumption 3. In particular, we shall prove that the lth-order moment of $(Df)^{-1}$ exists, which will imply the validity of assumption 2 (a).

First, we introduce some notation. By c we denote various positive constants whose precise values are inessential to us. By M_∞ we denote the set of random variables on $(\mathfrak{X}, \Sigma, \mathbf{P})$ that have moments of all orders.

By (16.13) and (16.14), we have

$$Df(X) = (f^{(1)}(X), K(X))_{H(X)} = \sum_{x \in X} \frac{\langle 2Au + b^*, \theta(x)\rangle^2}{|x|^{1+1/\alpha}}\rho(|x|).$$

The properties of the function ρ imply the following lower bound for Df:

$$Df \geq cR^2, \qquad \text{where} \quad c > 0,$$

$$R^2 = \sum_{|x| \leq 2} \langle 2Au + b^*, \theta(x)\rangle^2 + \sum_{|x|>2} \frac{\langle 2Au + b^*, \theta(x)\rangle^2}{|x|^{2+2/\alpha}}. \tag{16.15}$$

Now let us obtain an upper bound for the higher derivatives (in the direction of the field K) of the functions f and $\operatorname{tr} K^{(1)}$.

LEMMA 16.3. *For any $j = 1, 2, \dots$ and $\xi \in M_\infty$, the following conditions hold:*
1) $|D^j f| \le \xi R^2 (1+R)^{j-1}$,
2) $|D^j \operatorname{tr} K^{(1)}| \le \xi (1+R)^{j+1}$.

PROOF. The statement is obvious for $j = 1$. For $j = 2$, we have

$$\begin{aligned} D^2 f &= D(Df) \\ &= 2 \sum_{x,y \in X} \frac{\langle 2Au + b^*, \theta(x)\rangle}{|x|^{1+1/\alpha}} \rho(|x|) \langle 2A\theta(y), \theta(x)\rangle \frac{\langle 2Au + b^*, \theta(y)\rangle}{|y|^{1+1/\alpha}} \rho(|y|) \\ &\quad - \alpha^2 \sum_{x \in X} \langle 2Au + b^*, \theta(x)\rangle^3 \varkappa'(|x|) \rho(|x|) = A_1 + A_2, \\ \varkappa(z) &= \frac{\rho(z)}{z^{1+1/\alpha}}. \end{aligned}$$

Let us estimate the first term:

$$\begin{aligned} |A_1| &\le c \Big(\sum_{x \in X} \frac{|\langle 2Au + b^*, \theta(x)\rangle|}{|x|^{1+1/\alpha}} \rho(|x|) \Big)^2 \\ &\le c \Big(\Big(\sum_{|x| \le 2} |\langle 2Au + b^*, \theta(x)\rangle| \Big)^2 + \Big(\sum_{|x|>2} \frac{\langle 2Au + b^*, \theta(x)\rangle}{|x|^{2+2/\alpha}} \Big)^2 \Big) \\ &\le c \Big(\operatorname{card}(X \cap S \times (0,2]) \sum_{|x| \le 2} \langle 2Au + b^*, \theta(x)\rangle^2 \\ &\qquad + \sum_{|x|>2} \frac{1}{|x|^{2+2/\alpha}} \sum_{|x|>2} \frac{\langle 2Au + b^*, \theta(x)\rangle^2}{|x|^{2+2/\alpha}} \Big) \le \xi R^2, \qquad \xi \in M_\infty. \end{aligned}$$

Let us estimate the second term. Since $\varkappa'([0,1]) = 0$, we have

$$\begin{aligned} |A_2| &\le c \Big(\sum_{1 \le |x| \le 2} |\langle 2Au + b^*, \theta(x)\rangle|^3 + \sum_{|x|>2} |\langle 2Au + b^*, \theta(x)\rangle|^3 \frac{1}{|x|^{4+3/\alpha}} \Big) \\ &\le c \Big(\max_{\{x | 1 \le |x| \le 2\}} |\langle 2Au + b^*, \theta(x)\rangle| \sum_{1 \le |x| \le 2} \langle 2Au + b^*, \theta(x)\rangle^2 \\ &\qquad + \max_{\{x | |x|>2\}} \frac{|\langle 2Au + b^*, \theta(x)\rangle|}{|x|^{1+1/\alpha}} \sum_{|x|>2} \frac{\langle 2Au + b^*, \theta(x)\rangle^2}{|x|^{2+2/\alpha}} \Big) \le \xi R^2. \end{aligned}$$

Thus, we have obtained the desired estimate for $j = 2$. For $j > 2$, the proof is similar. We differentiate the function Df along K appropriately many times and estimate each term by one of the two methods. Namely, if the sum over $x \in X$ contains the first power of the expression $\langle 2Au + b^*, \theta(x)\rangle$, then we use the Hölder inequality. If this expression is raised to a power $j > 2$, then we replace the factor $|\langle 2Au + b^*, \theta(x)\rangle|^{j-2}$ (for $1 \le |x| \le 2$) or $|\langle 2Au + b^*, \theta(x)\rangle|^{j-2} |x|^{-(1+1/\alpha)(j-2)}$ (for $|x| > 2$) by its maximum over $\{x \mid 1 \le x \le 2\}$ or $\{x \mid |x| > 2\}$, respectively.

Now let us prove the second statement of Lemma 16.3. By (16.14), we have

$$K^{(1)}(X)[h,g] = \sum_{x,y\in X} \frac{\langle 2A\theta(y), \theta(x)\rangle}{|y|^{1+1/\alpha}} \rho(|x|)h(x)g(y) - \alpha^2 \sum_{x\in X} \langle 2Au + b^*, \theta(x)\rangle \rho'(|x|)h(x)g(y)$$

for $h, g \in \mathbf{H}(X)$. Hence,

$$\begin{aligned}\operatorname{tr} K^{(1)} &= 2\sum_{x\in X} \frac{\langle A\theta(x), \theta(x)\rangle}{|x|^{1+1/\alpha}} \rho(|x|) - \alpha^2 \sum_{x\in X} \langle 2Au + b^*, \theta(x)\rangle \rho'(|x|) \\ &\le \xi(1+R), \qquad \xi \in M_\infty.\end{aligned}$$

The estimates for the higher derivatives of $\operatorname{tr} K^{(1)}$ can be obtained just as in the proof of the first statement. □

Now let us verify assumption 3 of Theorem 16.1. We need an upper bound for products of the form

$$\begin{aligned}\Big|(Df)^{-l-\sum_{j=1}^l k_j}(D^2 f)^{k_1}\dots(D^{l+1}f)^{k_l} \\ \times (\operatorname{tr} K^{(1)})^{\beta_1}(D \operatorname{tr} K^{(1)})^{\beta_2}\dots(D^{i-1}\operatorname{tr} K^{(1)})^{\beta_i}\Big|,\end{aligned} \tag{16.16}$$

where

$$\sum_{j=1}^{l} jk_j = l - i, \quad \sum_{j=1}^{i} j\beta_j = i, \qquad i = 0, 1, \dots, l.$$

By (16.15) and Lemma 16.3, we obtain the upper bound

$$\xi_1 R^{-2((l+\sum k_j)} R^{2\sum k_j}(1+R)^{\sum jk_j}(1+R)^{\sum j\beta_j} = \xi_1 R^{-2l}(1+R)^l, \qquad \xi_1 \in M_\infty,$$

for (16.16) for all $i = 0, 1, \dots, l$.

It follows from this estimate that the existence of the $2(l+\delta)$th-order moment of R^{-1} for some $\delta > 0$ is sufficient for the expectations in assumption 3 of Theorem 16.1 to be finite.

LEMMA 16.4. *For some $\delta > 0$, we have $\mathsf{E}R^{-2(l+\delta)} < \infty$.*

PROOF. Consider the subset $S_0 \subset S$ of θ such that $\pi(U_\theta) > 0$ for any neighborhood U_θ of θ. Since $\mathbf{B}_0$ is the affine support of μ, it follows that $\mathbf{B}_0$ coincides with the closure of the linear span S_0 up to a translation by some vector.

Furthermore, the condition $\dim A_0\mathbf{B}_0 > 2l$ implies that there exist $m = 2l+1$ linearly independent vectors $\theta_1^0, \dots, \theta_m^0 \in S_0$ and m linearly independent vectors $\widetilde{\theta}_1^0, \dots, \widetilde{\theta}_m^0 \in S_0$ such that the matrix $(A\widetilde{\theta}_i^0, \theta_i^0)_{i,j=1}^m$ is nondegenerate. The sets $\{\theta_i^0\}$ and $\{\widetilde{\theta}_i^0\}$ need not consist of different vectors. Suppose that the number of coincidences is equal to $m - m_0$, $0 \le m_0 \le m$ (that is, $m - m_0 = \operatorname{card}\{i \mid \theta_i^0 = \widetilde{\theta}_j^0 \text{ for some } j\}$).

We renumber the vectors in both sets so that $\theta_i^0 = \widetilde{\theta}_i^0$ for all $i = m_0 + 1, \dots, m$. Next we choose disjoint neighborhoods $\widetilde{U}_1, \dots, \widetilde{U}_m, U_1, \dots, U_{m_0}, U_{m_0+1} = \widetilde{U}_{m_0+1}, \dots, U_m = \widetilde{U}_m$ of $\widetilde{\theta}_1, \dots, \widetilde{\theta}_m, \theta_1, \dots, \theta_{m_0}, \theta_{m_0+1} = \widetilde{\theta}_{m_0+1}, \dots, \theta_m = \widetilde{\theta}_m$ so that the matrix $(A\widetilde{\theta}_i, \theta_j)_{i,j=1}^m$ is uniformly nondegenerate with respect to all $\widetilde{\theta}_i \in \widetilde{U}_i$

and $\theta_i \in U_i$. We also note that $\pi(U_i)\pi(\widetilde{U}_i) > 0$ for all i. Furthermore, for $\widetilde{n}_1 \geq 2, \dots, \widetilde{n}_m \geq 2, n_1 \geq 2, \dots, n_{m_0} \geq 2$, by $\widetilde{F}(\widetilde{n}_1, \dots, \widetilde{n}_m)$ (respectively, by $F(n_1, \dots, n_{m_0})$) we denote the set of configurations X for which the point x of the configuration $X \cap (\widetilde{U}_i \times [2, \infty))$, $i = 1, \dots, m$, (respectively, of $X \cap (U_i \times [2, \infty))$, $i = 1, \dots, m_0$) with the least real coordinate lies in $\widetilde{U}_i \times (\widetilde{n}_i, \widetilde{n}_i + 1]$ (respectively, in $U_i \times (n_i, n_i + 1]$). Note that the events $\widetilde{F}(\widetilde{n}_1, \dots, \widetilde{n}_m)$ and $F(n_1, \dots, n_{m_0})$ are independent for any sets of numbers $(\widetilde{n}_1, \dots, \widetilde{n}_m)$ and $(n_1, \dots, n_{m_0})$.

We have

$$\mathsf{E}R^{-2(l+\delta)} \leq \mathsf{E}\Big(\sum_{|x|>2} \frac{\langle 2Au + b^*, \theta(x)\rangle^2}{|x|^{2+2/\delta}}\Big)^{-(l+\delta)}$$

$$= \sum_{\substack{\widetilde{n}_1, \dots, \widetilde{n}_m \\ n_1, \dots, n_{m_0}}} \mathsf{E}\Big(\sum_{|x|>2} \frac{\langle 2Au + b^*, \theta(x)\rangle^2}{|x|^{2+2/\delta}}\Big)^{-(l+\delta)} \mathbf{1}_{\widetilde{F}(\widetilde{n}_1, \dots, \widetilde{n}_m)} \mathbf{1}_{F(n_1, \dots, n_{m_0})}.$$

Let us separately estimate each term in the last sum. To this end, we represent $\mathbf{1}_{F(\widetilde{n}_1, \dots, \widetilde{n}_m)}$ as the product of two independent indicators $\mathbf{1}_1 \cdot \mathbf{1}_2$, where $\mathbf{1}_1$ is the indicator of the event $\{X \cap (\bigcup_{j=1}^m \widetilde{U}_j \times (2, n_j]) = \varnothing\}$, and $\mathbf{1}_2$ is the indicator of the event $\{X \cap \widetilde{U}_j \times (\widetilde{n}_j, \widetilde{n}_j + 1] \neq \varnothing \text{ for all } j = 1, \dots, m\}$.

We have

$$\Delta_{\widetilde{n}_1, \dots, \widetilde{n}_{m_0}}^{n_1, \dots, n_{m_0}} = \mathsf{E}\Big(\sum_{|x|>2} \frac{\langle 2Au + b^*, \theta(x)\rangle^2}{|x|^{2+2/\alpha}}\Big)^{-(l+\delta)} \mathbf{1}_{\widetilde{F}(\widetilde{n}_1, \dots, \widetilde{n}_m)} \mathbf{1}_{F(n_1, \dots, n_{m_0})}$$

$$\leq \mathsf{E}\Big(\sum_{x \in X \cap \mathcal{D}} \frac{\langle 2Au + b^*, \theta(x)\rangle^2}{|x|^{2+2/\alpha}}\Big)^{-(l+\delta)} \mathbf{1}_1 \mathbf{1}_2 \mathbf{1}_{F(n_1, \dots, n_{m_0})},$$

where

$$\mathcal{D} = \Big(\bigcup_{j=1}^{m_0} U_j \times (n_j, n_j + 1]\Big) \cap \Big(\bigcup_{j=m_0+1}^{m} \widetilde{U}_j \times (\widetilde{n}_j, \widetilde{n}_j + 1]\Big).$$

In turn, the last summand does not exceed

$$\begin{aligned} &\Big(\prod_{j=1}^{m_0} (n_j + 1) \prod_{j=1}^{m} (\widetilde{n}_j + 1)\Big)^{(2+2/\alpha)(l+\delta)} \\ &\qquad \times \mathsf{E}\Big(\sum_{x \in X \cap \mathcal{D}} \langle 2Au + b^*, \theta(x)\rangle^2\Big)^{-(l+\delta)} \mathbf{1}_1 \mathbf{1}_2 \mathbf{1}_{F(n_1, \dots, n_{m_0})}. \end{aligned} \tag{16.17}$$

We represent the last mean value as the integral of the conditional mathematical expectation under the condition that the configurations are fixed outside the set $\bigcup_{j=1}^m \widetilde{U}_j \times (\widetilde{n}_j, \widetilde{n}_j + 1]$, in each set the number of points of a configuration is equal to $\widetilde{U}_j \times (\widetilde{n}_j, \widetilde{n}_j + 1]$, and finally, in each of these sets all points of a configuration except one are fixed and the first coordinate of this exceptional point is also fixed, that is, $\theta \in S$. Since the values of a Poisson measure on disjoint sets are independent and $\Pi(d\theta, dr) = \pi(d\theta)\, dr$, it follows that the free coordinates that correspond to the sets $\widetilde{U}_j \times (\widetilde{n}_j, \widetilde{n}_j + 1]$ are uniformly distributed on $(\widetilde{n}_j, \widetilde{n}_j + 1]$ and are independent of each other.

Now we rewrite the conditional expectation as the integral over the distribution and obtain an upper bound for the integrand by taking only one (arbitrary) term in each sum over $x \in X \cap (U_j \times (n_j, n_j + 1])$ as well as in the sum over $x \in X \cap (\widetilde{U}_j \times (\widetilde{n}_j, \widetilde{n}_j + 1])$. We obtain

$$\Delta^{n_1,\dots,n_{m_0}}_{\widetilde{n}_1,\dots,\widetilde{n}_m} \leq \Big(\prod_{j=1}^{m_0}(n_j+1)\prod_{j=1}^{m}(\widetilde{n}_j+1)\Big)^{(2+2/\alpha)(l+\delta)} \mathsf{E}1_1\mathbf{1}_{F(n_1,\dots,n_{m_0})}\mathbf{1}_2 \times \int_{\prod_{j=1}^m(\widetilde{n}_j,\widetilde{n}_j+1]} \frac{dr_1\dots dr_m}{(\langle 2Au+b^*,\theta_1\rangle^2+\dots+\langle 2Au+b^*,\theta_m\rangle^2)^{l+\delta}}. \tag{16.18}$$

Note that by construction we have $\theta_i \in U_i$, $i = 1,\dots,m$. Let us estimate the inner integral by a constant independent of the conditions. By (16.12), u can be represented in the form

$$u = \frac{1}{\alpha}\Big(\frac{\widetilde{\theta}_1}{r_1^{1/\alpha}}+\dots+\frac{\widetilde{\theta}_m}{r_m^{1/\alpha}}\Big)+u_1, \tag{16.19}$$

where $\widetilde{\theta}_i \in \widetilde{U}_i$ corresponds to the unique point $x = (\widetilde{\theta}_i, r_i) \in \widetilde{U}_i \times (\widetilde{n}_i, \widetilde{n}_i + 1]$ whose second coordinate is free. In this representation, the vector u_1 depends only on the assumptions.

Now we substitute (16.19) into the inner integral in (16.18) and perform the change of variables $r_j = s_j^{-\alpha}$. We obtain

$$\begin{aligned}\int_{\prod_{j=1}^m(\widetilde{n}_j,\widetilde{n}_j+1]} &\big(\langle 2Au+b^*,\theta_1\rangle^2+\dots+\langle 2Au+b^*,\theta_m\rangle^2\big)^{-(l+\delta)}\,dr_1\dots dr_m \\ \leq \alpha^m\prod_{j=1}^m(\widetilde{n}_j+1)^{1+1/\alpha} &\int_{\prod_{j=1}^m(\widetilde{n}_j^{-1/\alpha},(\widetilde{n}_j+1)^{-1/\alpha}]} \Big(\Big\langle\frac{2}{\alpha}A\Big(\sum_{j=1}^m s_j\widetilde{\theta}_j\Big)+2Au_1+b^*,\theta_1\Big\rangle^2 \\ &+\dots+\Big\langle\frac{2}{\alpha}A\Big(\sum_{j=1}^m s_j\widetilde{\theta}_j\Big)+2Au_1+b^*,\theta_m\Big\rangle^2\Big)^{-(l+\delta)} ds_1\dots ds_m.\end{aligned} \tag{16.20}$$

Let us perform another change of variables,

$$t_i = \frac{2}{\alpha}\sum_{j=1}^m\langle A\widetilde{\theta}_j,\theta_i\rangle s_j+\langle 2Au_1+b^*,\theta_i\rangle.$$

Since the matrix $\langle A\widetilde{\theta}_j, \theta_i\rangle_{i,j=1}^m$ is uniformly nondegenerate, it follows that the Jacobian of this change of variables is bounded above by a constant that depends only on α. Hence, the right-hand side of (16.20) has the upper bound

$$c_\alpha\int_{[-c_\alpha,c_\alpha]}\frac{dt_1\dots dt_m}{(\sum_{i=1}^m t_i^2)^{l+\delta}}.$$

Since $m = 2l + 1$, it follows that the last integral is finite for $\delta < 1/2$.

Summarizing, we obtain the upper bound

$$\begin{aligned}\Delta^{n_1,\dots,n_{m_0}}_{\tilde n_1,\dots,\tilde n_m} &\le c_\alpha\Big(\prod_{j=1}^{m_0}(n_j+1)\prod_{j=1}^{m}(\tilde n_j+1)\Big)^{(2+2/\alpha)(l+\delta)}\\ &\quad\times\prod_{j=1}^{m}(\tilde n_j+1)^{1+1/\alpha}\int \mathbf{1}_1\,d\mathbf{P}\cdot\mathbf{P}\{F(n_1,\dots,n_{m_0})\}\end{aligned}\tag{16.21}$$

for $\Delta^{n_1,\dots,n_{m_0}}_{\tilde n_1,\dots,\tilde n_m}$. On the other hand, we have the estimates

$$\int \mathbf{1}_1\,d\mathbf{P} \le c\exp\Big\{-\sum_{j=1}^{m}\tilde n_j\pi(\widetilde U_j)\Big\},$$

$$\mathbf{P}\{F(n_1,\dots,n_{m_0})\} \le c\exp\Big\{-\sum_{j=1}^{m_0}n_j\pi(U_j)\Big\}$$

for $\int \mathbf{1}_1\,d\mathbf{P}$ and $\mathbf{P}\{F(n_1,\dots,n_{m_0})\}$. These estimates together with (16.21) readily imply that the series $\Delta^{n_1,\dots,n_{m_0}}_{\tilde n_1,\dots,\tilde n_m}$ converges. This completes the proof of Lemma 16.4 and hence of Theorem 16.2 □

§17. Distribution of the norm of a stable vector

Suppose that $(\mathbf{B},\|\cdot\|)$ is a separable Banach space and $u\in\mathbf{B}$ is a random element of $\mathbf{B}$ whose distribution is stable with exponent α, $0<\alpha<2$. We set $F(x)=\mathbf{P}\{\|u\|^\rho<x\}$ for $\rho>0$. The aim of the present section is to study the smoothness of F. The methods introduced in §6 allow us to obtain sufficient conditions for the existence and boundedness of the derivatives of F in a wide class of spaces.

First, let us describe the conditions to be imposed on $\mathbf{B}$. We consider spaces equipped with a uniformly convex differentiable norm whose derivatives satisfy the conditions of boundedness (12.13) and power-law uniform convexity (12.8). We denote the value of the kth-order derivative at a point $b\in\mathbf{B}$ by $\nabla^{(k)}(b)[b_1,\dots,b_k]$, and the norm of the derivative treated as a polylinear form by $\|\nabla^{(k)}(b)\|$.

Let us state the main result of this section

THEOREM 17.1. *Suppose that μ is the distribution of a stable vector u in a Banach space $\mathbf{B}$ in which the norm satisfies the condition of mth-order power-law uniform convexity and the derivatives of order $1,2,\dots,l+1$ are bounded. If the support of μ is infinite-dimensional, then for any $\rho>0$ the distribution of the random variable $\|u\|^\rho$ belongs to the class $\overline{\mathbf{W}}{}^l_1$.*

PROOF. Let us use the representation (16.12) of a stable random vector in terms of a stochastic integral over a random Poisson measure. We define a function $f\colon\mathfrak{X}\to\mathbf{R}^1$ by setting $f(X)=\|u(X)\|^\rho$. By (14.5), we have

$$(f^{(1)}(X))(x) = -\frac{\rho}{\alpha^2|x|^{1+1/\alpha}}(\nabla^{(1)}(u)[\theta(x)])\|u\|^{\rho-1}\tag{17.1}$$

(here, as before, $|x|=r(x)$, that is, $x=(\theta(x),|x|)$).

The remaining part of the proof is quite similar to the proof of Theorem 16.2. First, we need to choose a vector field along which f varies sufficiently rapidly.

To this end, we choose and fix some monotone infinitely differentiable function $\varkappa\colon \mathbf{R}^1 \to \mathbf{R}_+$ such that $\varkappa((-\infty, 0]) = 0$ and $\varkappa([1,\infty)) = 1$. Using $\varkappa$, we introduce smoothing functions $\varphi_1(x) = \varkappa(2-x)$, $\varphi_0 = 1 - \varphi_1$, $\varphi_3 = \varkappa(x-3)$, and $\Psi(x) = \varkappa(x-99)$. In the sequel, we need two random vectors related to the vector u as follows:

$$u_1 = \alpha^{-1} \sum \frac{\theta(x)}{|x|^{1/\alpha}} \varphi_1(|x|), \qquad u_0 = u - u_1. \tag{17.2}$$

Then it is easy to see that the number of summands in the definition of u_1 is finite, the random variable $\|u_0\|$ has moments of any order, and the vector u_0 can be represented in the form

$$u_0 = b_0 + \int \frac{\theta(x)}{|x|^{1/\alpha}} \varphi_0(|x|) \widetilde{\nu}(dx),$$

where b_0 is some nonrandom vector.

Now we consider the partition of unity (I_1, I_2) in the configuration space $\mathfrak{X}$, where $I_1 = \Psi(\|u_1\|/(\|u_0\|+1))$ and $I_2 = 1 - I_1$.

LEMMA 17.1. 1. *The inequality*

$$\langle \nabla^{(1)}(u), u_1 \rangle \geq \frac{1}{2}\|u\| \tag{17.3}$$

is satisfied on the set $\{I_1 > 0\}$.

2. *On the set* $\{I_2 > 0\}$, *the random variables* $\|u\|$ *and* $\|u_1\|$ *have moments of any order.*

PROOF. Statement 2 readily follows from the fact that $\|u_0\|$ has moments of all orders. Let us prove (17.3). On the set $I_1 > 0$, we have

$$\|u_1\| \geq 99(\|u_0\| + 1) \geq 99\|u_0\|,$$

and hence,

$$\|u\| = \|u_1 + u_0\| \geq \|u_1\| - \|u_0\| \geq 98\|u_0\|.$$

Thus,

$$\begin{aligned}\langle \nabla^{(1)}(u), u_1 \rangle &= \langle \nabla^{(1)}(u), u \rangle - \langle \nabla^{(1)}(u), u_0 \rangle \\ &\geq \|u\| - |\langle \nabla^{(1)}(u), u_0 \rangle| \geq \|u\| - \|u_0\| \geq \frac{1}{2}\|u\|. \quad \square\end{aligned}$$

Let us introduce two vector fields. The "radial" vector field is defined by the formula

$$K_1(x) = -|x|\varphi_1(|x|) \qquad (x = (\theta(x), |x|) \in X). \tag{17.4}$$

This field vanishes for $|x| > 2$ and hence has only finitely many nonzero coordinates for all $X \in \mathfrak{X}$.

The "gradient" vector field is a truncated, normalized and smoothed gradient of the function $X \mapsto \|u(X)\|$. This field is constructed as follows. Let $\theta_1^0, \dots, \theta_q^0 \in S$ be linearly independent unit vectors that belong to the support of the spectral measure. The number q must be sufficiently large (we estimate this number in the sequel). Let $U_1, \dots, U_q$ be small neighborhoods of the vectors θ_i^0 such that their spectral measures are positive and the following linear independence property is satisfied:

$$\inf_{\{c_i\in\mathbf{R}^1,\ \sum c_i^2=1,\ \theta_i\in U_i\}} \Big\| \sum_{i=1}^{q} c_i\theta_i \Big\| > 0 \tag{17.5}$$

(the existence of such vectors follows from the fact that the linear support of μ is finite-dimensional).

We set

$$(K_2(X))(x) = -\nabla^{(1)}(u(X))[\theta]|x|^{-1-1/\alpha}\varphi_3(|x|)\mathbf{1}_{\{\theta\in\bigcup_{i=1}^q U_i\}} \tag{17.6}$$

for $x = (\theta(x), |x|) \in X$. Finally, we define the desired vector field by the formula

$$K(X) = K_1(X)I_1 + K_2(X)I_2.$$

The field K is constructed according to the following idea. Although the derivative of f along K will be sufficiently far from the zero, for large values of $\|u\|$ we fail to obtain good estimates for the higher derivatives of f along K_2. Therefore, for large values of $\|u\|$, we must use the field K instead of K_2; in a sense, K is almost linear and does not present problems in estimating the higher derivatives.

We need to prove that the distribution of the random variable $\|u\|^\rho$ belongs to the class $\overline{\mathbf{W}}_1^l$. By Theorem 16.1, it suffices to prove that the expectations

$$\begin{aligned} \mathsf{E}|Df|^{-(l+\sum_{j=1}^l k_j)}&|D^2 f|^{k_1}\dots|D^{l+1}f|^{k_l} \\ &\times |\operatorname{tr} K^{(1)}|^{\beta_1}|D\operatorname{tr} K^{(1)}|^{\beta_2}\dots|D^{i-1}\operatorname{tr} K^{(1)}|^{\beta_i} \end{aligned} \tag{17.7}$$

are finite for all possible sets of nonnegative numbers $\{k_j\}$ and $\{\beta_j\}$ such that $\sum_j j\beta_j = i$ and $\sum_j jk_j = l - i$, $i = 0, 1, \dots, l$. Just as above, by c we denote various positive constants whose values are of no importance for our estimates, and by M_∞ we denote the set of random variables on $(\mathfrak{X}, \Sigma, \mathbf{P})$ that have moments of all orders.

Now let us estimate the higher derivatives of f.

LEMMA 17.2. 1. *For any* $j = 0, 1, \dots, l+1$, *we have*

$$|D^j f| \le \xi(1 + \|u_1\|^\rho), \qquad \xi \in M_\infty. \tag{17.8}$$

2. *For any* $j = 0, 1, \dots, l$, *we have*

$$|D^j(\operatorname{tr} K^{(1)})| \in M_\infty. \tag{17.9}$$

3.

$$\|u\|^{-1} \in M_\infty, \quad \|u_0\|^{-1} \in M_\infty. \tag{17.10}$$

PROOF. Statements 1 and 2 can be proved in the same way as Lemma 16.1. Statement 3 follows from the fact that the support of the spectral measure is finite-dimensional. □

Now to prove that the expectations (17.7) are finite, it suffices to obtain a lower bound for $|Df|$. We have

$$
\begin{aligned}
Df &= \frac{\rho}{\alpha^2}\|u\|^{\rho-1}\Bigg(\sum_{\{x\in X\,|\,|x|\le 2\}} \frac{(\nabla^{(1)}(u)[\theta(x)])\varphi_1(|x|)}{|x|^{1/\alpha}} I_1 \\
&\qquad + \sum_{\{x\in X\,|\,|x|>2,\theta\in\bigcup_i U_i\}} \frac{(\nabla^{(1)}(u)[\theta(x)])^2}{|x|^{2+2/\alpha}}\varphi_3(|x|)I_2\Bigg) \\
&= \frac{\rho}{\alpha^2}\|u\|^{\rho-1}\Bigg(\alpha\langle\nabla^{(1)}(u),u_1\rangle I_1 \\
&\qquad + \sum_{\{x\in X\,|\,|x|>2,\theta\in\bigcup_i U_i\}} \frac{\nabla^{(1)}(u)[\theta(x)]^2}{|x|^{2+2/\alpha}}\varphi_3(|x|)I_2\Bigg) \\
&\ge \frac{\rho}{2\alpha}\|u\|^{\rho}I_1 + \frac{\rho}{\alpha^2}\|u\|^{\rho-1}\sum_{\{x\in X\,|\,|x|>3,\theta\in\bigcup_i U_i\}} \frac{(\nabla^{(1)}(u)[\theta(x)])^2}{|x|^{2+2/\alpha}} I_2
\end{aligned}
\tag{17.11}
$$

(at the last step, we have used statement 1 of Lemma 17.1).

Let us rewrite (17.7) as the sum of integrals over the domains $\{I_1 \ge 1/2\}$ and $\{I_1 < 1/2\}$.

By Lemma 17.2 and (17.11), the integral over the first domain has the upper bound

$$
\int_{\{I_1\ge 1/2\}} \mathbf{P}(dX)\xi[1+\|u_1\|^{\rho}]^{\sum k_j}\Big/\|u\|^{(l+\sum k_j)\rho},
$$

where $\xi \in M_\infty$ and $\xi \ge 0$. The last integral is finite, since $\|u_1\| \le 2\|u\|$ and $\|u\| \ge \|u_1\| - \|u_0\| \ge \|u_1\| - \|u_1\|/99 \ge 99/2$ in the domain in question.

Now let us consider the integrals over the domain $\{I_1 < 1/2\}$. To estimate the expression $|D^2 f|^{k_1}\dots|D^{l+1}f|^{k_l}|\operatorname{tr} K^{(1)}|^{\beta_1}\dots|D^{i-1}\operatorname{tr} K^{(1)}|^{\beta_i}$, we again use Lemma 17.2 and take into account the fact that the factor $[1+\|u_1\|^{\rho}]^{\sum k_j}$ can be included in ξ in the domain considered. Moreover, to obtain a lower bound for Df, we need to retain only the second summand in (17.11). We have

$$
\begin{aligned}
&\int_{\{I_1<1/2\}} |Df|^{-(l+\sum k_j)}|D^2 f|^{k_1}\dots|D^{l+1}f|^{k_l} \\
&\qquad \times|\operatorname{tr} K^{(1)}|^{\beta_1}\dots|D^{i-1}\operatorname{tr} K^{(1)}|^{\beta_i}\,\mathbf{P}(dX) \\
&\qquad\qquad \le \int_{\{I_1<1/2\}} \mathbf{P}(dX)\xi\Big/(\|u\|^{\rho-1}Z(X))^{l+\sum k_j},
\end{aligned}
\tag{17.12}
$$

where

$$
Z(X) = \sum_{\{x\,|\,|x|>3,\theta\in\bigcup_i U_i\}} |\nabla^{(1)}(u)[\theta(x)]|^2|x|^{-2(1+1/\alpha)}.
$$

Since in the domain of integration any power of $\|u\|$ belongs to M_∞, we see that the existence of the corresponding moment of Z^{-1} is sufficient for the convergence of the integral (17.14). Taking the maximum of $\sum k_j$, we see that it is necessary to estimate the moment of order $2(1+\delta)l$, $\delta > 0$.

LEMMA 17.3. *For any $\delta > 0$, we have $\mathsf{E}Z^{-2l(1+\delta)} < \infty$.*

PROOF. For positive integers $n_1, \dots, n_q$, by $\mathcal{D}(n_1, \dots, n_q)$ we denote the set of configurations $X \in \mathfrak{X}$ such that the point $x \in X \cap (U_i \times [3, \infty))$ with the minimal real coordinate lies in $U_i \times (n_i, n_i + 1]$ for all $i = 1, \dots, q$. We divide each set $\mathcal{D}(n_1, \dots, n_q)$ into q-dimensional layers. Then we fix the part of the configuration that lies outside the set $\bigcup_i U_i \times (n_i, n_i+1]$ and all but one points of the configuration in each of the sets $U_i \times (n_i, n_i + 1]$. Finally, we fix the first coordinate $\theta_i \in U_i$ of each of the q points $(\theta_i, |x_i|) \in U_i \times (n_i, n_i + 1]$ that remain free. Now only $|x_i|$ remains free in each element of the partition.

Since disjoint points are independent with respect to $\mathbf{P}$, it follows that the conditional measure that arises on an element of the partition is just the Lebesgue measure on $\Delta_{n_1} \times \cdots \times \Delta_{n_q}$, where $\Delta_{n_i} = (n_i, n_i + 1]$, and the conditional moment is bounded above by the integral

$$\int_{\Delta_{n_1} \times \cdots \times \Delta_{n_q}} dr_1 \dots dr_q \Big(\sum_{i=1}^{q} |\nabla^{(1)}(u)[\theta_i]|^2 r_i^{-2-2/\alpha} \Big)^{-2(1+\delta)l}. \tag{17.13}$$

In addition to (17.13), we have the following expression for u:

$$u(X) = u(r_1, \dots, r_q) = b + \frac{1}{\alpha} \Big(\frac{\theta_1}{r_1^{1/\alpha}} + \cdots + \frac{\theta_q}{r_q^{1/\alpha}} \Big),$$

where the vector $b \in \mathbf{B}$ is constant within any element of the partition. We need an estimate of (17.13) which does not depend on b too strongly. In (17.13), we perform the change of variables $r_i = \alpha^{-\alpha} s_i^\alpha$ and arrive at the integral

$$c \int_{\overline{\Delta}_{n_1} \times \cdots \times \overline{\Delta}_{n_q}} ds_1 \dots ds_q \Big(\sum_{i=1}^{q} \Big| \nabla^{(1)} \Big(b + \sum_{i=1}^{q} s_i \theta_i \Big) [\theta_i] \Big|^2 s_i^{(2+2/\alpha)\alpha} \Big) \prod_{i=1}^{q} s_i^{-(1+\alpha)}.$$

We replace the inessential factor of the form $(\min_i s_i)^{-c}$ by its upper bound, namely, by the function $\max_i (n_i + 1)^c$, which has all moments, eliminate it from the integral, extend the domain of integration to the cube $[0, \alpha^{-1}]^q$, and obtain

$$\int_{[0,c]^q} ds_1 \dots ds_q \Big(\sum_{i=1}^{q} \Big| \nabla^{(1)} \Big(b + \sum_{i=1}^{q} s_i \theta_i \Big) [\theta_i] \Big|^2 \Big)^{-2(1+\delta)l}, \tag{17.14}$$

where $c = \alpha^{-1}$.

In this integral, we pass to polar coordinates centered at the point of the cube where the convex function $\|b + \sum_{i=1}^q s_i \theta_i\|$ attains its minimum. By replacing the vector b and by extending the domain of integration, we can assume that this minimum is attained at $s_i = 0$.

Now let us estimate the integrand in (17.14). We set $s_i^0 = s_i / \big(\sum_{j=1}^q s_j^2 \big)^{1/2}$. By the linearity of the form $\nabla^{(1)}$ and the Hölder inequality, we obtain

$$\begin{aligned} \Big| \nabla^{(1)} \Big(b + \sum_{i=1}^{q} s_i \theta_i \Big) \Big[\sum_{i=1}^{q} s_i^0 \theta_i \Big] \Big|^2 &= \Big| \sum_{i=1}^{q} s_i^0 \nabla^{(1)} \Big(b + \sum_{i=1}^{q} s_i \theta_i \Big) [\theta_i] \Big|^2 \\ &\le \sum_{i=1}^{q} \Big| \nabla^{(1)} \Big(b + \sum_{i=1}^{q} s_i \theta_i \Big) [\theta_i] \Big|^2. \end{aligned} \tag{17.15}$$

Suppose that $s^0 = \{s_i^0\} \in \mathbf{R}^d$, $\gamma^0 = \sum s_i^0 \theta_i$, and $\gamma = \gamma^0 / \|\gamma^0\|$ is the unit vector in $\mathbf{B}$. The choice of the sets U_i implies that the vectors θ_i are uniformly linearly independent, that is, $\|\gamma^0\|$ is uniformly bounded away from zero and infinity. By passing to polar coordinates in (17.14) and substituting the estimate (17.15), we obtain

$$\int ds^0 \int_0^c dy \big|\nabla^{(1)}(b + y\gamma)[\gamma]\big|^{-4(1+\delta)l} y^{q-1} \mathbf{1}_{\{\nabla^{(1)}(b)[\gamma] \geq 0\}}. \tag{17.16}$$

To estimate the one-dimensional integral, we can use the following lemma, which follows from the homogeneity of the norm and the definition of power-law uniform convexity.

LEMMA 17.4. *Suppose that the space* $(\mathbf{B}, \|\cdot\|)$ *satisfies the condition of power-law uniform convexity* (12.8). *Let* $k, A, \zeta > 0$, $b \in \mathbf{B}$, $\gamma \in S \subset \mathbf{B}$, *and* $\nabla^{(1)}(b)[\gamma] > 0$. *Then*

$$\int_0^A \big|\nabla^{(1)}(b + y\gamma)[\gamma]\big|^{-k} y^{km+\zeta}\, dy \leq c\Big(\max\Big\{\frac{A}{\sigma}, \|b\|\Big\}\Big)^{km+\zeta+1}. \tag{17.17}$$

PROOF. We perform the change of variables $z = y\|b\|^{-1}$ on the left-hand side in (17.17) and set $b_0 = b\|b\|^{-1}$. Since the gradient $\nabla^{(1)}$ is constant along the radial rays, we have

$$\int_0^A \big|\nabla^{(1)}(b + y\gamma)[\gamma]\big|^{-k} y^{km+\zeta}\, dy = \int_0^{A/\|b\|} \big|\nabla^{(1)}(b_0 + z\gamma)[\gamma]\big|^{-k} (z\|b\|)^{km+\zeta} \|b\|\, dz.$$

We estimate the integral over $[0, \sigma]$ by using the uniform convexity condition:

$$\int_0^\sigma \big|\nabla^{(1)}(b_0 + z\gamma)[\gamma]\big|^{-k} (z)^{km+\zeta}\, dz \cdot \|b\|^{km+\zeta+1} \leq c\|b\|^{km+\zeta+1}.$$

In the domain $[\sigma, A\|b\|^{-1}]$, since the derivative of a convex function is monotone, we obtain the estimate

$$\int_\sigma^{A\|b\|^{-1}} \big|\nabla^{(1)}(b_0 + z\gamma)[\gamma]\big|^{-k} z^{km+\zeta}\, dz \cdot \|b\|^{km+\zeta+1}$$
$$\leq c\sigma^{-mk}\|b\|^{km+\zeta+1}\Big(\frac{A}{\|b\|}\Big)^{km+\zeta+1} \leq c\Big(\frac{A}{\sigma}\Big)^{km+\zeta+1}. \quad \square$$

Let us return to the proof of Lemma 17.3. We need to apply (17.17) to the integral (17.16); therefore, we need to ensure that $\zeta = q - 1 - 4(1+\delta)lm > 0$. This can be achieved by choosing a sufficiently large q. Since the estimate (17.17) is uniform in γ, we obtain a majorant of the form $c(1 + \|b\|)^c$ for (17.16). Furthermore, since the vector b has been chosen so as to minimize the norm, we can pass from conditional moments to unconditional moments, thus obtaining

$$\int \mathbf{P}(dX)(Z(X))^{-2(1+\delta)l} \leq c \int_{\{x \in \mathfrak{X} | I_1 \leq 1/2\}} \mathbf{P}(dX)(1 + \|u(X)\|)^c \xi < \infty.$$

This implies that the integrals (17.12) are finite. The proofs of Lemma 17.3 and Theorem 17.1 are complete. $\square$

Now let us discuss some consequences of Theorem 17.1.

1. As follows from the properties of distributions of the class $\overline{\mathbf{W}}_1^l$, under the assumptions of the theorem, the function F has l bounded derivatives and the lth derivative is a function of bounded variation.

2. In fact, the assumptions of the theorem are also satisfied for the distribution of the random variable $\|u + b\|^\rho$, $b \in \mathbf{B}$, since $u + b$ is also a stable vector.

3. For a Hilbert space and the spaces $\mathbf{L}^p$ (where p is an even integer) the theorem allows us to claim that the distribution of the norm belongs to $\overline{\mathbf{W}}_1^l$ for any $l > 0$, that is, the distribution function is infinitely differentiable. For other spaces $\mathbf{L}^p$, $p > 2$, the theorem implies that $[p - 1]$ derivatives of the distribution function are bounded. (Here $[s]$ denotes the integral part of s for noninteger s and the integral part of $s - 1$ for integer s).

4. Here we consider only stable vectors with exponents $\alpha \in (0, 2)$. For Gaussian vectors ($\alpha = 2$), a result similar to Theorem 17.1 was given in §12.

5. Since ρ is arbitrary in the statement of the theorem, we see that in a neighborhood of zero the distribution of the norm not only is bounded, but decreases faster than any power-law function.

6. The statement of the theorem remains valid if instead of the assumption that the support is infinite-dimensional, we only require that its dimension exceeds some critical value $n = n(l, m, \rho)$.

7. For an arbitrary Banach space $\mathbf{B}$, the function F may have a more complicated structure. For example, in a space with a nonsmooth norm, there exists a stable vector such that F' is unbounded for $\rho = 1$. This means that the distribution of the norm does not belong to any class $\overline{\mathbf{W}}_1^l$. Still, some facts can be proved for the space with unsmooth norm. If $0 < \alpha < 1$ and the distribution of u is strongly stable, then F' is bounded [**65**].

CHAPTER 5

Local Limit Theorems

§18. Strong convergence theorems

We now proceed to studying the strong convergence of distributions of stochastic functionals. The problem can be stated as follows. Suppose that a sequence $(\mathbf{P}_n)$ of probability measures on a complete separable metric space $(\mathbf{X}, \mathfrak{B}_{\mathbf{X}})$ weakly converges to some measure $\mathbf{P}_\infty$. Let $f\colon \mathbf{X} \to \mathbf{R}^1$ be a functional. The question is whether $\mathbf{P}_n f^{-1}$ converges in variation to $\mathbf{P}_\infty f^{-1}$. If the distributions $\mathbf{P}_n f^{-1}$ and $\mathbf{P}_\infty f^{-1}$ are absolutely continuous, then (see Problem 2.4) the convergence in variation is equivalent to the convergence of the densities $d\mathbf{P}_n f^{-1}/d\lambda$ to $d\mathbf{P}_\infty f^{-1}/d\lambda$ in the metric of $\mathbf{L}^1(\mathbf{R}^1)$. Thus, we essentially deal with local limit theorems for distributions of stochastic functionals.

If the measures $\mathbf{P}_n$ themselves converge in variation to $\mathbf{P}_\infty$, then the answer to our question is obvious, since in this case the convergence $\mathbf{P}_n f^{-1} \overset{\text{var}}{\to} \mathbf{P}_\infty f^{-1}$ holds for any functional f (see Theorem 2.5). However, this situation rarely occurs in practice. In a majority of the interesting cases, the measures $\mathbf{P}_n$ are singular with respect to $\mathbf{P}_\infty$ for each n (for example, this is the case if $\mathbf{P}_n$ is the distribution of random open polygons constructed from sums of independent identically distributed random variables in the ordinary way and $\mathbf{P}_\infty$ is the Wiener measure). It is just the mutual singularity of $\mathbf{P}_n$ and $\mathbf{P}_\infty$ that causes difficulties in solving the problem.

Preliminary results. For a measure $\mathbf{P}_n$ and a measurable partition Γ of the space $\mathbf{X}$ we denote the quotient measure and the system of conditional measures of $\mathbf{P}_n$ with respect to Γ by $\mathbf{P}_{n,\Gamma}$ and $\{\mathbf{P}_{n,\gamma}, \gamma \in \mathbf{X}/\Gamma\}$.

THEOREM 18.1. *Let $\mathbf{P}_n \Rightarrow \mathbf{P}_\infty$. Suppose that the following assumptions are satisfied for the functional f and some partition Γ:*

(i) $\int_{\mathbf{X}/\Gamma} \|\mathbf{P}_{n,\gamma} f^{-1} - \mathbf{P}_{\infty,\gamma} f^{-1}\| \mathbf{P}_{n,\Gamma}(d\gamma) \to 0, \quad n \to \infty;$

(ii) *the mapping $\gamma \mapsto \mathbf{P}_{\infty,\gamma} f^{-1}$ of $\mathbf{X}/\Gamma$ into $\mathbf{Z}(\mathbf{R}^1)$ is $\mathbf{P}_{\infty,\Gamma}$-almost everywhere continuous.*

Then $\mathbf{P}_n f^{-1} \overset{\text{var}}{\to} \mathbf{P}_\infty f^{-1}$.

REMARK 1. Since $\|\mathbf{P}_{n,\gamma} f^{-1} - \mathbf{P}_{\infty,\gamma} f^{-1}\| \le 2$, it follows that (i) is equivalent to the following assumption: $\mathbf{P}_{n,\Gamma}\{\gamma \mid \|\mathbf{P}_{n,\gamma} f^{-1} - \mathbf{P}_{\infty,\gamma} f^{-1}\| > \varepsilon\} \to 0$ as $n \to \infty$ for all $\varepsilon > 0$. We also note that $\|\mathbf{P}_{n,\gamma} f^{-1} - \mathbf{P}_{\infty,\gamma} f^{-1}\| \le \|\mathbf{P}_{n,\gamma} - \mathbf{P}_{\infty,\gamma}\|$. Therefore, (i) is necessarily satisfied if the conditional distributions for the measures $\mathbf{P}_n$ tend in variation to the corresponding conditional distributions for $\mathbf{P}_\infty$.

REMARK 2. The structure of the range of f is irrelevant. In this theorem, we can assume that f is a measurable mapping of $\mathbf{X}$ into an arbitrary measurable space $(\mathbf{E}, \mathfrak{E})$ (in this case, (ii) will contain the space $\mathbf{Z}(\mathbf{E})$ instead of $\mathbf{Z}(\mathbf{R}^1)$). In particular, a result similar to Theorem 18.1 is valid for functionals with values in $\mathbf{R}^d$.

PROOF. We define a measure $\mathbf{Q}_n$ by setting

$$\mathbf{Q}_n = \int_{\mathbf{X}/\Gamma} \mathbf{P}_{\infty,\gamma} f^{-1} \mathbf{P}_{n,\Gamma}\,(d\gamma).$$

By (3.4) and (2.3), we have

$$\|\mathbf{P}_n f^{-1} - \mathbf{P}_\infty f^{-1}\| \le \int_{\mathbf{X}/\Gamma} \|\mathbf{P}_{n,\gamma} f^{-1} - \mathbf{P}_{\infty,\gamma} f^{-1}\| \mathbf{P}_{n,\Gamma}\,(d\gamma) + \|\mathbf{Q}_n - \mathbf{P}_\infty f^{-1}\|.$$

By (ii), the first term on the right-hand side tends to zero as $n \to \infty$. Let us consider the second term:

$$\|\mathbf{Q}_n - \mathbf{P}_\infty f^{-1}\| = \left\| \int_{\mathbf{X}/\Gamma} \mathbf{P}_{\infty,\gamma} f^{-1} \mathbf{P}_{n,\Gamma}\,(d\gamma) - \int_{\mathbf{X}/\Gamma} \mathbf{P}_{\infty,\gamma} f^{-1} \mathbf{P}_{\infty,\Gamma}\,(d\gamma) \right\|.$$

Condition (ii) permits us to use Theorem 2.3 and thus obtain the desired convergence $\|\mathbf{Q}_n - \mathbf{P}_\infty f^{-1}\| \to 0$. □

The following result shows how Theorem 18.1 can be applied.

THEOREM 18.2. *Let (μ_n) and (ν_n) be two sequences of probability measures on $\mathbf{R}^d$. Suppose that $\mu_n \Rightarrow \mu$ and $\nu_n \overset{\text{var}}{\to} \nu$. If ν is absolutely continuous, then*

$$\mu_n * \nu_n \overset{\text{var}}{\to} \mu * \nu.$$

PROOF. We set $\mathbf{X} = \mathbf{R}^d \times \mathbf{R}^d$, $\mathbf{P}_n = \mu_n \times \nu_n$, and $\mathbf{P}_\infty = \mu \times \nu$, and define $f\colon \mathbf{X} \to \mathbf{R}^d$ by $f(x,y) = x + y$. Obviously, $\mu_n * \nu_n = \mathbf{P}_n f^{-1}$ and $\mu * \nu = \mathbf{P}_\infty f^{-1}$, so that it is natural to use Theorem 18.1 (with regard to Remark 2). Let Γ be the partition of $\mathbf{X}$ into the d-dimensional manifolds $\gamma = \{x\} \times \mathbf{R}^d$, $x \in \mathbf{R}^d$. Since the measure $\mathbf{P}_{n,\gamma} f^{-1}$ (respectively, $\mathbf{P}_{\infty,\gamma} f^{-1}$) is just the measure ν_n (respectively, ν) translated by x, we have $\|\mathbf{P}_{n,\gamma} f^{-1} - \mathbf{P}_{\infty,\gamma} f^{-1}\| = \|\nu_n - \nu\|$. Hence, condition (i) of Theorem 18.1 is satisfied. By assumption, we obtain $\nu \ll \lambda^d$. We denote the density $d\nu/d\lambda^d$ by p. Let $\gamma_1 = \{x_1\} \times \mathbf{R}^d$ and $\gamma_2 = \{x_2\} \times \mathbf{R}^d$. Then

$$\|\mathbf{P}_{\infty,\gamma_1} f^{-1} - \mathbf{P}_{\infty,\gamma_2} f^{-1}\| = \int_{\mathbf{R}^d} |p(z - x_1) - p(z - x_2)| \lambda^d\,(dz).$$

Since the translation operator is continuous in $\mathbf{L}^1(\mathbf{R}^d)$, it follows that this expression tends to zero as $x_2 \to x_1$. Thus, the mapping $\gamma \to \mathbf{P}_{\infty,\gamma} f^{-1}$ is everywhere continuous, and hence, condition (ii) of Theorem 18.1 is satisfied. □

Unfortunately, conditional distributions of weakly convergent measures seldom approach each other in variation. This fact hinders the straightforward application of Theorem 18.1 and forces one to seek other ways of solving the problem.

However, before stating the main result, Theorem 18.4, we present one more generalization of Theorem 18.1 to be used in the sequel.

THEOREM 18.3. *Let $\mathbf{P}_n \Rightarrow \mathbf{P}_\infty$, and let f_n, $n \in \overline{\mathbf{N}}$, be a sequence of functionals. Suppose that some partition Γ satisfies the following conditions:*

(i) $\displaystyle\int_{\mathbf{X}/\Gamma} \|\mathbf{P}_{n,\gamma} f_n^{-1} - \mathbf{P}_{\infty,\gamma} f_\infty^{-1}\| \mathbf{P}_{n,\Gamma}\,(d\gamma) \to 0, \quad n \to \infty;$

(ii) *the mapping $\gamma \mapsto \mathbf{P}_{\infty,\gamma} f_\infty^{-1}$ of $\mathbf{X}/\Gamma$ into $\mathbf{Z}(\mathbf{R}^1)$ is $\mathbf{P}_{\infty,\Gamma}$-almost everywhere continuous.*

Then $\mathbf{P}_n f_n^{-1} \stackrel{\text{var}}{\to} \mathbf{P}_\infty f_\infty^{-1}$.

PROOF. The proof virtually coincides with that of Theorem 18.1, and we leave it to the reader as an exercise. □

The superstructure method. This method was introduced in §5, where it was used to study the absolute continuity of distributions of stochastic functionals. In what follows, we use the superstructure method for studying the strong convergence. Although Theorem 18.4 looks rather cumbersome at first glance, it is particularly convenient in applications.

THEOREM 18.4. *Consider a sequence of probability measures* $(\mathbf{P}_n, n \in \overline{\mathbf{N}})$, *defined on the* σ*-algebra* $\mathfrak{B}_\mathbf{X}$ *of Borel subsets of a complete separable metric space* $(\mathbf{X}, \rho)$. *Suppose that* $\mathbf{P}_n \Rightarrow \mathbf{P}_\infty$ *and for* $\mathbf{P}_\infty$*-almost all* x *there exists an open ball* V *centered at* x, *a number* $\varepsilon > 0$, *and a family* $(G_{n,c}, n \in \overline{\mathbf{N}}, c \in (0, \varepsilon])$ *of measurable transformations of* $\mathbf{X}$ *such that*

(i) *for each* $c \in (0, \varepsilon)$, *one has* $G_{n,c} \to G_{\infty,c}$ *in the measure* $\mathbf{P}_n$ *as* $n \to \infty$;

(ii) *for each* $c \in (0, \varepsilon)$, *the mapping* $G_{\infty,c}$ *is* $\mathbf{P}_\infty$*-almost everywhere continuous, and moreover,*

$$d(S, c) = \sup_{z \in S} \rho(z, G_{\infty,c} z) \to 0 \qquad as \quad c \to 0$$

for each open ball S;

(iii) $\lim_{c \to 0} \overline{\lim}_n \|\mathbf{P}_n G_{n,c}^{-1} - \mathbf{P}_n\| = 0$;

(iv) *for each* $\delta \in (0, \varepsilon)$, *one has*

$$\int_V \|\lambda_{[0,\delta]} \varphi_{n,z}^{-1} - \lambda_{[0,\delta]} \varphi_{\infty,z}^{-1}\| \mathbf{P}_n\,(dz) \to 0 \qquad as \quad n \to \infty,$$

where $\varphi_{n,z}(c) = f(G_{n,c} z)$, $n \in \overline{\mathbf{N}}$, *and* $c \in (0, \varepsilon]$;

(v) *for each* $\delta \in (0, \varepsilon)$, *the mapping* $z \mapsto \lambda_{[0,\delta]} \varphi_{\infty,z}^{-1}$ *of* V *into* $\mathbf{Z}(\mathbf{R}^1)$ *is* $\mathbf{P}_\infty$*-almost everywhere continuous.*

Then $\mathbf{P}_n f^{-1} \stackrel{\text{var}}{\to} \mathbf{P}_\infty f^{-1}$.

Before proving this theorem, let us discuss conditions (i)–(v).

Condition (i) needs no clarification. Condition (ii) says that the mappings $G_{n,c}$ are close to the identity mapping for small c on bounded sets.

Condition (iii) shows that the perturbations of $\mathbf{P}_n$ by the mappings $G_{n,c}$ are uniformly small for small c. Although this condition does not cover the case $n = \infty$, we have

$$\lim_{c \to 0} \|\mathbf{P}_\infty G_{\infty,c}^{-1} - \mathbf{P}_\infty\| = 0, \tag{18.1}$$

which follows from (iii) by virtue of (i) and (ii) (see the reasoning after Lemma 18.1). Here the assumption that the mapping $G_{\infty,c}$ is continuous is important. However, once (18.1) is established, we can drop this assumption.

According to Proposition 2.1, condition (iv) is satisfied if for $\mathbf{P}_\infty$-almost all z the convergence $z_n \to z$ implies

$$\|\lambda_{[0,\delta]} \varphi_{n,z_n}^{-1} - \lambda_{[0,\delta]} \varphi_{\infty,z_n}^{-1}\| \to 0$$

as $n \to \infty$. To verify this relation as well as assumption (v), we can use the results of §4 concerning the transfer of one-dimensional measures.

PROOF. Let X_0 be the set of x for which there exist an open ball V, a number $\varepsilon > 0$, and a family $(G_{n,c})$ with the desired properties. By assumption, $\mathbf{P}_\infty\{X_0\} = 1$. Decreasing the radius of the ball V if necessary, we can assume that $\mathbf{P}_\infty\{\partial V\} = 0$. The family $(V_x,\ x \in X_0)$ is a cover of X_0. Using the separability, we can choose a countable subcover $(V_k,\ k \in \mathbf{N})$, where V_k is a ball centered at x_k. Let ε_k and $(G_{n,c}^{(k)})$ be, respectively, the number and the family of mappings corresponding to x_k. We can assume that $\varepsilon_k \to 0$. We fix a $\beta > 0$ and choose M so large that

$$\mathbf{P}_n\Big\{\Big(\bigcup_{j=1}^{M} V_j\Big)^{\complement}\Big\} < \beta, \qquad n \in \overline{\mathbf{N}}.$$

Setting $W = \bigcup_{j=1}^{M} V_j$ and $U_k = V_k \setminus \bigcup_{j=1}^{k-1} V_j$, we note that

$$\begin{aligned}\|\mathbf{P}_n f^{-1} - \mathbf{P}_\infty f^{-1}\| &\le 2\beta + \|\mathbf{P}_{n,W} f^{-1} - \mathbf{P}_{\infty,W} f^{-1}\| \\ &\le 2\beta + \sum_{j=1}^{M} \|\mathbf{P}_{n,U_j} f^{-1} - \mathbf{P}_{\infty,U_j} f^{-1}\|.\end{aligned}$$

Therefore, to prove the theorem, it suffices to prove that each term in this sum tends to zero as $n \to \infty$. We fix k and omit the index k in the sequel: $U = U_k$, $G_{n,c} = G_{n,c}^{(k)}$, etc. Let us show that

$$\|\mathbf{P}_{n,U} f^{-1} - \mathbf{P}_{\infty,U} f^{-1}\| \to 0, \qquad n \to \infty. \tag{18.2}$$

To this end, we use the superstructure method. Let $\delta \in (0, \varepsilon_M)$. We set $I = (0, \delta)$ and $\mu = \delta^{-1}\lambda_{[0,\delta]}$, and consider the measures $\mathbf{Q}_n = \mu \times \mathbf{P}_{n,U}$ and the functions $F_n(c, x) = f(G_{n,c}x)$, $n \in \overline{\mathbf{N}}$, on the direct product $I \times U$. Obviously, we have

$$\|\mathbf{P}_{n,U} f^{-1} - \mathbf{P}_{\infty,U} f^{-1}\| \le \Delta_1 + \Delta_2 + \Delta_3,$$

where

$$\Delta_1 = \|\mathbf{P}_{n,U} f^{-1} - \mathbf{Q}_n F_n^{-1}\|, \qquad \Delta_2 = \|\mathbf{Q}_n F_n^{-1} - \mathbf{Q}_\infty F_\infty^{-1}\|,$$
$$\Delta_3 = \|\mathbf{Q}_\infty F_\infty^{-1} - \mathbf{P}_{\infty,U} f^{-1}\|.$$

To estimate Δ_1 and Δ_3, we need the following two lemmas.

LEMMA 18.1. *Under conditions* (i) *and* (ii) *of Theorem* 18.4, *for all* $c \in (0, \varepsilon]$ *we have the weak convergence* $\mathbf{P}_n G_{n,c}^{-1} \Rightarrow \mathbf{P}_\infty G_{\infty,c}^{-1}$ *as* $n \to \infty$.

PROOF. Let h be a uniformly continuous function bounded on $\mathbf{X}$. We have

$$\beta_n = \left| \int_{\mathbf{X}} h\, d\mathbf{P}_n G_{n,c}^{-1} - \int_{\mathbf{X}} h\, d\mathbf{P}_\infty G_{\infty,c}^{-1} \right| \le I_1 + I_2,$$

where

$$I_1 = \int_{\mathbf{X}} |h(G_{n,c}x) - h(G_{\infty,c}x)| \mathbf{P}_n\,(dx),$$
$$I_2 = \left| \int_{\mathbf{X}} h(G_{\infty,c}x) \mathbf{P}_n\,(dx) - \int_{\mathbf{X}} h(G_{\infty,c}x) \mathbf{P}_\infty\,(dx) \right|.$$

Since the mapping $G_{\infty,c}$ is $\mathbf{P}_\infty$-almost everywhere continuous, it follows that the function $x \to h(G_{\infty,c}x)$ is $\mathbf{P}_\infty$-almost everywhere continuous and bounded. Therefore, by Proposition 2.1 we have $I_2 \to 0$ as $n \to \infty$. The integral I_1 can be estimated as follows. Suppose that $\alpha > 0$ and $L = \sup_{z\in X} |h(z)|$; then

$$I_1 \leq 2L\mathbf{P}_n\{x \mid \rho(G_{n,c}x, G_{\infty,c}x) \geq \alpha\} + \omega_h(\alpha).$$

Hence, $\overline{\lim}_n \beta_n \leq \omega_h(\alpha)$. Since α is arbitrarily small and h is continuous, we see that $\overline{\lim}_n \beta_n = 0$, which implies the desired weak convergence. □

Note that the weak convergence together with Theorem 2.7 yields the inequality

$$\|\mathbf{P}_\infty G_{\infty,c}^{-1} - \mathbf{P}_\infty\| \leq \underline{\lim}_n \|\mathbf{P}_n G_{n,c}^{-1} - \mathbf{P}_n\|,$$

which shows that under the assumptions of Theorem 18.4 one has

$$\lim_{c\to 0} \|\mathbf{P}_\infty G_{\infty,c}^{-1} - \mathbf{P}_\infty\| = 0. \tag{18.3}$$

LEMMA 18.2[1]. *Suppose that* $\mathbf{P}_n \Rightarrow \mathbf{P}_\infty$ *and conditions* (i)–(iii) *of Theorem* 18.4 *are satisfied for some family* $(G_{n,c},\ n \in \overline{\mathbf{N}},\ c \in (0,\delta))$ *of transformations of the space* $\mathbf{X}$. *Then*

$$\lim_{c\to 0} \overline{\lim_n} \|\mathbf{P}_{n,U} G_{n,c}^{-1} - \mathbf{P}_{n,U}\| = 0. \tag{18.4}$$

for any set U *such that* $\mathbf{P}_\infty\{\partial U\} = 0$.

PROOF. We have

$$\begin{aligned}|\nu(A \cap G^{-1}(B)) - \nu(A\cap B) &\leq |\nu(A\cap B) - \nu(G^{-1}(A\cap B))| \\ &\quad + |\nu(A\cap G^{-1}(B)) - \nu(G^{-1}(A)\cap G^{-1}(B))|\end{aligned}$$

for any measure ν on $\mathfrak{B}_\mathbf{X}$, any transformation G of the space $\mathbf{X}$, and any $A, B \in \mathfrak{B}_\mathbf{X}$.

By setting $\nu = \mathbf{P}_n$, $G = G_{n,c}$, and $A = U$ in this inequality and by taking the supremum over $B \times \mathfrak{B}_\mathbf{X}$, we obtain

$$\|\mathbf{P}_{n,U} G_{n,c}^{-1} - \mathbf{P}_{n,U}\| \leq \|\mathbf{P}_n G_{n,c}^{-1} - \mathbf{P}_n\| + \mathbf{P}_n\{U \Delta G_{n,c}^{-1}(U)\}. \tag{18.5}$$

The estimate of the first term is obvious. Let us consider the second term. For any $\varepsilon > 0$, we have

$$U\Delta G_{n,c}^{-1}(U) \subset \{x \mid \rho(x, G_{n,c}x) > \varepsilon\} \cup \{U^\varepsilon \cap (U^\complement)^\varepsilon\}.$$

Let us choose a sufficiently large ball S such that $\mathbf{P}_\infty\{S^\complement\} < \varepsilon$ and $\mathbf{P}_\infty\{\partial S\} = 0$, and use the fact that

$$\begin{aligned}&\{x \mid \rho(x, G_{n,c}x) > \varepsilon\} \\ &\qquad \subset S^\complement \cup \{x \mid \rho(G_{\infty,c}x, G_{n,c}x) > \varepsilon/2\} \cup \{x \in S \mid \rho(x, G_{\infty,c}x) > \varepsilon/2\}.\end{aligned}$$

By (ii), the last set is empty for sufficiently small c. Therefore,

$$\begin{aligned}\mathbf{P}_n\{U\Delta G_{n,c}^{-1}(U)\} \leq \mathbf{P}_n\{S^\complement\} + \mathbf{P}_n\{x \mid \rho(G_{\infty,c}x, G_{n,c}x) > \varepsilon/2\} \\ + \mathbf{P}_n\{U^\varepsilon \cap (U^\complement)^\varepsilon\}.\end{aligned} \tag{18.6}$$

[1] Cf. Lemma 5.1.

Let us pass to the limit as $n \to \infty$. The weak convergence and the choice of S imply that $\lim_n \mathbf{P}_n\{S^{\complement}\} = \mathbf{P}\{S^{\complement}\} < \varepsilon$. By (i), the second term in (18.6) converges to zero. Therefore,

$$\begin{aligned} \overline{\lim_n} \mathbf{P}_n\{U \Delta G_{n,c}^{-1}(U)\} &\le \varepsilon + \overline{\lim_n} \mathbf{P}_n\{U^\varepsilon \cap (U^{\complement})^\varepsilon\} \\ &\le \varepsilon + \mathbf{P}\{(U^\varepsilon \cap (U^{\complement})^\varepsilon)^-\}. \end{aligned} \tag{18.7}$$

By letting ε and c tend to zero, we obtain

$$\lim_{c\to 0} \overline{\lim_n} \mathbf{P}_n\{U \Delta G_{n,c}^{-1}(U)\} \le \varepsilon + \mathbf{P}\{\partial U\} \to 0. \tag{18.8}$$

Inequalities (18.8) and (18.5) imply (18.4). □

Now we can estimate Δ_1. We have

$$\Delta_1 = \left\| \int_I \mathbf{P}_{n,U} f^{-1}\, d\mu - \int_I \mathbf{P}_{n,U} G_{n,c}^{-1} f^{-1}\, d\mu \right\| \le \int_I \|\mathbf{P}_{n,U} - \mathbf{P}_{n,U} G_{n,c}^{-1}\|\, d\mu.$$

By Fatou's lemma and Lemma 18.2, the last inequality readily implies that

$$\lim_{\delta\to 0} \overline{\lim_n} \Delta_1 = 0.$$

Just as in estimating Δ_1, we obtain the inequality

$$\Delta_3 \le \int_I \|\mathbf{P}_{\infty,U} - \mathbf{P}_{\infty,U} G_{\infty,c}^{-1}\|\, d\mu.$$

By (18.3), the sequence of measures $\widetilde{\mathbf{P}}_n = \mathbf{P}_\infty$ and the family of mappings $\widetilde{G}_{n,c} = G_{\infty,c}$, $n \in \overline{\mathbf{N}}$, satisfy the assumptions of Lemma 18.2. By this lemma, we have $\lim_{\delta\to 0} \Delta_3 = 0$.

It remains to estimate Δ_2. We consider the partition Γ of the space $I \times U$ into the fibers $I \times \{z\}$, $z \in U$, and apply Theorem 18.3 to the measures $\mathbf{Q}_n$ and the functionals F_n. Conditions (iv) and (v) of Theorem 18.4 imply the validity of conditions (i) and (ii) of Theorem 18.3. Therefore, $\Delta_2 \to 0$ as $n \to \infty$.

Thus,

$$\overline{\lim_n} \|\mathbf{P}_{n,U} f^{-1} - \mathbf{P}_{\infty,U} f^{-1}\| \le \overline{\lim_n} \Delta_1 + \Delta_3.$$

The left-hand side is independent of δ, whereas the right-hand side tends to zero as $\delta \to 0$. Hence, we arrive at (18.2). □

PROBLEM 18.1. It would be interesting to obtain an analog of Theorem 18.2 for the case of an infinite-dimensional linear topological space. The above proof cannot be generalized to this case. Why?

PROBLEM 18.2. State and prove an analog of Theorem 18.4 for the strong convergence of the distributions $\mathbf{P}_n f_n^{-1}$ to $\mathbf{P}_\infty f_\infty^{-1}$.

§19. Strong convergence of distributions of Gaussian functionals

In this section, we consider sequences $(\mathbf{P}_n)$ of Gaussian measures in a separable Banach space $\mathbf{X}$ and show that the weak convergence $\mathbf{P}_n \Rightarrow \mathbf{P}_\infty$ implies the strong convergence of $\mathbf{P}_n f^{-1}$ to $\mathbf{P}_\infty f^{-1}$ for a wide class of functionals without any additional assumptions.

The classes $\mathcal{M}_{\mathbf{P}}$ and $\mathcal{M}_{\mathbf{P}}^{(a)}$. First, we define two classes of functionals and prove some properties of these classes. In this section, $\mathbf{P}$ is an arbitrary probability measure, so that the definition is meaningful not only for Gaussian measures.

Sets of the form $\Delta = \{x + cl, c \in [0, a]\}$, $x, l \in \mathbf{X}$, will be called *segments parallel to a vector* l. We say that segments $\Delta_n = \{x_n + cl_n, c \in [0, a_n]\}$ *converge* to Δ if $x_n \to x$, $l_n \to l$, and $a_n \to a$. For a functional f and a segment Δ, we define a function f_Δ by setting

$$f_\Delta(c) = f(x + cl), \qquad c \in [0, a_n].$$

Recall that $\mathbf{H_P}$ denotes the set of admissible directions for the measure $\mathbf{P}$.

We say that a functional f *belongs to the class* $\mathcal{M}_{\mathbf{P}}$ if for $\mathbf{P}$-almost each x there exist a vector $l \in \mathbf{H_P}$ and an open ball V centered at x with the following properties: for $\mathbf{P}$-almost all $y \in V$ and any segment $\Delta \subset V$, $\Delta = \{y + cl, c \in [0, a]\}$, the convergence $\Delta_n \to \Delta$ implies the convergence

$$\lambda f_{\Delta_n}^{-1} \overset{\text{var}}{\to} \lambda f_\Delta^{-1}. \tag{19.1}$$

By $\mathcal{M}_{\mathbf{P}}^{(a)}$ we denote the class of $f \in \mathcal{M}_{\mathbf{P}}$ such that, in addition to (19.1), we have

$$\lambda f_\Delta^{-1} \ll \lambda. \tag{19.2}$$

These classes are quite broad: Problem 19.1 shows that the class $\mathcal{M}_{\mathbf{P}}^{(a)}$ is everywhere dense with respect to the uniform norm in the class of all $\mathbf{P}$-almost everywhere continuous functionals. Problem 19.2 shows that the distributions of functionals from the class $\mathcal{M}_{\mathbf{P}}^{(a)}$ are absolutely continuous.

Let us consider smooth functionals.

THEOREM 19.1. *Suppose that the following conditions are satisfied for* $\mathbf{P}$-*almost all* x:

(i) *the derivative* $Df(x)$ *is weakly continuous at* x;

(ii) $Df(x)(\mathbf{H_P}) \neq \{0\}$.

Then $f \in \mathcal{M}_{\mathbf{P}}^{(a)}$.

PROOF. Let X_0 be the set of all points x for which conditions (i) and (ii) are satisfied. Let $x \in X_0$. Condition (ii) means that there exists an $l_0 \in \mathbf{H_P}$ such that

$$\langle l_0, Df(x)\rangle = D_{l_0} f(x) = \delta \neq 0.$$

Without loss of generality, we can assume that $\delta > 0$. Since the derivative is continuous at x, we can find an open ball V centered at x such that

$$D_l f(y) \geq \delta/2$$

for all $y \in V$ and all vectors l close to l_0. Let us take some segment $\Delta = \{y + cl_0, c \in [0, a]\}$ in V such that $y \in X_0$. We intend to show that (19.1) is satisfied as $\Delta_n \to \Delta$.

Let $\Delta_n = \{y_n + cl_n, c \in [0, a_n]\}$. Since $a_n \to a$, without loss of generality we can assume that all functions f_{Δ_n} are defined on the same interval $[0, a]$. It is easy to verify that the f_{Δ_n} uniformly converge to f_Δ on $[0, a]$. Since $f'_{\Delta_n}(c) = D_{l_n} f(y_n + cl_n)$ and the derivative of f is continuous at y, it follows that the functions f'_{Δ_n} also converge uniformly to f'_Δ on $[0, a]$. By the choice of V, the function f'_Δ is bounded away from zero on the entire interval $[0, a]$. Therefore, (19.1) follows from

Theorem 4.5. Theorem 4.2 implies that λf_{Δ}^{-1} is absolutely continuous. The proof is complete. □

Let us apply this result to integral functionals. Let $\mathbf{X} = \mathbf{C}(T)$, where T is a metric compact set and $(\xi_t, t \in T)$ is a continuous Gaussian process with correlation function $K(s,t)$ and distribution $\mathbf{P}$.

We consider integral functionals of the form

$$f(x) = \int_T q(x(t))\mu\,(dt), \tag{19.3}$$

where μ is a finite measure on $\mathfrak{B}_T$ and q is a measurable function on $\mathbf{R}^1$.

Consider the following nondegeneracy condition:

$$\int_{T\times T} K(s,t)\varphi(t)\varphi(s)\mu\,(dt)\mu\,(ds) > 0, \qquad \varphi \neq 0, \quad \varphi \in \mathbf{L}^1(d\mu). \tag{19.4}$$

This condition means that any mixture

$$\int_T \xi_t\varphi(t)\mu\,(dt),$$

where $\varphi \neq 0$ and $\varphi \in \mathbf{L}^1(d\mu)$, has a nondegenerate distribution. A condition of this type was used in §9 when we studied conditions under which integral functionals are absolutely continuous.

THEOREM 19.2. *Suppose that* (19.4) *is satisfied, q is locally Lipschitzian, and the derivative q' is almost everywhere continuous and nonzero. Then $f \in \mathcal{M}_{\mathbf{P}}^{(a)}$.*

PROOF. We have already considered the functional f in Proposition 9.2, and proved that

$$D_l f(x) = \int_T l(t)q'(x(t))\mu(dt).$$

Moreover, we have verified that $D_l f$ converges weakly at almost all $x \in \mathbf{X}$ and that for each of these x there exists an $l \in \mathbf{H_P}$ such that $D_l f(x) \neq 0$. Thus, all assumptions of Theorem 19.1 are satisfied, and so $f \in \mathcal{M}_{\mathbf{P}}^{(a)}$. □

The condition imposed on q can be weakened for some specific processes; see Problem 19.3.

Let us consider sufficient conditions under which convex functionals belong to the class $\mathcal{M}_{\mathbf{P}}^{(a)}$.

We assume that the measure $\mathbf{P}$ has a barycenter $a_{\mathbf{P}}$ and f is a continuous convex functional. Set

$$\alpha_0 = \inf_{x \in \operatorname{supp}(\mathbf{P})} f(x).$$

THEOREM 19.3. *Suppose that the following conditions are satisfied:*

(i) $(a_{\mathbf{P}} + \mathbf{H_P})^- = \operatorname{supp}(\mathbf{P})$;

(ii) $\mathbf{P}\{x \mid f(x) = \alpha_0\} = 0$.

Then $f \in \mathcal{M}_{\mathbf{P}}^{(a)}$.

PROOF. Let $x \in \mathrm{supp}(\mathbf{P})$ be a point such that $f(x) > \alpha_0$. Since f is convex, it follows that there exist an $l \in \mathbf{H_P}$ and an open ball V centered at x such that for any segment $\Delta \subset V$ parallel to l, the function f_Δ is strictly monotone and convex. Therefore, $\lambda f_\Delta^{-1} \ll \lambda$. Furthermore, if $\Delta_n \to \Delta$, then the sequence f_{Δ_n} converges to f_Δ pointwise. Thus, (19.1) follows from Theorem 4.4. □

COROLLARY. *Let* $\mathbf{P}$ *be a Gaussian measure that is not concentrated at one point and let* $f(x) = \|x\|$. *Then the following assertions hold:*

(i) *if* $a_\mathbf{P} \in \mathbf{H_P}$, *then* $f \in \mathcal{M}_\mathbf{P}^{(a)}$;

(ii) *if* $(\mathbf{X}, \|\cdot\|)$ *is strictly convex, then* $f \in \mathcal{M}_\mathbf{P}^{(a)}$.

PROOF. Condition (i) of Theorem 19.3 is satisfied since $\mathbf{P}$ is Gaussian (see Proposition 7.4). Condition (ii) of Theorem 19.3 holds by Proposition 12.1. □

REMARK. For a Gaussian measure $\mathbf{P}$, a convex functional f belongs to $\mathcal{M}_\mathbf{P}^{(a)}$ if and only if $\mathbf{P}f^{-1} \ll \lambda$.

The main theorem and its applications. Now we are in a position to state the main result of this section.

THEOREM 19.4. *Suppose that* $(\mathbf{P}_n, n \in \overline{\mathbf{N}})$ *is a sequence of Gaussian measures,* $\mathbf{P}_n \Rightarrow \mathbf{P}_\infty$, *and* $f \in \mathcal{M}_{\mathbf{P}_\infty}$. *Then* $\mathbf{P}_n f^{-1} \overset{\text{var}}{\to} \mathbf{P}_\infty f^{-1}$.

This theorem combined with the result of Problem 19.2 implies the following functional local limit theorem.

THEOREM 19.5. *Suppose that* $\mathbf{P}_n \Rightarrow \mathbf{P}_\infty$ *and* $f \in \mathcal{M}_\mathbf{P}^{(a)}$. *Then the density of the absolutely continuous component of the distribution* $\mathbf{P}_n f^{-1}$ *converges to the density of the distribution* $\mathbf{P}_\infty f^{-1}$ *in the* $\mathbf{L}^1$*-metric.*

Note some applications of Theorems 19.4 and 19.5.

Let (ξ_k) be a stationary Gaussian sequence with $\mathsf{E}\xi_k = 0$. We set $S_n = \sum_1^n \xi_k$ and consider a continuous random open polygon $\xi_n(t)$, $t \in [0,1]$, with vertices $(k/n,\ S_k/\sqrt{\mathsf{D}S_n})$, $k = 0, \dots, n$. By $\mathbf{P}_n$ we denote the distribution in $\mathbf{C}[0,1]$ corresponding to the Gaussian process $(\xi_n(t), t \in [0,1])$. We assume that the variance of the sums S_n regularly varies as n increases, that is,

$$\mathsf{D}S_n = n^\gamma L(n),$$

where $L(\cdot)$ is a slowly varying function and $\gamma \in (0,2)$. In this case, as was shown in [**29**], $\mathbf{P}_n \Rightarrow \mathbf{P}_\infty$, where $\mathbf{P}_\infty$ is the distribution in $\mathbf{C}[0,1]$ of the semistable Gaussian process w_γ considered in Example 7.7. In this case, we can apply Theorems 19.4 and 19.5, thus obtaining local limit theorems for all $f \in \mathcal{M}_{\mathbf{P}_\infty}^{(a)}$ simultaneously.

Suppose that $\mathbf{P}$ is a Gaussian measure, $f(x) = \|x\|$, and $\mathbf{X}_n$ is an increasing system of subspaces of $\mathbf{X}$ such that $\mathcal{L}(\bigcup_n \mathbf{X}_n)^- = \mathbf{X}$. Next, let $\pi_n \colon \mathbf{X} \to \mathbf{X}_n$ be a system of projections such that $\pi_n x \to x$ for $\mathbf{P}$-almost all x. Then we have the weak convergence $\mathbf{P}_n = \mathbf{P}\pi_n^{-1} \Rightarrow \mathbf{P}$. If $\mathbf{P}f^{-1} \ll \lambda$, then, as we have seen, $f \in \mathcal{M}_\mathbf{P}^{(a)}$, and by Theorem 19.5, the distribution density of the functional $\|\pi_n(\cdot)\|$ converges to the distribution density of $\mathbf{P}f^{-1}$ in $\mathbf{L}^1$.

The limit behavior of conditional distributions. When one applies Theorem 18.4 to Gaussian measures, the perturbing transformations are naturally chosen to be translations along admissible directions. Indeed, suppose that $\mathbf{P}$ is a

Gaussian measure, $l \in \mathbf{H_P}$, and G_c is the translation mapping in the direction l, $G_c x = x + cl$. To estimate $\|\mathbf{P} - \mathbf{P}G_c^{-1}\|$, we consider the partition Γ of the space $\mathbf{X}$ into lines parallel to l. By (2.3), we have

$$\|\mathbf{P} - \mathbf{P}G_c^{-1}\| \le \int_{\mathbf{X}/\Gamma} \|\mathbf{P}_\gamma - \mathbf{P}_\gamma G_c^{-1}\| \mathbf{P}_\Gamma\,(d\gamma).$$

We know that the conditional measures $\mathbf{P}_\gamma$ are one-dimensional Gaussian measures with variance $\sigma^2_{\mathbf{P}}(l)$ independent of γ. If we denote the density of the measure $\mathbf{P}_\gamma$ by $p(t)$, then the density of the measure $\mathbf{P}_\gamma G_c^{-1}$ is $p(t-c)$. Therefore,

$$\|\mathbf{P}_\gamma - \mathbf{P}_\gamma G_c^{-1}\| = \int_{\mathbf{R}^1} |p(t) - p(t-c)|\, dc \le \frac{2c}{\sqrt{2\pi}\sigma_{\mathbf{P}}(l)}. \tag{19.5}$$

(This inequality becomes obvious if we draw two Gaussian densities with variance $\sigma^2_{\mathbf{P}}(l)$ translated by c with respect to each other.)

Thus,

$$\|\mathbf{P} - \mathbf{P}G_c^{-1}\| \le \sqrt{\frac{2}{\pi}} \frac{c}{\sigma_{\mathbf{P}}(l)}. \tag{19.6}$$

It is natural to try to use this inequality to derive condition (iii) of Theorem 18.4. However, one can show that using the same direction for all measures $\mathbf{P}_n$ is not sufficient.[2] It is necessary to carry out a more detailed analysis.

The following theorem provides a clue to the use of Theorem 18.4 in the Gaussian case.

THEOREM 19.6. *Let $(\mathbf{P}_n, n \in \overline{\mathbf{N}})$ be a sequence of Gaussian measures in a separable Banach space $\mathbf{X}$. Suppose that $\mathbf{P}_n \Rightarrow \mathbf{P}_\infty$. Then for each $l_\infty \in \mathbf{H}_{\mathbf{P}_\infty}$ there exists a sequence (l_n), $l_n \in \mathbf{H}_{\mathbf{P}_n}$, such that $\|l_n - l_\infty\| \to 0$ and $\sigma_{\mathbf{P}_n}(l_n) \to \sigma_{\mathbf{P}_\infty}(l_\infty)$ as $n \to \infty$.*

PROOF. By Ball_1^* we denote the closed unit ball in $\mathbf{X}^*$. Let Ψ_n be the characteristic functional of the measure $\mathbf{P}_n$.

Since the spaces $\mathbf{H}_{\mathbf{P}_n}$ are invariant under translations of the measures $\mathbf{P}_n$ and since $a_n \to a_\infty$ by the result of Problem 11.6, we can assume that the measures $\mathbf{P}_n$ have zero barycenters a_n.

Let $K_n \colon \mathbf{X}^* \to \mathbf{H}_{\mathbf{P}_n}$ be the correlation operator. We set

$$\beta_n = \sup_{x^* \in \mathrm{Ball}_1^*} \|K_n(x^*) - K_\infty(x^*)\|.$$

LEMMA 19.1. *$\beta_n \to 0$ as $n \to \infty$.*

PROOF OF THE LEMMA. Since the means are zero, we have

$$\Psi_n(x^*) = \exp\bigl\{-\tfrac{1}{2}|K_n(x^*)|_n^2\bigr\},$$

where $|\cdot|_n$ is the norm in $\mathbf{H}_{\mathbf{P}_n}$. By Theorem 2.4, $\Psi_n(x^*) \to \Psi_\infty(x^*)$ as $n \to \infty$ uniformly with respect to $x^* \in \mathrm{Ball}_1^*$. Therefore,

$$\sup_{x^* \in \mathrm{Ball}_1^*} \bigl| |K_n(x^*)|_n - |K_\infty(x^*)|_\infty \bigr| \to 0, \quad n \to \infty. \tag{19.7}$$

[2] See Problem 19.4.

Noting that $(x,y)_n = \frac{1}{2}(|x|_n^2 + |y|_n^2 - |x-y|_n^2)$, from (19.7) we obtain

$$\delta_n = \sup \left|(K_n(x^*), K_n(y^*))_n - (K_\infty(x^*), K_\infty(y^*))_\infty\right| \to 0, \qquad n \to \infty.$$

By the definition of K_n, we have $(l, K_n(y^*))_n = \langle l, y^* \rangle$ for any $l \in \mathbf{H}_{\mathbf{P}_n}$ and $y^* \in \mathbf{X}^*$. Therefore, $\delta_n = \beta_n$, and so $\beta_n \to 0$. □

Obviously, in Lemma 19.1 the unit ball can be replaced by a ball of any other radius. Thus, if $l_\infty \in \mathbf{H}_{\mathbf{P}_\infty}$ has the form $l_\infty = K_\infty(x^*)$, then, by setting $l_n = K_n(x^*)$, we obtain the desired sequence. Indeed, the convergence of l_n to l_∞ in the norm of $\mathbf{X}$ follows from Lemma 19.1, and the convergence of conditional variances follows from (19.7), since $\sigma^2_{\mathbf{P}_n}(l_n) = |K_n(x^*)|_n^{-2}$. For an arbitrary $l_\infty \in \mathbf{H}_{\mathbf{P}_\infty}$, $l_\infty \neq 0$, we must use the fact that $K_\infty(\mathbf{X}^*)$ is everywhere dense in $\mathbf{H}_{\mathbf{P}_\infty}$ both with respect to the norm of $\mathbf{H}_{\mathbf{P}_\infty}$ and with respect to the norm of $\mathbf{X}$. Hence, there exists a sequence $(l_\infty^{(m)})$, $l_\infty^{(m)} = K_\infty(x_m^*)$, such that $\|l_\infty^{(m)} - l_\infty\| \to 0$ and $|l_\infty^{(m)}|_\infty \to |l_\infty|_\infty$. We have already proved that for each m there exists a sequence $l_n^{(m)}$ such that $\|l_n^{(m)} - l_\infty^{(m)}\| \to 0$ and $|l_n^{(m)}|_n \to |l_\infty^{(m)}|_\infty$. Now the desired sequence approximating l_∞ can be constructed by the diagonal process. □

PROOF OF THEOREM 19.4. Let X_0 be the set of $\mathbf{P}_\infty$-full measure that arises in the assumption that $f \in \mathcal{M}_{\mathbf{P}_\infty}$. Suppose that $x \in X_0$; then l_∞ and W are, respectively, the $\mathbf{P}_\infty$-admissible direction and the open ball corresponding to the point x. Let V be the open ball of radius ε centered at x, where ε is half the radius of W. Using Theorem 19.6, we can find elements l_n such that they determine $\mathbf{P}_n$-admissible directions convergent to l_∞ and such that the variances σ_n^2 of the conditional measures for $\mathbf{P}_n$ on the lines parallel to l_n converge to the variances σ_∞^2 of the conditional measures for $\mathbf{P}_\infty$ on the lines parallel to l_∞.

Set $G_{n,c}x = x + cl_n$, $c \geq 0$, $n \in \overline{\mathbf{N}}$. Let us show that the neighborhood V, the number ε, and the family of mappings $G_{n,c}$ satisfy the assumptions of Theorem 18.4.

Assumptions (i) and (ii) are obviously satisfied.

To prove that assumption (iii) is satisfied, we consider the partition Γ_n of $\mathbf{X}$ into lines parallel to l_n. By (19.6), we have

$$\|\mathbf{P}_n G_{n,c}^{-1} - \mathbf{P}_n\| \leq \sqrt{\frac{2}{\pi}} \frac{c}{\sigma_n}.$$

Since $\sigma_n^2 \to \sigma_\infty^2$, we see that this inequality proves assumption (iii).

Now let us verify assumption (iv). Note that for $\delta \in (0, \varepsilon)$ and $z \in V$, the measure $\lambda_{[0,\delta]}\varphi_{n,z}^{-1}$ coincides with $\lambda f_{\Delta_n}^{-1}$, where $\Delta_n = \{z + cl_n, c \in [0,\delta]\}$. Moreover, the segment Δ_n is parallel to l_n and lies entirely in the ball W. Since $f \in \mathcal{M}_{\mathbf{P}_\infty}$, it follows from the choice of l_∞ and W that the function

$$\delta_n(z) = \|\lambda_{[0,\delta]}\varphi_{n,z}^{-1} - \lambda_{[0,\delta]}\varphi_{\infty,z}^{-1}\|, \qquad z \in V,$$

has the following property: it follows from the convergence $z_n \to z$ that $\delta_n(z_n) \to 0$. This, together with the remark following Theorem 18.4, means that assumption (iv) is satisfied.

Assumption (v) can be verified similarly to assumption (iv).

Thus, all assumptions of Theorem 18.4 are satisfied, and Theorem 19.4 is thereby proved. □

PROBLEM 19.1. Suppose that $\mathbf{H_P} \neq \{0\}$. Then for any $\mathbf{P}$-almost everywhere continuous functional f and any $\varepsilon > 0$, there exists a functional $f_\varepsilon \in \mathcal{M}_{\mathbf{P}}^{(a)}$ such that $\sup_{x\in\mathbf{X}} |f(x) - f_\varepsilon(x)| < \varepsilon$. (*Hint*: in a neighborhood of a point of continuity, approximate f by a linear functional that grows in the direction of $l \in \mathbf{H_P}$.)

PROBLEM 19.2. Suppose that $f \in \mathcal{M}_{\mathbf{P}}^{(a)}$. Then $\mathbf{P}f^{-1} \ll \lambda$. (*Hint*: let x be a point of the set of $\mathbf{P}$-full measure occurring in the definition of the class $\mathcal{M}_{\mathbf{P}}^{(a)}$; let V and l be the corresponding ball and admissible direction. Consider the partition Γ of the space $\mathbf{X}$ into lines parallel to l, and use (3.4) and (19.2) to prove that $\mathbf{P}_V f^{-1} \ll \lambda$.)

PROBLEM 19.3. Let $\mathbf{W}$ be the Wiener measure on the space $\mathbf{X} = \mathbf{C}[0,1]$, and let $f(x) = \int_0^1 q(x(t))\,d(t)$. Prove that $f \in \mathcal{M}_{\mathbf{W}}^{(a)}$ provided that for each $\varepsilon > 0$ there exists an open interval $J \subset (-\varepsilon, \varepsilon)$ on which the derivative q' is continuous and does not vanish. (*Hint*: use the fact that for any interval $[a,b] \subset [0,1]$ there exists a function $l \in \mathbf{H_W}$ that is equal to zero outside (a,b) and positive on (a,b).)

PROBLEM 19.4. Let $\mathbf{W}$ be a Wiener measure on the space $\mathbf{X} = \mathbf{C}[0,1]$. By Λ we denote the set of all nondecreasing automorphisms of the interval $[0,1]$ whose derivatives are bounded away from zero and infinity. To each $u \in \Lambda$, we assign a bounded invertible linear operator T_u by setting

$$T_u x(s) = x(u(s)), \qquad s \in [0,1], \quad x \in \mathbf{X}.$$

Let us choose automorphisms $u_n \in \Lambda$ so that

1) u_n converges to $u_\infty(t) \equiv t$ uniformly on $[0,1]$;
2) for all $E \in \mathfrak{B}_{[0,1]}$ and $\lambda(E) > 0$, one has

$$\int_E \frac{1}{u_n'(s)}\,ds \to \infty$$

as $n \to \infty$.

Prove that the sequence of Gaussian measures $\mathbf{P}_n = \mathbf{W}T_n^{-1}$ has the following properties:

1) $\mathbf{P}_n \Rightarrow \mathbf{W}$;
2) the correlation operators $K_{\mathbf{P}_n}$ converge strongly to $K_\mathbf{W}$;
3) for each n, $\mathbf{H}_{\mathbf{P}_n}$ coincides (as a set) with $\mathbf{H_W}$;
4) for each $l \in \mathbf{H_W}$, the variance $\sigma_n^2(l)$ of the conditional distributions for $\mathbf{P}_n$ on lines parallel to l tends to zero as $n \to \infty$.

(*Hint*: verify that $\sigma_n^2(l) = \left(\int_0^1 |l'(v_n(s))|^2 |v_n'(s)|^2\,ds\right)^{-1}$, where v_n is the inverse of u_n.)

§20. The local invariance principle

The classical Donsker–Prokhorov invariance principle [**4,** Ch. 2, §10] states that the distributions $\mathbf{P}_n$ of continuous random open polygons in $\mathbf{C}[0,1]$ constructed in the standard way from sums S_n of independent identically distributed random variables with finite variance converge weakly to a Wiener measure $\mathbf{W}$. In other words, for any $\mathbf{W}$-almost everywhere continuous functional f, the distributions $\mathbf{P}_n f^{-1}$ converge weakly to the distribution $\mathbf{W}f^{-1}$. At the same time, for some specific functionals (for example, $f(x) = \sup x(t)$ or $f(x) = \sup |x(t)|$), it is known that, under natural additional restrictions on the distribution S_1, strong convergence takes place.

In this section, we prove that the convergence $\mathbf{P}_n f^{-1} \overset{\text{var}}{\to} \mathbf{W} f^{-1}$ holds for all $f \in \mathcal{M}_{\mathbf{W}}$. A similar result in the case of attraction to a stable law will be obtained in §21.

Suppose that $\{\xi_n, n \in \mathbf{N}\}$ is a sequence of identically distributed random variables with zero mean and finite variance, $S_n = \xi_1 + \cdots + \xi_n$ are the corresponding partial sums, $\sigma_n^2 = \mathsf{D} S_n$ are their variances, and $\mathbf{P}_n$ are the distributions in the space $\mathbf{X} = \mathbf{C}[0,1]$ of the random elements

$$\zeta_n(t) = \sigma_n^{-1}\big(S_{[nt]} + (nt - [nt])\xi_{[nt]+1}\big), \qquad t \in [0,1]. \tag{20.1}$$

To guarantee the strong convergence of distributions for a wide class of functionals, we need to impose additional restrictions on the joint distribution $\mathcal{P}$ of the variables ξ_n. We assume that $\mathcal{P}$ has an absolutely continuous density p such that $p'/p \in \mathbf{L}^2(d\mathcal{P})$. In other words, we assume that the Fisher information is finite,

$$I_p = \int_{\mathbf{R}^1} \left(\frac{p'}{p}\right)^2 d\mathcal{P} = \int_{\mathbf{R}^1} \frac{(p')^2}{p}\, d\lambda < \infty. \tag{20.2}$$

This means that the density p possesses a certain smoothness. Indeed, the weaker condition $p'/p \in \mathbf{L}^1(d\mathcal{P})$ is equivalent to the assumption that p is a function of bounded variation.

By analogy with the case of weak convergence, the following theorem is called the *local invariance principle*, or the *functional local limit theorem*.

THEOREM 20.1. *Suppose that ξ_n, $n = 1, 2, \ldots$, are independent identically distributed random variables with zero mean and finite nonzero variance σ^2. Suppose that the distribution density p of the random variables ξ_n is absolutely continuous and satisfies $I_p < \infty$.*

Then

$$\mathbf{P}_n f^{-1} \overset{\text{var}}{\to} \mathbf{W} f^{-1} \tag{20.3}$$

for any functional $f \in \mathcal{M}_{\mathbf{W}}$.

PROOF. Since $f \in \mathcal{M}_{\mathbf{W}}$, we can choose a set X_0, a point $x \in X_0$, balls U and V, a number ε, and a vector l_∞ just as in the proof of Theorem 19.4. We set $l_n = \pi_n(l_\infty)$, where π_n is the self-mapping of $\mathbf{X}$ that assigns the continuous piecewise linear open polygon with vertices $\big(k/n, x(k/n)\big)$, $k = 0, 1, \ldots, n$, to each function x. Let $G_{n,c} x = x + c l_n$, $c > 0$, and $n \in \mathbf{N}^+$. Let us verify that the ball V, the number ε, and the family $(G_{n,c})$ satisfy the assumptions of Theorem 18.4.

Assumptions (i) and (ii) can be readily verified. Assumptions (iv) and (v) can be verified by the method used in the proof of Theorem 19.4, since they are related only to the fact that $f \in \mathcal{M}_{\mathbf{W}}$. It remains to prove (iii).

Let us consider the mapping $J_n \colon \mathbf{X} \to \mathbf{R}^n$ given by

$$J_n(x) = \sqrt{n}\Big(x\Big(\frac{1}{n}\Big), x\Big(\frac{2}{n}\Big) - x\Big(\frac{1}{n}\Big), \ldots, x(1) - x\Big(\frac{n-1}{n}\Big)\Big).$$

Let $\mathcal{P}_n = \mathbf{P}_n J_n^{-1}$ and $\mathbf{l}_n = J_n(l_n) = (l_n^{(1)}, \ldots, l_n^{(n)})$.

For $c > 0$ and all $n \in \mathbf{N}$, we define mappings $\overline{G}_{n,c} \colon \mathbf{R}^n \to \mathbf{R}^n$ by setting

$$\overline{G}_{n,c}\mathbf{x} = \mathbf{x} + c\mathbf{l}_n.$$

Since the subspace $E_n = \pi_n(\mathbf{X})$ is the support of the measure $\mathbf{P}_n$ and J_n is an isomorphism of E_n onto $\mathbf{R}^n$, we have

$$\Delta_n(c) = \|\mathbf{P}_n G_{n,c}^{-1} - \mathbf{P}_n\| = \|\mathcal{P}_n \overline{G}_{n,c}^{-1} - \mathcal{P}_n\|.$$

Since the measure $\mathcal{P}_n$ is absolutely continuous and its density is equal to $\prod_1^n p(x_i)$, we obtain

$$\Delta_n(c) = \int_{\mathbf{R}^n} \left| \prod_1^n p(x_i) - \prod_1^n p(x_i - c l_n^{(i)}) \right| d\mathbf{x}.$$

LEMMA 20.1. *Suppose that p is absolutely continuous. Then*

$$\Delta_n(c) \le c I_p^{1/2} \|l'_\infty\|_{\mathbf{L}^2_{[0,1]}}. \tag{20.4}$$

PROOF. First, note that $\Delta_n(c) = \mathsf{E}|1 - Z_n(c)|$, where

$$Z_n(c) = \prod_1^n \frac{p(\xi_i - c l_n^{(i)})}{p(\xi_i)}.$$

If $l_\infty(t) \equiv t$, then $l_n^{(i)} = 1/\sqrt{n}$ and $Z_n(c)$ is just the likelihood ratio in the estimation problem for the translation parameter. This relation allows us to prove (20.4) by using a specific method of the asymptotic estimation theory.

Let

$$Q(s, \mathbf{x}) = \prod_1^n p(x_i - s l_n^{(i)}).$$

Obviously,

$$\frac{\partial Q(s,\mathbf{x})}{\partial s} = Q(s,\mathbf{x}) \sum_{i=1}^n (-l_n^{(i)}) \frac{p'(x_i - s l_n^{(i)})}{p(x_i - s l_n^{(i)})}.$$

Therefore,

$$\begin{aligned} \Delta_n(c) &= \int_{\mathbf{R}^n} \left| \int_0^c \frac{\partial Q(s,\mathbf{x})}{\partial s} \, ds \right| d\mathbf{x} \\ &\le \int_0^c \left(\int_{\mathbf{R}^n} \left| \sum_1^n (-l_n^{(i)}) \frac{p'(x_i - s l_n^{(i)})}{p(x_i - s l_n^{(i)})} \right|^2 Q(s,\mathbf{x}) \, d\mathbf{x} \right)^{1/2} ds. \end{aligned} \tag{20.5}$$

For a given s, the functions

$$\gamma_i(\mathbf{x}) = \frac{p'(x_i - s l_n^{(i)})}{p(x_i - s l_n^{(i)})},$$

treated as random variables on the probability space $(\mathbf{R}^n, \mathfrak{B}^n, \mathbf{Q})$, where $\mathbf{Q}$ is the measure with density $Q(s, \mathbf{x})$, are independent and have zero means. Their variance is

$$\begin{aligned} \mathsf{D}_{\mathbf{Q}} \gamma_i &= \int_{\mathbf{R}^n} \left(\frac{p'(x_i - s l_n^{(i)})}{p(x_i - s l_n^{(i)})} \right)^2 Q(s,\mathbf{x}) \, d\mathbf{x} \\ &= \int_{\mathbf{R}^1} \left(\frac{p'(x_i - s l_n^{(i)})}{p(x_i - s l_n^{(i)})} \right)^2 p(x_i - s l_n^{(i)}) \, dx_i = I_p. \end{aligned}$$

Therefore,

$$\int_{\mathbf{R}^n}\left|\sum_1^n(-l_n^{(i)})\frac{p'(x_i-sl_n^{(i)})}{p(x_i-sl_n^{(i)})}\right|^2 Q(s,\mathbf{x})\,d\mathbf{x} \tag{20.6}$$
$$=\mathsf{D}_{\mathbf{Q}}\Bigl(\sum_1^n l_n^{(i)}\gamma_i\Bigr)=\sum_1^n(l_n^{(i)})^2\mathsf{D}_Q\gamma_i=I_p\sum_1^n(l_n^{(i)})^2.$$

Since

$$\sum_1^n(l_n^{(i)})^2=n\sum_1^n\Bigl(l\Bigl(\frac{i+1}{n}\Bigr)-l\Bigl(\frac{i}{n}\Bigr)\Bigr)^2=\frac{1}{n}\sum_1^n\Bigl(n\int_{i/n}^{(i+1)/n}l'(s)\,ds\Bigr)^2$$
$$\le\sum_1^n\int_{i/n}^{(i+1)/n}(l'(s))^2\,ds=\|l'_\infty\|^2_{\mathbf{L}^2[0,1]},$$

we see that relations (20.5) and (20.6) prove the lemma. □

This lemma readily implies assumption (iii). An application of Theorem 18.4 completes the proof of Theorem 20.1. □

Along with continuous random open polygons ζ_n, one often deals with processes of the form

$$\widehat{\zeta}_n(t)=\sigma_n^{-1}S_{[nt]},\qquad t\in[0,1], \tag{20.7}$$

whose trajectories are right continuous step-functions. The space $\mathbf{C}[0,1]$ cannot be used to study weak convergence of such processes and hence is usually replaced by the space $\mathbf{D}[0,1]$ of functions without second-order discontinuities, equipped with the Skorokhod topology. However, in this situation it is easier to use a different space, which is more convenient for our goals.

Let $\mathfrak{D}$ be the σ-algebra of Borel subsets of $\mathbf{D}[0,1]$, and let $\mathfrak{U}$ be the σ-algebra of subsets of $\mathbf{D}[0,1]$ generated by the uniform topology. Let $\mathbf{P}_n$ be the probability measures on $\mathfrak{D}$ corresponding to $\widehat{\zeta}_n$, and let $\mathbf{W}$ be a Wiener measure on $\mathfrak{D}$. The measures $\mathbf{P}_n$ and $\mathbf{W}$ can be continued from $\mathfrak{D}$ to $\mathfrak{U}$ (see [**4**, Ch. 3, §18]. We use the same notation $\mathbf{P}_n$ and $\mathbf{W}$ for these continuations. If the variables (ξ_n) that determine $\widehat{\zeta}_n$ are independent and identically distributed and have zero mean and finite variance, then, by analogy with the invariance principle in $\mathbf{C}[0,1]$, we have the weak convergence $\mathbf{P}_n\overset{\mathfrak{U}}{\Rightarrow}\mathbf{W}$.

By $\mathbf{E}[0,1]$ we denote the uniform closure of the set of functions $f\in\mathbf{D}[0,1]$ that have finitely many jumps at rational points. The set $\mathbf{E}[0,1]$ equipped with the uniform norm is a separable Banach space; the σ-algebra of Borel subsets of $\mathbf{E}[0,1]$ coincides with the σ-algebra $\mathbf{E}[0,1]\cap\mathfrak{U}$. By $\mathbf{W}_{\mathbf{E}}$ we denote the restriction of the Wiener measure $\mathbf{W}$ to this σ-algebra.

Since $\mathbf{E}[0,1]$ is the support of the measures $\mathbf{P}_n$ and $\mathbf{W}$, we can literally repeat the proof of Theorem 20.1 and obtain the following statement.

THEOREM 20.2. *Let $\mathbf{P}_n$, $n=1,2,\dots$, be distributions in $\mathbf{E}[0,1]$ corresponding to the processes $\widehat{\zeta}_n$ constructed according to (20.7) from independent identically distributed random variables ξ_k. Suppose that $\mathsf{E}\xi_k=0$, $\mathsf{D}\xi_k=\sigma^2<\infty$, and the* *density p of the distribution of ξ_k is absolutely continuous. If $I_p < \infty$ and f is a functional on the space* $\mathbf{E}[0,1]$ *belonging to the class* $\mathcal{M}_{\mathbf{W}_\mathbf{E}}$, *then* $\mathbf{P}_n f^{-1} \overset{\text{var}}{\to} \mathbf{W} f^{-1}$.

One can readily see that for any functional $f \in \mathcal{M}_\mathbf{W}$ on the space $\mathbf{D}[0,1]$, the restriction of f to $\mathbf{E}[0,1]$ belongs to the class $\mathcal{M}_{\mathbf{W}_\mathbf{E}}$. Therefore, Theorem 20.2 implies the following statement.

THEOREM 20.3. *The convergence* $\mathbf{P}_n f^{-1} \overset{\text{var}}{\to} \mathbf{W} f^{-1}$ *holds under the assumptions of Theorem* 20.2 *for any functional* $f \in \mathcal{M}_\mathbf{W}$ *on the space* $\mathbf{D}[0,1]$.

We conclude this section by giving some problems that deal with some other analogs and generalizations of the local invariance principle for independent random variables with finite dispersion (see also the "Bibliographical notes").

PROBLEM 20.1. Consider independent random variables (ξ_n) uniformly distributed on $[-1,1]$. Let $\mathbf{P}_n$ be the distributions in $\mathbf{C}[0,1]$ of random open polygons ζ_n of the form (1), and let $G_{n,c}$ be the group of translations along $l_n(t) = l_\infty(t) = t$. Then for any $c > 0$ one has

$$\overline{\lim_n} \|\mathbf{P}_n G_{n,c}^{-1} - \mathbf{P}_n\| = 2.$$

This shows that the condition in Lemma 20.1 stating that the Fisher information is finite is essential.

PROBLEM 20.2 (a generalization of Lemma 20.1). Consider independent random variables $\xi_1, \dots, \xi_n$ with distribution densities $p_1, \dots, p_n$. Let $I_1, \dots, I_n$ be the corresponding Fisher information. Then for any real numbers $a_1, \dots, a_n$ we have the inequality

$$\int_{\mathbf{R}^n} \Big| \prod_1^n p_i(x_i) - \prod_1^n p_i(x_i - a_i) \Big| \, d\bar{x} \le \Big(\sum_1^n I_k a_k^2 \Big)^{1/2}.$$

PROBLEM 20.3 (P. Breier). Suppose that $(\xi_{n,i}, i = 1, \dots, n; n \in \mathbf{N})$ is a sequence of series of random variables that are independent in each series. Suppose that $F_{n,i}$ is the distribution of $\xi_{n,i}$; $\mathsf{E}\xi_{n,i} = 0$; $\mathsf{D}\xi_{n,i} = \sigma_{n,i}^2$; $S_n = \sum_1^n \xi_{n,i}$; $\sigma_n^2 = \mathsf{D}S_n$; $t_{n,k} = \sigma_n^{-2} \sum_1^k \sigma_{n,i}^2$.

Let $\mathbf{P}_n$ be the distributions in $\mathbf{C}[0,1]$ of continuous piecewise linear random open polygons ζ_n with vertices at the points $(t_{n,k}, \sigma_n^{-1} S_k)$. By $I_{n,i}$ we denote the Fisher information of the distribution $F_{n,i}$. If

$$\overline{\lim_n} \sum_{i=1}^n I_{n,i} \sigma_{n,i}^2 < \infty$$

and $f \in \mathcal{M}_\mathbf{W}$, then $\mathbf{P}_n f^{-1} \overset{\text{var}}{\to} \mathbf{W} f^{-1}$. (*Hint*: to verify assumption (iii) of Theorem 18.4, use the result of Problem 20.2.)

PROBLEM 20.4. From the previous statement, derive the following version of the local limit theorem for the series.

If the variables $(\xi_{n,i})$ satisfy the assumptions of Problem 20.3, then the distribution densities of the normalized sums $\sigma_n^{-1} S_n$ converge in the metric of $\mathbf{L}^1(\mathbf{R}^1)$ to the distribution density of the standard Gaussian law.

PROBLEM 20.5 (a bound for the convergence rate). Suppose that $\xi_1, \dots, \xi_n$ are independent identically distributed random variables, $\mathsf{E}\xi_k = 0$, $\mathsf{D}\xi_k = 1$, and $\mathsf{E}|\xi_k|^3 < \infty$. Let $\mathbf{P}_n$ be the distribution in $\mathbf{D}[0,1]$ of the random element $\widehat{\zeta}_n$, constructed from $\xi_1, \dots, \xi_n$. Let $\mathbf{W}$ be a Wiener measure on $\mathbf{D}[0,1]$. We set

$$\psi_1(t) = \left(t \ln \frac{e}{t}\right)^{1/2}, \qquad \psi_2(t) = t^{1/2} |\ln t|^{-1}, \qquad t \in [0,1];$$
$$H_i(L) = \{x \in \mathbf{D}[0,1] \mid \omega_x(t) \le L\psi_i(t) \ \forall t \in [0,1]\}, \qquad i = 1, 2.$$

Consider the functionals

$$f_\varepsilon(x) = \sup_{[\varepsilon,1]} \{x(t) + h(t)\}, \qquad g_\varepsilon(x) = \sup_{[\varepsilon,1]} |x(t) + h(t)|,$$

where $x \in \mathbf{D}[0,1]$, h is a given function from $\mathbf{D}[0,1]$, and $\varepsilon \in [0, 1/2]$.

Then the following statements hold.

1) (Davydov [**34**]) Suppose that the absolutely continuous component of the distribution of the sum $\sum_1^{n_0} \xi_k$ is not zero for some n_0. Then there exists a constant C_1, depending only on L and $\mathsf{E}|\xi_k|^3$, such that

$$\sup_{h \in H_1(L)} \|\mathbf{P}_n f_\varepsilon^{-1} - \mathbf{W} f_\varepsilon^{-1}\| \le C_1 \varepsilon^{-3/8} (\ln n)^{3/8} n^{-1/8}$$

for all $n \ge \varepsilon^{-1}$, $\varepsilon > 0$.

2) (Lifshits) Suppose that for some n_0 the distribution of the sum $\sum_1^{n_0} \xi_k$ has a bounded density. Then there exist constants C_2 and C_3, depending only on L and $\mathsf{E}|\xi_k|^3$, such that

$$\sup_{h \in H_2(L)} \|\mathbf{P}_n f_0^{-1} - \mathbf{W} f_0^{-1}\| \le C_2 n^{-1/8} (\ln n)^6 \exp(C_3 \sqrt{\ln n}).$$

The same estimates hold for the functionals g_ε.

§21. The local invariance principle in the case of attraction to a stable law

Now we assume that independent identically distributed variables (ξ_n) lie in the domain of attraction of a stable law F_α with exponent $\alpha \in (0, 2)$. This means that the normalized sums $B_n^{-1} S_n$, where $S_n = \sum_1^n \xi_k$, $B_n = n^{1/\alpha} L(n)$, and L is a slowly varying function that weakly converges to F_α. In this case, Skorokhod proved that the random processes

$$\widehat{\zeta}_n(t) = B_n^{-1} S_{[nt]}, \qquad t \in [0,1], \tag{21.1}$$

weakly converge in the space $\mathbf{D}[0,1]$ to a stable process v_α for which $\mathcal{P}_{v_\alpha(1)} = F_\alpha$. Thus, he obtained an analog of the Donsker–Prokhorov invariance principle. Note that in this case the continuous open polygons of the type (20.1) (with σ_n replaced by B_n) do not weakly converge in $\mathbf{C}[0,1]$, since the trajectories of the process v_α are discontinuous with probability 1. However, the main difficulty that arises when we try to obtain an analog of the local invariance principle in this case is related not to the passage from the space of continuous functions to the Skorokhod space $\mathbf{D}[0,1]$, but to the fact that the set of admissible translations of the measure $\mathbf{Q}_\alpha$, the distribution of the process v_α, is trivial. Therefore, the translation mappings used in the proof of Theorem 20.1 prove useless, and we have to seek other families of mappings.

First of all, we list some well-known properties of the processes v_α.

The characteristic function of the variable $v_\alpha(t)$ has the form

$$\mathsf{E}\exp\{isv_\alpha(t)\} = \exp\left\{ias + t\int_{-\infty}^{\infty}\left(\exp\{ius\} - 1 - \frac{ius}{1+u^2}\right)\pi_\alpha(u)\,du\right\}, \tag{21.2}$$

where $c_1, c_2 \geq 0$, $c_1, c_2 \neq 0$, and $\pi_\alpha(u) = |u|^{-1-\alpha}\bigl(c_1\mathbf{1}_{(0,\infty)}(u) + c_2\mathbf{1}_{(-\infty,0)}(u)\bigr)$ is the density of the Lévy measure of the distribution F_α.

We shall consider the case of strong stability; this means that the constant a in (21.2) is zero.

With probability 1, for any $\varepsilon > 0$ the trajectories of the process v_α have only finitely many jumps of magnitude $> \varepsilon$. At the same time, with probability 1 the process v_α has at least one jump in any subinterval of $[0,1]$. Moreover, if the constant c_1 in (21.2) is not zero, then there is a jump of positive sign; a similar statement holds in the case $c_2 \neq 0$.

If $c_2 = 0$ and $c_1 \neq 0$, then

$$\operatorname{supp}\mathbf{Q}_\alpha = \mathbf{X}_+ = \bigl\{x \in \mathbf{D}[0,1] \mid x(0) = 0,\ x \text{ does not decrease}\bigr\}.$$

If $c_1 = 0$ and $c_2 \neq 0$, then

$$\operatorname{supp}\mathbf{Q}_\alpha = \mathbf{X}_- = \bigl\{x \in \mathbf{D}[0,1] \mid x(0) = 0,\ x \text{ does not increase}\bigr\}.$$

If $c_1 \neq 0$ and $c_2 \neq 0$, then

$$\operatorname{supp}\mathbf{Q}_\alpha = \mathbf{X}_0 = \bigl\{x \in \mathbf{D}[0,1] \mid x(0) = 0\bigr\}.$$

These and some other properties of stable processes are described in [**94**].

In this section, we consider the following families $\{G_c,\ c \in \mathbf{R}_+\}$ of transformations of the space $\mathbf{X} = \mathbf{D}[0,1]$:

$$G_c x = x + c l_x, \tag{21.3}$$

where $(l_x, x \in \mathbf{X})$ is a set of vectors from $\mathbf{X}$, called a *local direction field.* Obviously, if $l_x = l$ for all x, then we arrive at the familiar translation mappings.

It turns out that the direction field (l_x) can be chosen so that the measure $\mathbf{Q}_\alpha$ is quasi-invariant with respect to the mapping G_c, that is,

$$\mathbf{Q}_\alpha G_c^{-1} \sim \mathbf{Q}_\alpha$$

for all $c \geq 0$.

In this case, the local direction field (l_x) is said to *admissible* for $\mathbf{Q}_\alpha$.

The set of such fields is rather wide, and this allows us to apply the general results of §18 for the derivation of the local invariance principle.

To describe the local direction fields, we introduce some auxiliary notation.

We denote the value of the jump of a function x at a point s by $\delta_x(s)$; that is, $\delta_x(s) = x(s) - x(s-)$.

For $\varepsilon > 0$, we set

$$\begin{aligned}
J_x^+(\varepsilon) &= \{s \in [0,1] \mid \delta_x(s) > \varepsilon\},\\
J_x^-(\varepsilon) &= \{s \in [0,1] \mid \delta_x(s) < -\varepsilon\},\\
J_x(\varepsilon) &= J_x^+(\varepsilon) \cup J_x^-(\varepsilon).
\end{aligned}$$

As was already noted, card $J_x(\varepsilon) < \infty$.

Suppose that $m \in \mathbf{N}$, $\Delta_1, \dots, \Delta_m$ are disjoint open subintervals of the interval $[0,1]$, and $\tau_1, \dots, \tau_m$ are real numbers. These objects and the number ε are called the *parameters* of the field.

For $s, t \in [0,1]$, we set

$$\varphi_s(t) = \sum_{i=1}^{m} \tau_i \mathbf{1}_{\Delta_i}(s) \mathbf{1}_{[s,\infty)}(t),$$

and for $\varepsilon > 0$ we define the local direction field $(l_x, x \in \mathbf{X})$ by setting

$$l_x = \sum_{s \in J_x^+(\varepsilon)} \varphi_s^+ - \sum_{s \in J_x^-(\varepsilon)} \varphi_s^-; \tag{21.4}$$

here φ_s^+ and φ_s^- are the positive and the negative part of φ_s, respectively.

Relation (21.3) defines the mappings G_c corresponding to this field.

Let us describe the action of G_c on the function x informally. Suppose that x has a jump at some point $s \in \Delta_i$, so that $\delta_x(s) > \varepsilon$ and $\tau_i > 0$. Then at this point the function $G_c x$ also has the jump $\delta_x(s) + c\tau_i$. The functions x and $G_c x$ have equal increments between neighboring jumps.

This relation implies that the family $(G_c, c \in \mathbf{R}_+)$ is a semigroup, that is, $G_{c_1+c_2} x = G_{c_2}(G_{c_1} x)$ for all $c_1, c_2 > 0$. Note that the trajectories of the semigroup (G_c) are half-lines, that is, rays (which are however not parallel to each other). That is why the local direction fields in question are said to be *radial*. The semi-group (G_c) is a very special case of the set of semigroups of transformations, described by Lifshits in connection with functionals of random processes with independent increments. He obtained a result (see Theorem 9.2 in [**40**]), which readily implies the following statement.

THEOREM 21.1. *The radial field $\{l_x\}$ is admissible for the measure $\mathbf{Q}_\alpha$. The conditional measures for $\mathbf{Q}_\alpha$ on the orbits of the semigroup (G_c) exist and have densities.*

REMARK. Precise expressions for these densities are given in [**40,** §9]; we do not need them in the sequel.

The following statement shows that the set of radial fields is sufficiently large.

LEMMA 21.1. *Suppose that $c_1 \cdot c_2 \neq 0$ in (21.2). Then for $\mathbf{Q}_\alpha$-almost each x and any open ball U in X_0, there exist an open ball V_x centered at x and parameters ε, N, $\{\Delta_1, \dots, \Delta_N\}$, and $\{\tau_1, \dots, \tau_N\}$ such that the radial field $\{l_x\}$ defined by these parameters has the following property: for each $y \in V_x$ the vector l_y belongs to U.*

If $c_1 \neq 0$ and $c_2 = 0$ (or $c_1 = 0$ and $c_2 \neq 0$), then a similar statement remains valid with $\mathbf{X}_0$ replaced by $\mathbf{X}_+$ (respectively, by $\mathbf{X}_-$). The hint to Problem 21.1 outlines the proof of this lemma.

We say that a functional $f\colon \mathbf{X} \to \mathbf{R}^1$ belongs to the class $\mathcal{K}_{\mathbf{Q}_\alpha}$ if for $\mathbf{Q}_\alpha$-almost each x there exist an open ball V_x centered at x and a radial field $\{l_y, y \in \mathbf{X}\}$ such that for $\mathbf{Q}_\alpha$-almost each $y \in V_x$ and for any segment $\Delta = \{y + cl_y,\ c \in [0,a]\}$ lying in V_x the convergence of segments Δ_n to Δ implies that $\lambda_{\Delta_n} f_{\Delta_n}^{-1} \overset{\text{var}}{\to} \lambda_\Delta f_\Delta^{-1}$.

This definition is close to the definition of the classes $\mathcal{M}_{\mathbf{P}}$; therefore, it is not surprising that the class $\mathcal{K}_{\mathbf{Q}_\alpha}$ has similar properties (see Problems 21.2 and 21.3).

Now we can state the main result of this section.

THEOREM 21.2. *Suppose that* (ξ_n) *are independent identically distributed random variables that lie in the domain of attraction of a strongly stable law with exponent* α, $0 < \alpha < 2$ (*the constant* a *in representation* (21.2) *is equal to zero*). *Let* $\mathbf{P}_n$ *be the distribution in the space* $\mathbf{D}[0,1]$ *of the random processes* (21.1) *constructed from the variables* ξ_n. *Suppose that the distribution* F *of the variables* ξ_n *has density* q *which is an absolutely continuous function. Let* $A_q(u)$ *be the total variation of* q *on the set* $\mathbf{R}^1 \setminus [-u, u]$,

$$A_q(u) = \int_{|s|>u} |q'(s)|\, ds, \tag{21.5}$$

and let

$$\sup nB_n A_q(B_n) < \infty, \tag{21.6}$$

where the B_n *are the normalizing constants in* (21.1).

Then for any $f \in \mathcal{K}_{\mathbf{Q}_\alpha}$ *we have the convergence in variation*

$$\mathbf{P}_n f^{-1} \overset{\text{var}}{\to} \mathbf{Q}_\alpha f^{-1}. \tag{21.7}$$

Condition (21.6) provides some regularity of the density q. This condition plays the same role as the condition stating that the Fisher information is finite in Theorem 20.4.

Before proving the theorem, we formulate the following corollary.

COROLLARY 1. *Suppose that under the assumptions of Theorem* 21.2 *the distribution of the variables* ξ_n *is stable, that is,* $F = F_\alpha$. *Then* (21.7) *holds for any* $f \in \mathcal{K}_{\mathbf{Q}_\alpha}$.

The proof follows from the fact that in this case $|q'(s)| \sim b_1 s^{-2-\alpha}$ as $s \to \infty$ and $|q'(s)| \sim b_2 s^{-2-\alpha}$ as $s \to -\infty$ for some choice of the constants b_1 and b_2 [**51**]. Therefore, $A_q(u) \sim Cu^{-1-\alpha}$, and inequality (21.6) holds since $B_n \sim Cn^{1/\alpha}$.

PROOF OF THE THEOREM. Just as in the proof of the local invariance principle for random variables with finite dispersion, we use Theorem 18.4.

By $\mathbf{X}_0$ we denote the set of $\mathbf{Q}_\alpha$-full measure of points x for which there exist an open ball V_x and an admissible radial field $\{l_y, y \in \mathbf{X}\}$ that possess the properties listed in the definition of the class $\mathcal{K}_{\mathbf{Q}_\alpha}$.

For each $n \in \mathbf{N}^+$, we set

$$G_{n,c} y = G_c y = y + c l_y, \qquad y \in \mathbf{X}.$$

Let us show that all assumptions of Theorem 18.4 are satisfied.

Assumptions (i) and (iv) are obvious. Assumption (v) readily follows from the fact that $f \in \mathcal{K}_{\mathbf{Q}_\alpha}$.

Since on the orbits of the semigroup (G_c) there exist densities of the conditional measures for the measure $\mathbf{Q}_\alpha$, we have

$$\|\mathbf{Q}_\alpha G_c^{-1} - \mathbf{Q}_\alpha\| \to 0, \qquad c \to 0 \tag{21.8}$$

(see Problem 21.4). Therefore, we need not verify that the mappings G_c are continuous $\mathbf{Q}_\alpha$-almost everywhere (see the discussion of the assumptions of Theorem 18.4). The verification of the second part of assumption (ii) does not pose any problems.

Indeed, let $\varepsilon > 0$. Let $M_\varepsilon(x)$ be the number of jumps of x with magnitude $\geq \varepsilon$. Then

$$\sup_{t\in[0,1]} |l_x(t)| \leq T M_\varepsilon(x),$$

where $T = \max|\tau_i|$, and therefore,

$$\mathbf{P}_n\{x \mid \rho(x, G_c x) > \varepsilon\} = \mathbf{P}_n\{x \mid c \sup_{t\in[0,1]} |l_x(t)| > \varepsilon\} \leq \mathbf{P}_n\Big\{x \mid M_\varepsilon(x) \geq \frac{\varepsilon}{cT}\Big\}.$$

Since for any $k \in \mathbf{N}$ the set $\{x \mid M_\varepsilon(x) \geq k\}$ is closed (see Problem 21.6), we have

$$\overline{\lim_n} \mathbf{P}_n\{x \mid \rho(x, G_c x) > \varepsilon\} \leq \mathbf{P}_\infty\Big\{x \mid M_\varepsilon(x) \geq \Big[\frac{\varepsilon}{cT}\Big]\Big\}.$$

Since for $\mathbf{Q}_\alpha$-almost all x the number of jumps whose absolute values are bounded away from zero by some positive number is finite, it follows that the right-hand side of the above inequalities tends to zero as $c \to 0$, and the desired statement is thereby proved.

Thus, it remains to verify that (iii) is satisfied under our assumptions. This can be performed by arguments close to those used in the proof of Lemma 20.1.

By $\mathbf{E}_n$ we denote the subspace of functions that are constant on the intervals $[(k-1)/n, k/n)$. Obviously, $\mathbf{P}_n(\mathbf{E}_n) = 1$. Let $J_n : \mathbf{E}_n \to \mathbf{R}^n$ be the mapping defined by the formula $J_n(x) = \mathbf{x} = (x_1, \ldots, x_n)$, $x_k = x(k/n)$. We set $\widetilde{\mathbf{P}}_n = \mathbf{P}_n J_n^{-1}$ and $\widetilde{G}_c = J_n \circ G_c \circ J_n^{-1}$. Then

$$\delta_n(c) = \|\mathbf{P}_n - \mathbf{P}_n G_c^{-1}\| = \|\widetilde{\mathbf{P}}_n - \widetilde{\mathbf{P}}_n \widetilde{G}_c^{-1}\| = \int_{\mathbf{R}^n} |Q(c, \mathbf{x}) - Q(0, \mathbf{x})|\, d\mathbf{x},$$

where $Q(s, \mathbf{x})$ is the density of the distribution $\widetilde{\mathbf{P}}_n G_s^{-1}$.

It follows from the construction of the transformation G_c that $Q(s, \mathbf{x})$ has the form

$$Q(s, \mathbf{x}) = \prod_1^n q\big(x_i - sB_n w_i(x_i)\big),$$

where

$$w_i(t) = \begin{cases} \tau_j & \text{if } |t| > \varepsilon B_n, \quad i/n \in \Delta_j\,, \text{ and } \tau_j t \geq 0, \\ 0 & \text{otherwise.} \end{cases}$$

Hence,

$$\frac{\partial Q(s, \mathbf{x})}{\partial s} = Q(s, \mathbf{x}) \sum_1^n (-sB_n w_i(x_i)) \frac{q'(x_i - sB_n w_i(x_i))}{q(x_i - sB_n w_i(x_i))},$$

and therefore,

$$\begin{aligned} \delta_n(c) &= \int_{\mathbf{R}^n} \left| \int_0^c \frac{\partial Q(s, \mathbf{x})}{\partial s}\, ds \right| d\mathbf{x} \\ &\leq \int_0^c \int_{\mathbf{R}^n} \left(\sum_1^n sB_n |w_i(x_i)| \frac{|q'(x_i - sB_n w_i(x_i))|}{q(x_i - sB_n w_i(x_i))} \right) Q(s, \mathbf{x})\, d\mathbf{x} ds \\ &= B_n \int_0^c s \left(\sum_1^n \int_{\mathbf{R}^1} |w_i(x_i)|\, |q'(x_i - sB_n w_i(x_i))|\, dx_i \right) ds. \end{aligned}$$

Since $|w_i(x_i)| \le T\mathbf{1}_{\{|x_i|>\varepsilon B_n\}}$, we have

$$\delta_n(c) \le TB_n \int_0^c s\Big(\sum_1^n \int_{|t|>\varepsilon B_n} |q'(t - sB_n w_i(x_i))|\,dt\Big)\,ds$$
$$\le TnB_n \int_0^c s \int_{|t|>\frac{\varepsilon}{2}B_n} |q'(u)|\,du ds = \frac{c^2}{2} T_n B_n A_q\Big(\frac{\varepsilon}{2}B_n\Big)$$

for $c \le \varepsilon/2T$.

For any positive integer k we have $B_n \sim k^{1/\alpha} B_{[n/k]}$; therefore, the inequalities

$$2k^{1/\alpha} B_{[n/k]} \ge B_n \ge \frac{1}{2} k^{1/\alpha} B_{[n/k]}$$

hold for sufficiently large n. Let us choose k so that $k > (4/\varepsilon)^\alpha$. Then

$$nB_nA_q\Big(\frac{\varepsilon}{2}B_n\Big) \le 2nk^{1/\alpha} B_{[n/k]} A_q\Big(\frac{\varepsilon}{4} k^{1/\alpha} B_{[n/k]}\Big)$$
$$\le 2k^{1/\alpha} \frac{n}{[n/k]} \Big[\frac{n}{k}\Big] B_{[n/k]} A_q(B_{[n/k]}).$$

Hence,

$$\overline{\lim_n}\, nB_nA_q\Big(\frac{\varepsilon}{2}B_n\Big) = C < \infty,$$

whence we obtain

$$\overline{\lim_n}\, \delta_n(c) \le CTc^2.$$

This inequality means that assumption (iii) of Theorem 18.4 is satisfied. □

PROBLEM 21.1. Prove Lemma 21.1. (*Hint*: suppose that $c_1 \ne 0$ and $c_2 \ne 0$. The other cases can be considered in a similar way. Without loss of generality, we can assume that U is a ball centered at l, where l has the form

$$l(t) = \sum_1^M \alpha_j \mathbf{1}_{[t_j,1]}(t);$$

here M, α_j, t_j, $j = 1, \dots, M$, are given numbers, and $t_j \in [0,1]$. Suppose that $\delta > 0$ is so small that the intervals $I_j = (t_j - \delta, t_j + \delta)$ are disjoint and for any $u_j \in I_j$ the function $\widetilde{l}(t) = \sum_1^M \alpha_j \mathbf{1}_{[u_j,1]}$ lies in U. Suppose that on each interval I_j the function x has at least one jump of the same sign as α_j (say, at a point $w_j \in I_j$). The set of such x is of full $\mathbf{Q}_\alpha$-measure. Note that for an ε sufficiently close from the left to $\min_j |\delta_x(w_j)|$, there exist subintervals I_j' centered at w_j and an open ball V_x centered at x such that each function $y \in V_x$ has exactly one discontinuity point in I_j' at which the value of the jump is greater than ε. Then the field $\{l_y\}$ with parameters ε, $N = M$, $\Delta_j = I_j$, and $\tau_j = \alpha_j$, $j = 1, \dots, N$, is the desired field.)

PROBLEM 21.2. A functional $f\colon \mathbf{X} \to \mathbf{R}^1$ is said to belong to the class $\mathcal{K}^a_{\mathbf{Q}_\alpha} \subset \mathcal{K}_{\mathbf{Q}_\alpha}$ if $f \in \mathcal{K}_{\mathbf{Q}_\alpha}$ and $\lambda_\Delta f_\Delta^{-1} \ll \lambda$. Prove the following analogs of Problems 19.1 and 19.2:

a) the class $\mathcal{K}^a_{\mathbf{Q}_\alpha}$ is everywhere dense in the class of all $\mathbf{Q}_\alpha$-almost everywhere continuous functionals;
b) if $f \in \mathcal{K}^a_{\mathbf{Q}_\alpha}$, then $\mathbf{Q}_\alpha f^{-1} \ll \lambda$.

PROBLEM 21.3. Suppose that for $\mathbf{Q}_\alpha$-almost each $x \in \mathbf{X}$ the following conditions hold:

a) there exists an l such that $D_l f(x) = 0$;

b) the mapping $(h, y) \to D_h f(y)$ is continuous at the point (l, x) (in the direct product topology).

Then $f \in \mathcal{K}^a_{\mathbf{Q}_\alpha}$. (*Hint*: see the proof of Theorem 19.1.)

PROBLEM 21.4. Suppose that $f(x) = \int_0^1 h(x(t))\,dt$, h is locally Lipschitzian, and h' is continuous and nonzero almost everywhere on $\mathbf{R}^1$. Then $f \in \mathcal{K}^a_{\mathbf{Q}_\alpha}$. (*Hint*: use the result of the preceding problem.)

PROBLEM 21.5. Prove relation (21.8). (*Hint*: use the total probability formula (3.4) with respect to the partition Γ into orbits of the semigroup (G_c).)

PROBLEM 21.6. Suppose that $M_x(\varepsilon) = \operatorname{card} J_x(\varepsilon)$, $k \in \mathbf{N}$. Prove that the set $\{x \mid M_x\varepsilon \geq k\}$ is closed.

§22. The infinite-dimensional local limit theorem

Suppose that ξ is a random element of a separable Banach space $(\mathbf{X}, \|\cdot\|)$ and $\mathbf{P}$ is the distribution of ξ on the Borel σ-algebra $\mathfrak{B}_\mathbf{X}$ of the space $\mathbf{X}$. Suppose that $\xi_1, \xi_2, \ldots$ are independent copies of ξ, $S_n = n^{-1/2}(\xi_1 + \cdots + \xi_n)$, and $\mathbf{P}_n$ is the distribution of S_n. Suppose that the central limit theorem (CLT) holds for $\mathbf{P}_n$, that is, the $\mathbf{P}_n$ converge weakly to a nondegenerate Gaussian measure $\mathbf{P}_\infty$.

If $\mathbf{X}$ is finite-dimensional, then we understand the local limit theorem for the sequence (ξ_n) as the statement that the densities $d\mathbf{P}_n/d\lambda$ converge to $d\mathbf{P}_\infty/d\lambda$, where λ is the Lebesgue measure. In the infinite-dimensional case there is no analog of the Lebesgue measure. Therefore, the problem of how to interpret the local limit theorem is nontrivial.

If we admit that for any given measure μ the densities $d\mathbf{P}_n/d\mu$ converge to $d\mathbf{P}_\infty/d\mu$ in mean with respect to μ, then, as is well known, this means that the measures $\mathbf{P}_n$ converge in variation to $\mathbf{P}_\infty$. This means that there is no dependence on the auxiliary measure, and we must study convergence in variation. In this connection, it is interesting to know whether the following analog of Prokhorov's theorem [**51**] is valid in the infinite-dimensional case.

THEOREM A. *The convergence* $\mathbf{P}_n \overset{\text{var}}{\to} \mathbf{P}_\infty$ *holds if and only if for some* n_0 *the measure* $\mathbf{P}_{n_0}$ *has a nonzero component that is absolutely continuous with respect to* $\mathbf{P}_\infty$.

Results of this type are interesting because in this case we have the convergence

$$\mathbf{P}_n f^{-1} \overset{\text{var}}{\to} \mathbf{P}_\infty f^{-1} \tag{22.1}$$

for any functional f on $\mathbf{X}$.

In this section we prove an auxiliary result; namely, we obtain the conditions under which (22.1) is satisfied for a certain class of functionals f. Note that for all n the measures $\mathbf{P}_n$ can be singular with respect to $\mathbf{P}_\infty$, and hence, our result is not covered by Theorem A.

To give the precise statement, we need some notions and notation.

Just as in the preceding sections, by

$$I_F = \int_{\mathbf{R}^1} \frac{(q')^2}{q}\,d\lambda$$

we denote the Fisher information for a probability distribution F on $\mathbf{R}^1$ with density q.

For each $l \in \mathbf{X}$, by Γ_l we denote the partition of the space $\mathbf{X}$ into lines $\gamma_y = \{y + cl, c \in \mathbf{R}^1\}$ parallel to l. By $\{\alpha_y\}$ and $\varkappa$, we denote, respectively, the system of conditional measures and the quotient measure for the measure $\mathbf{P}$ with respect to the partition Γ_l (naturally, they are independent of l). We set $I_{l,y} = I_{\alpha_y}$.

Suppose that $\mathcal{K}$ is a subset of $\mathbf{X}$ and $\mathbf{Q}$ is a measure on $\mathfrak{B}_{\mathbf{X}}$. We say that a functional f belongs to the class $\mathcal{M}_{\mathbf{Q}}(\mathcal{K})$ if for $\mathbf{Q}$-almost each $x \in \mathbf{X}$ there exist an open ball U and a vector $l \in \mathcal{K}$ such that for any segment Δ that lies in U and is parallel to l the condition $\Delta_n \to \Delta$ implies that

$$\lambda_{\Delta_n} f^{-1} \overset{\text{var}}{\to} \lambda_\Delta f^{-1}.$$

Note that the class $\mathcal{M}_{\mathbf{Q}}$ introduced earlier in this book is just $\mathcal{M}_{\mathbf{Q}}(\mathbf{H_Q})$, where $\mathbf{H_Q}$ is the space of admissible translations of the measure $\mathbf{Q}$. In the sequel, by $\mathcal{K}$ we denote the set

$$\mathcal{K} = \{l \in \mathbf{X} \mid M_l = \underset{\mathbf{P}_{\Gamma_l}}{\text{ess sup}} I_{l,y} < \infty\}.$$

THEOREM 22.1. *Suppose that the central limit theorem holds, that is,* $\mathbf{P}_n \Rightarrow \mathbf{P}_\infty$. *Then* (22.1) *is satisfied for any* $f \in \mathcal{M}_{\mathbf{P}_\infty}(\mathcal{K})$.

Note that specific conditions under which the central limit theorem can be used are not essential to us (we refer the reader to [**84**]).

Let us consider another example in which Theorem 22.1 can be used.

Suppose that $\mathbf{X}$ is a Banach sequence space with strictly convex norm and $f(x) = \|x\|$. Suppose that the first component ζ_1 of a random element $\xi = (\zeta_1, \zeta_2, \dots)$ is independent of the other components $(\zeta_2, \zeta_3, \dots)$. In other words, the distribution $\mathbf{P}$ of this element can be viewed as a product measure. We also assume that the distribution of ζ_1 has finite Fisher information I_1.

COROLLARY. *If the central limit theorem is satisfies for* ξ, *then the distribution of the variable* $\|S_n\|$ *converges in variation to the distribution of the norm of the Gaussian random element* S_∞ *with distribution* $\mathbf{P}_\infty$.

PROOF. In this case, the class $\mathcal{K}$ necessarily contains the vector $l = (1, 0, 0, \dots)$, since $I_{l,y} = I_1$ for $\mathbf{P}_{\Gamma_l}$-almost all y. Then we have $f \in \mathcal{M}_{\mathbf{P}_\infty}(\mathcal{K})$, thanks to the strong convexity and Theorem 4.4. □

REMARKS. 1) The condition of strong convexity is, in fact, too strong. For the corollary to hold, it suffices to require that for $x \neq 0$ the function $c \to \|x + cl\|$, $c \in \mathbf{R}^1$, attain its minimum only at one point.

2) The measures $\mathbf{P}_n$ can be singular with respect to $\mathbf{P}_\infty$ for all n. Indeed, it suffices to assume that ζ_2 is independent of $(\zeta_3, \zeta_4, \dots)$ and the distribution of the variable ζ_2 is discrete.

PROOF OF THEOREM 22.1. We need the following auxiliary fact.

Let us consider the partition Γ_l of the space $\mathbf{X}$ into lines parallel to l. We identify the quotient space $\mathbf{X}/\Gamma_l$ with some subspace $\mathbf{Y}$ that does not contain l. By $\mathbf{Y}^n$ we denote the direct product $\prod_1^n \mathbf{Y}$.

LEMMA 22.1. *Let $\mathbf{Q}_1, \ldots, \mathbf{Q}_n$ be a system of probability measures on $\mathfrak{B}_{\mathbf{X}}$. Let $\{\alpha_y^{(k)}, y \in \mathbf{Y}\}$ and $\varkappa_k$ be the system of conditional measures and the quotient measure for $\mathbf{Q}_k$ with respect to the partition Γ_l. Suppose that $\{\alpha_y, y \in \mathbf{Y}\}$ and $\varkappa$ are the system of conditional measures and the quotient measure for the convolution $\mathbf{Q} = \mathbf{Q}_1 * \cdots * \mathbf{Q}_n$ with respect to Γ_l. Then the following assertions hold:*

(i) $\varkappa = \varkappa_1 * \cdots * \varkappa_n$;

(ii) $\alpha_y = \displaystyle\int_{E_{\mathbf{y}}} \alpha_{y_1}^{(1)} * \cdots * \alpha_{y_n}^{(n)} \overline{\varkappa}_{\mathbf{y}}\,(d\mathbf{y})$,

where $E_{\mathbf{y}} = \{\mathbf{y} \in \mathbf{Y}^n \mid y_1 + \cdots + y_n = \mathbf{y}\}$ and $\overline{\varkappa}_y$ are the conditional measures for $\varkappa_1 \times \cdots \times \varkappa_n$ on the fibers $E_{\mathbf{y}}$.

PROOF. Suppose that h is a nonnegative bounded measurable function on $\mathbf{X}$ and the mapping $\sigma\colon \mathbf{X}^n \to \mathbf{X}$ is defined by $\sigma(x_1, \ldots, x_n) = x_1 + \cdots + x_n$. Then $\mathbf{Q} = (\mathbf{Q}_1 \times \cdots \times \mathbf{Q}_n)\sigma^{-1}$; therefore

$$\begin{aligned}\int_{\mathbf{X}} h\,d\mathbf{Q} &= \int_{\mathbf{X}^n} h(x_1 + \cdots + x_n)\mathbf{Q}_1 \times \cdots \times \mathbf{Q}_n\,(d\mathbf{x}) \\ &= \int_{\mathbf{Y}^n} \biggl[\int_{\gamma_{y_1} \times \cdots \times \gamma_{y_n}} h\bigl((y_1 + c_1 l) + \cdots + (y_n + c_n l)\bigr) \\ &\qquad \times \alpha_{y_1}^{(1)} \times \cdots \times \alpha_{y_n}^{(n)}(d\mathbf{c})\biggr] \varkappa_1 \times \cdots \times \varkappa_n\,(d\mathbf{y});\end{aligned} \tag{22.2}$$

here we use the partition Γ_l on each copy of $\mathbf{X}$. The partition of the space $\mathbf{Y}^n$ into the fibers $E_{\mathbf{y}}$ can be considered as the partition into preimages of points under the mapping $\sigma\colon \mathbf{Y}^n \to \mathbf{Y}$. Thus, it is natural to identify the quotient space with $\mathbf{Y}$. Obviously, for $\varkappa_1 \times \cdots \times \varkappa_n$ the quotient measure with respect to this partition is equal to $\varkappa = \varkappa_1 * \cdots * \varkappa_n$. Therefore, we can transform the integrals in (22.2) as follows:

$$\begin{aligned}\int_{\mathbf{X}} h\,d\mathbf{Q} &= \int_{\mathbf{Y}} \int_{E_{\mathbf{y}}} \int_{\gamma_{y_1} \times \cdots \times \gamma_{y_n}} h\bigl(y + (c_1 + \cdots + c_n)l\bigr) \\ &\qquad \times \alpha_{y_1}^{(1)} \times \cdots \times \alpha_{y_n}^{(n)}(d\mathbf{c})\overline{\varkappa}_{\mathbf{y}}\,(d\mathbf{y})\varkappa\,(d\mathbf{y}) \\ &= \int_{\mathbf{Y}} \int_{\gamma_y} h(\mathbf{y} + cl)\alpha_{\mathbf{y}}(dc)\varkappa(d\mathbf{y}),\end{aligned} \tag{22.3}$$

where $\alpha_{\mathbf{y}} = \int_{E_{\mathbf{y}}} \alpha_{y_1}^{(1)} * \cdots * \alpha_{y_n}^{(n)} \overline{\varkappa}_y\,(d\mathbf{y})$.

Relation (22.3) readily proves our lemma. □

Now we return to the proof of the theorem. Let us verify that under our assumptions all five conditions of Theorem 18.4 are satisfied.

Let $x \in \mathbf{X}_0$ be the set of $\mathbf{P}_\infty$-full measure mentioned in the definition of the inclusion $f \in \mathcal{M}_{\mathbf{P}_\infty}(\mathcal{K})$, let U be the corresponding open ball, and let l be the corresponding vector from $\mathcal{K}$.

We define a mapping $G_{n,c} = G_{\infty,c} = G_c$ by setting

$$G_c z = z + cl, \qquad c \in [0, 1].$$

We can assume that U is a ball centered at the point x. Suppose that r is its radius. Then we can take V to be the open ball of radius $r/2$ centered at x and set $\varepsilon = r\|l\|/2$.

Let us show that the above mappings, the ball V, and the number ε satisfy all the necessary assumptions.

Assumptions (i) and (ii) of Theorem 18.4 are automatically satisfied. Assumptions (iv) and (v) can br readily derived from the fact that $f \in \mathcal{M}_{\mathbf{P}_\infty}(\mathcal{K})$ and from Theorem 4.5. It is more difficult to verify assumption (iii).

By $\widehat{\mathbf{P}}_n$ we denote the distribution of the sum $\xi_1 + \cdots + \xi_n$. Obviously, we have

$$\|\mathbf{P}_n - \mathbf{P}_n G_c^{-1}\| = \|\widehat{\mathbf{P}}_n - \widehat{\mathbf{P}}_n G_{c\sqrt{n}}^{-1}\|.$$

Suppose that $\{\nu_y, y \in \mathbf{Y}\}$ and ν are the system of conditional measures and the quotient measure for $\widehat{\mathbf{P}}_n$ with respect to the partition Γ_l; next, suppose that $\{\widetilde{\nu}_y, y \in \mathbf{Y}\}$ and $\widetilde{\nu}$ are similar objects for $\widehat{\mathbf{P}}_n G_{c\sqrt{n}}^{-1}$. Obviously, $\nu = \widetilde{\nu}$. Therefore,

$$\|\mathbf{P}_n - \mathbf{P}_n G_c^{-1}\| = \int_{\mathbf{Y}} \|\nu_y - \widetilde{\nu}_y\| \nu\,(dy).$$

By applying Lemma 22.1 to the measures $\mathbf{Q}_k = \mathbf{P}$, $k = 1, \ldots, n$, and then to the measures $\widetilde{\mathbf{Q}}_k = \mathbf{P}_n G_{c/\sqrt{n}}^{-1}$, $k = 1, \ldots, n$, we obtain

$$\begin{aligned} \|\nu_y - \widetilde{\nu}_y\| &= \left\| \int_{E_y} \alpha_{x_1} * \cdots * \alpha_{x_n} \varkappa_y\,(d\mathbf{x}) - \int_{E_y} \widetilde{\alpha}_{x_1} * \cdots * \widetilde{\alpha}_{x_n} \varkappa_y\,(d\mathbf{x}) \right\| \\ &\leq \int_{E_y} \left\| \alpha_{x_1} * \cdots * \alpha_{x_n} - \widetilde{\alpha}_{x_1} * \cdots * \widetilde{\alpha}_{x_n} \right\| \varkappa_y\,(d\mathbf{x}); \end{aligned} \tag{22.4}$$

here $\{\alpha_y, y \in \mathbf{Y}\}$ and $\{\widetilde{\alpha}_y, y \in \mathbf{Y}\}$ are, respectively, the systems of conditional measures for $\mathbf{P}$ and $\mathbf{P}G_{c/\sqrt{n}}^{-1}$ with respect to Γ_l; $\varkappa$ is the corresponding quotient measure.

Since $l \in \mathcal{K}$, it follows that the measures α_x have finite Fisher information I_x, and $M_l = \operatorname{ess\,sup} I_x < \infty$. It follows from (22.4) and the estimate in Problem 20.2 (with $a_i = c\sqrt{n}$) that

$$\|\nu_y - \widetilde{\nu}_y\| \leq \int_{E_y} \left(\sum_1^n I_{x_j} c^2 n^{-1} \right)^{1/2} \varkappa_y\,(d\mathbf{x}) \leq c M_l^{1/2}.$$

The obtained estimate is uniform in y and n. Therefore,

$$\|\mathbf{P}_n - \mathbf{P}_n G_c^{-1}\| \leq c M_l^{1/2},$$

and assumption (iii) is thereby proved. Thus, the measures $\mathbf{P}_n$ and the functional f satisfy all assumptions of Theorem 18.4. Hence, $\mathbf{P}_n f^{-1} \overset{\text{var}}{\to} \mathbf{P}_\infty f^{-1}$. □

Bibliographical Notes

§**1.** The notions considered in this section are conventional. The problems discussed here are considered in more detail in the books [**4, 15, 16, 18, 24, 27, 44, 53, 88, 97**].

§**2.** A. D. Aleksandrov was the first to study weak convergence of measures. The theory of weak convergence was developed by Kolmogoroff, Prokhorov, Skorokhod, and Donsker. For references to the classical works in this field, see [**4**]. Theorem 2.2 is a special case of results obtained by Billingsley and Topsøe [**125**]. Apparently, the result of Theorem 2.3 has not been mentioned previously.

§**3.** Rokhlin constructed the theory of measurable partitions. Concerning the existence of conditional measures, see [**78,** §§45, 46].

§**4.** The main idea of the stratification method first appeared in Davydov [**30**]. Later on, this method was used for solving various problems in [**31–36, 56–58, 60–62, 89–101**]. A review of these papers can be found in [**40**]. The general scheme in terms of semigroups was proposed in [**60**]. The one-dimensional analog of Lemma 4.1, which is very important, was proved by Vershik and Skorokhod [**95**]. Our proof of this lemma is based on an idea proposed by Polishchuk.

Theorem 4.2 for absolutely continuous functions was proved by Geman and Horowitz [**137**]; the corresponding result for monotone functions was obtained by Zaretskii (see [**74**]). Wide generalizations of Theorems 4.2 and 4.3 [**41, 111**] are well known; their proofs go back to Banach's theorem on the indicatrix [**121, 74**]. Theorems 4.4–4.6 are taken from [**34, 40**]. The classical Theorems 4.7 and 4.8 are due to Gnedenko [**26; 5,** Ch. 4], but the proof of Theorem 4.8 in this book is new. It was given by Lifshits.

§**5.** This section is a slight modification of Davydov's paper [**170**].

§**6.** The method presented in this section was stated by Smorodina in [**104**]. The main ideas of this approach have common features with the methods used by Malliavin [**146**] and his numerous successors [**42, 20, 122, 123, 124, 160, 163**] to prove the smoothness of one-dimensional distributions of diffusion processes.

§**7.** The first examples and definitions of Gaussian measures were published by Wiener [**165**] and Kolmogorov [**141**]. The most important properties of these measures can be found in the monographs [**27, 44, 88, 97, 120**]; see also [**129, 138, 161**]. Cameron and Martin [**131**] were the first to consider a special case of the very important formula for the density of the translated Gaussian measure in terms of the original measure. Systems of conditional Gaussian measures that appear due to the partition into parallel lines, as well as the description of the space of admissible translations in terms of the factorization of the correlation vector (Theorem 7.3), were considered in [**40, 57**]; for the factorization, see also [**15, 16**]. The following classes of Gaussian measures and processes, which are most interesting to us, have been considered: measures in the Hilbert space, by Skorokhod [**97**]; stationary processes by Rozanov [**52, 88**]; semistable processes by Molchan and Golesov [**70**]; the white noise by Bochner [**126**]; and the multiparameter field by Yeh [**166, 167**]. Aronszajn studied reproducing kernels [**119, 116**].

§**8.** The material considered in this section goes back to [**104**]. The system of definitions follows in many aspects the cycle of Shigekawa's papers [**158, 159, 139**] in which he considered a special case of the Wiener measure. The book by Daletskii and Fomin [**44**] contains numerous integration by parts formulas for Gaussian measures, as well as an alternative approach to the definition of differentiability of a measure [**44,** Ch. 4]. Among the papers dealing with the properties of differentiable measures, we point out [**6, 7, 43, 72, 108, 109**], which are closely related to our monograph.

§**9.** We discuss the results about absolute continuity following [**40**], but these results were published earlier in [**30, 34, 37, 58, 99, 158**]. We consider the boundedness condition for the distribution density of functionals, which is very useful for estimating the rate of convergence in the invariance principle [**8, 9, 11; 84,** Ch. 5]. Theorem 9.1 was proved in [**40, 64**]. In [**157, 92, 10, 57, 93**], successively weakened versions of the boundedness condition for integral functionals of the Wiener process were found. Some results about the smoothness of functional distributions were obtained in [**102, 104**]; some closely related classes of functionals were considered in [**21**].

An important example of smooth functionals is the value of the diffusion process if we consider this value as a functional on the space of trajectories of the Wiener process that generates the diffusion. The analysis of studying this situation in relation to smooth solvability of differential equations led to the so-called "Malliavin calculus"; see the remarks to §6.

§**10.** Sudakov and Tsirel′son [**107**] and Borell [**128**] independently proved the isoperimetric property of half-spaces with respect to a Gaussian measure. Various kinds of Brunn–Minkowski inequality were obtained in [**127, 128, 135**], and some special cases of this inequality, in [**49, 118**]. In our monograph, we follow Ehrhard's approach [**135**], which allows us to obtain both the isoperimetric property and the Brunn–Minkowski inequality. For "non-Gaussian" generalizations, see [**5, 127**].

§**11.** Tsirel′son [**113**] presented a characterization of the distribution structure for convex functionals. The important Example 11.2 is taken from [**117**], Theorem 11.2 from [**168**], Theorem 11.3 from [**116**], and its proof from [**58**]. For the geometric characterization of the separation point, see [**40**]; in [**59**] it was shown how this point is related to degenerate functionals. The simplest results about the absolute continuity [**27**] and the boundedness of the distribution density [**150**] of a convex functional were pointed out long ago; some similar facts about non-Gaussian measures are proved in [**35, 36**].

More or less sharp exponential estimates for the measure of large deviations of a functional, which are close to Theorem 11.5, were obtained in [**71, 96, 127, 129, 136, 144**]. Freidlin [**17, 112**] used the statement with the functional action (Proposition 11.5); the statement with the norm of the correlation operator (Proposition 11.6) probably originates somewhere in mathematical folklore. For a review and a list of papers dealing with sharp exponential estimates for specific classes of Gaussian measures and functionals, see [**19, 85, 86, 98**]. The new Theorem 11.6 (Lifshits) appeared because of further development of the theory of large deviations [**2**].

§**12.** The absolute continuity of the norm distribution was considered in [**40**]. The idea of constructing the Gaussian measure with unbounded density of the norm distribution was proposed by Tsirel′son; then this idea was independently implemented by Paulauskas [**81, 84, 149**] and Lifshits [**40**]. Bentkus proposed Lemma 12.1, which is very useful for constructing a counterexample. Further consideration of examples with bad properties of the distribution of the norm can be found in [**75, 77, 151, 152**]. Theorem 12.1 belongs to Rhee and Talagrand [**153**]. This theorem generalizes earlier results due to Davydov [**40**] and Paulauskas [**80**] to the spaces $\mathbf{L}^p$ and $\mathbf{l}^p$, as well as the remarks from [**15, 143**] that pertain to Hilbert spaces. The dependence of the uniform estimate of the density on the correlation operator and the barycenter of the measure was studied in [**79, 80, 84, 110, 117**].

The entropy approach to estimating the density was proposed in [**62–64**]. This approach is based on the ideas of Dudley and Sudakov and on estimates obtained by Dmitrovskii [**46**]. Smorodina [**102, 103**] and Uglanov [**109**] proved the smoothness of the norm distribution for $\mathbf{L}^p$.

§§**13, 14.** The method used in these section appeared first in [**105, 106**].

§**15.** The main results of this section were obtained by Smorodina. These results are new; this is their first publication.

§**16.** The results of this section first appeared in [**105, 106**]. In [**123**], similar methods were used to study the smoothness of densities of one-dimensional distributions for locally infinitely divisible processes. The results about the boundedness of the distribution densities of functionals of processes with independent increments can be found in [**40, 61**].

§**17.** This result appeared first in [**106**]. Some other interesting results about the properties of the norm distribution for random vectors are proved in [**65, 130, 145, 154, 169**].

§**18.** Theorem 18.1, which is the simplest application of the stratification method for solving the strong convergence problem, was obtained in [**34**]. That paper also contains more difficult results on strong convergence.

Theorem 18.2, which illustrates the possibilities of Theorem 18.1, is a special case of a result obtained by Parthasarathy and Steerneman [**148**] by quite a different method. Note that this result can be substantially improved if we use Theorem 18.1.

Theorem 18.4 is an improved version of a theorem from [**38**].

The result of Problem 2 was also obtained in [**38**].

§**19.** Theorem 19.4 was proved by Davydov [**40**]. The superstructure method (Theorem 18.4), which appeared later, allowed us to simplify the proof considerably.

§**20.** The local invariance principle for independent identically distributed random variables (Theorems 20.1 and 20.2) was obtained by Davydov [**40**]. By using the superstructure method, just as in the Gaussian case, the proof can be substantially simplified and clarified.

The idea of using the spaces $E[0,1]$ for the consideration of discontinuous random open polygons is due to Borodin.

The condition that the Fisher information is finite seems too strong in Theorems 20.1 and 20.2. Recently, Davydov proved that if one considers a somewhat narrower functional class than $\mathcal{M}_{\mathbf{W}}$, then instead of the condition $I_F < \infty$ it suffices to require that the initial values of F have a distribution density. The proof of this result is rather complicated and cannot be presented in this monograph.

Various generalizations of Theorems 20.1 and 20.2 can be found in [**100, 101, 67, 68, 12, 13**].

Some other results of local invariance principle type for special classes of functionals can be found in the review [**40**].

§**21.** Theorem 21.2 was obtained by Breier.

§**22.** Theorem 22.1 was obtained in [**39**]. Bentkus [**171**] and later Zitikis [**172**] obtained local limit theorems in Banach spaces for some cases in which f is a homogeneous symmetric polynomial.

Bibliography

1. Ya. I. Belopol′skaya and Yu. L. Daletskii, *Stochastic equations and differential geometry*, "Vysshaya Shkola", Kiev, 1989; English transl., Kluwer, Dordrecht, 1990.
2. V. Yu. Bentkus, *Large deviations in Banach spaces*, Teor. Veroyatnost. i Primenen. **31** (1986), 710–716; English transl., Theory Probab. Appl. **31** (1986), 627–633.
3. V. Yu. Bentkus and D. Pap, *On the norm distribution for a stable random vector in a Hilbert space*, Litovsk. Mat. Sb. **26** (1986), 211–220; English transl. in Lithuanian Math. J. **26** (1986).
4. P. Billingsley, *Convergence of probability measures*, Wiley, New York, 1968.
5. D. G. Bobkov, *Upper functions and oscillations of Gaussian processes and related random processes*, Thesis, Leningrad State University, Leningrad, 1988. (Russian)
6. V. I. Bogachev, *Some results on differentiable measures*, Mat. Sb. **127** (1985), 336–351; English transl. in Math. USSR Sb. **55** (1986).
7. ———, *Subspaces of differentiability of smooth measures on infinite-dimensional spaces*, Dokl. Akad. Nauk SSSR **299** (1988), 18–22; English transl. in Soviet Math. Dokl. **37** (1988).
8. I. S. Borisov, *On the rate of convergence of distributions of functionals of integrable type*, Teor. Veroyatnost. i Primenen. **21** (1976), 293–308; English transl., Theory Probab. Appl. **21** (1976), 283–298.
9. ———, *On the rate of convergence in the "conditional" invariance principle*, Teor. Veroyatnost. i Primenen. **23** (1978), 67–79; English transl., Theory Probab. Appl. **34** (1978), 63–75.
10. ———, *On conditions for the existence of a bounded density for the distribution of additive functionals of some random processes*, Teor. Veroyatnost. i Primenen. **25** (1980), 464–475; English transl., Theory Probab. Appl. **25** (1980), 454–465.
11. A. A. Borovkov, *On the rate of convergence in the invariance principle*, Teor. Veroyatnost. i Primenen. **18** (1973), 217–234; English transl., Theory Probab. Appl. **18** (1973), 207–224.
12. P. Breier, *Local limit theorems for functionals in the conditional invariance principle*, Zap. Nauchn. Sem. LOMI **158** (1987), 32–38; English transl. in J. Soviet Math. **43** (1988), no. 6.
13. ———, *Local invariance principle for independent non-identically distributed random variables*, Rings and Modules. Limit Theorems in Probability Theory, vol. 2, Leningrad State Univ., Leningrad, 1988, pp. 143–149. (Russian)
14. ———, *Local invariance principle in the case of convergence to a stable law*, Manuscript. (Russian)
15. N. N. Vakhaniya, *Probability distributions in linear spaces*, "Metsniereba", Tbilisi, 1971; English transl., North-Holland, Amsterdam, 1981.
16. N. N. Vakhaniya, V. I. Tarieladze, and S. A. Chobanyan, *Probability distributions in Banach spaces*, "Nauka", Moscow, 1985; English transl., Reidel, Dordrecht, 1987.
17. A. D. Wentzel, *Theorems about the action functional for Gaussian random functions*, Teor. Veroyatnost. i Primenen. **17** (1972), 542–544; English transl., Theory Probab. Appl. **17** (1072), 515–518.
18. ———, *A course in the theory of random processes*, "Nauka", Moscow, 1975; German transl., *Theorie zufällige Prozesse*, Akademie-Verlag, Berlin, and Birkhäuser, Basel, 1979.
19. M. I. Freidlin and A. D. Wentzel, *Random perturbations of dynamical systems*, "Nauka", Moscow, 1979; English transl., Springer-Verlag, New York, 1984.
20. A. Yu. Veretennikov, *A probability approach to hyperellipticity*, Uspekhi Mat. Nauk **38** (1983), no. 3, 113–125; English transl. in Russian Math. Surveys **38** (1983).

21. T. R. Vinogradova, *On the accuracy of normal approximation on sets defined by a smooth function.* I, II, Teor. Veroyatnost. i Primenen. **30** (1985), 219–229, 554–557; English transl., Theory Probab. Appl. **30** (1985), 235–245, 590–593.
22. J. Galambos, *The asymptotic theory of extremal order statistics*, Wiley, New York, 1978.
23. W. Hoeffding, *On a theorem of V. M. Zolotarev*, Teor. Veroyatnost. i Primenen. **9** (1964), 96–99; English transl., Theory Probab. Appl. **9** (1964), 89–92.
24. I. I. Gikhman and A. V. Skorokhod, *Theory of random processes.* Vol. 1, "Nauka", Moscow, 1971; English transl., Springer-Verlag, New York, 1974.
25. I. I. Gikhman and A. V. Skorokhod, *Stochastic differential equations and their applications*, "Naukova Dumka", Kiev, 1982; English transl. of 1st ed., Springer-Verlag, New York, 1972.
26. B. V. Gnedenko, *A local limit theorem for densities*, Dokl. Akad. Nauk SSSR **95** (1954), 5–7. (Russian)
27. H.-H. Kuo, *Gaussian measures in Banach spaces*, Lecture Notes in Math., vol. 463, Springer-Verlag, New York, 1976.
28. V. M. Gol′dstein and Yu. G. Reshetnyak, *Introduction to the theory of functions with generalized derivatives and quasiconformal mappings*, "Nauka", Moscow, 1983; English transl., *Quasiconformal mappings and Sobolev spaces*, Kluwer, Dordrecht, 1990.
29. Yu. A. Davydov, *The invariance principle for stationary processes*, Teor. Veroyatnost. i Primenen. **15** (1970), 498–509; English transl., Theory Probab. Appl. **15** (1970), 487–498.
30. ———, *On absolute continuity of distributions of functionals of random processes*, Teor. Veroyatnost. i Primenen. **23** (1978), 228–229; English transl. in Theory Probab. Appl. **23** (1978).
31. ———, *On strong convergence of distributions of functionals of random processes*, Teor. Veroyatnost. i Primenen. **24** (1979), 429; English transl. in Theory Probab. Appl. **24** (1979).
32. ———, *Local limit theorems for functionals of supremum type in a lattice case*, Zap. Nauchn. Sem. LOMI **97** (1980), 51–61; English transl. in J. Soviet Math. **24** (1984), no. 5.
33. ———, *Local invariance principle.* I, II, Zap. Nauchn. Sem. LOMI **97** (1980), 45–50; **130** (1983), 69–77; English transl. in J. Soviet Math. **24** (1984), no. 5; **27** (1984), no. 6.
34. ———, *On strong convergence of distributions of functionals of random processes.* I, II, Teor. Veroyatnost. i Primenen. **25** (1980), 782–799; **26** (1981), 266–286; English transl., Theory Probab. Appl. **25** (1980), 772–788; **26** (1981), 258–278.
35. ———, *The structure of distributions of convex functionals*, Fourth Internat. Vilnius Conf. Probab. Theory and Math. Statist., Abstracts of Communications. Vol. I, Inst. Mat. i Kibernet. Akad. Nauk Litovsk. SSR, Vilnius, 1985, pp. 201–203. (Russian)
36. ———, *Distributions of convex functionals of random processes*, Rings and Moduli. Limit Theorems in Probability Theory, vol. 1, Leningrad State Univ., Leningrad, 1986, pp. 157–163. (Russian)
37. ———, *A remark on the absolute continuity of distributions of Gaussian functionals*, Teor. Veroyatnost. i Primenen. **33** (1988), 170–172; English transl., Theory Probab. Appl. **33** (1988), 158–160.
38. ———, *Local limit theorems for functionals of random processes*, Teor. Veroyatnost. i Primenen. **33** (1988), 788–794; English transl., Theory Probab. Appl. **33** (1988), 732–738.
39. ———, *A version of infinite-dimensional local limit theorem*, Zap. Nauchn. Sem. LOMI **177** (1989), 46–50; English transl. in J. Soviet Math. **61** (1992), no. 1.
40. Yu. A. Davydov and M. A. Lifshits, *Stratification method in some probability problems*, Itogi Nauki i Tekhniki: Teor. Veroyatnost. Mat. Statist. Teoret. Kibernet., vol. 22, VINITI, Moscow, 1984, pp. 61–157; English transl. in J. Soviet Math. **37** (1987), no. 1.
41. Yu. A. Davydov and A. L. Rozin, *On sojourn times for functions and random processes*, Teor. Veroyatnost. i Primenen. **23** (1978), 650–654; English transl., Theory Probab. Appl. **23** (1978), 633–637.
42. Yu. L. Daletskii, *Stochastic differential geometry*, Uspekhi Mat. Nauk **38** (1983), no. 3, 87–111; English transl. in Russian Math. Surveys **38** (1983).
43. Yu. L. Daletskii and B. D. Maryanin, *Smooth measures on infinite-dimensional manifolds*, Dokl. Akad. Nauk SSSR **285** (1985), 1297–1300; English transl. in Soviet Math. Dokl. **32** (1985).
44. Yu. L. Daletskii and S. V. Fomin, *Measures and differential equations on infinite-dimensional spaces*, "Nauka", Moscow, 1983; English transl., Kluwer, Dordrecht, 1991.
45. J. Dixmier, *Les C^*-algèbres et leurs représentations*, 2nd ed., Gauthier-Villars, Paris, 1969.

46. V. A. Dmitrovskii, *Estimates for the distribution of the maximum of a random Gaussian field*, Random Processes and Fields, Moscow State Univ., Moscow, 1979, pp. 22–31. (Russian)
47. ———, *Boundedness condition and estimates for the distribution of the maximum of random fields on arbitrary sets*, Dokl. Akad. Nauk SSSR **253** (1980), 271–274; English transl. in Soviet Math. Dokl. **22** (1980).
48. ———, *On the distribution of the maximum and local properties of Gaussian random fields*, Fifth Internat. Vilnius Conf. Probab. Theory and Math. Statist., Abstracts of Communications, vol. 1, Inst. Mat. i Kibernet. Akad. Nauk Litovsk. SSR, Vilnius, 1989, p. 196.
49. V. A. Zalgaller, *Compound volumes and the probabilities of convex domains under a multidimensional normal distribution*, Mat. Zametki **2** (1967), 97–104; English transl. in Math. Notes **2** (1967).
50. V. M. Zolotarev, *Concerning a certain probability problem*, Teor. Veroyatnost. i Primenen. **6** (1961), 219–222; English transl., Theory Probab. Appl. **6** (1961), 201–204.
51. I. A. Ibragimov and Yu. V. Linnik, *Independent and stationary sequences of random variables*, "Nauka", Moscow, 1965; English transl., Wolters–Noordhoff, Groningen, 1971.
52. I. A. Ibragimov and Yu. A. Rozanov, *Gaussian random processes*, "Nauka", Moscow, 1970; English transl., Springer-Verlag, New York, 1978.
53. K. Matthes, J. Kerstan, and J. Mecke, *Infinitely divisible point processes*, Akademie Verlag, Berlin, and Wiley, New York, 1978.
54. Yu. Ch. Kokaev, *The behavior of oscillations and conditional Gaussian distributions of linear functionals*, Zap. Nauchn. Sem. LOMI **119** (1982), 128–143; English transl. in J. Soviet Math. **27** (1984), no. 5.
55. A. N. Kolmogoroff, *Foundations of the theory of probability*, 2nd ed., "Nauka", Moscow, 1974; English transl. of 1st ed., Chelsea, New York, 1956.
56. M. A. Lifshits, *Some problems of the theory of random processes related to mappings of measures*, Thesis, Leningrad State University, Leningrad, 1981. (Russian)
57. ———, *The fiber method and its application to the study of functionals of stochastic processes*, Teor. Veroyatnost. i Primenen. **27** (1982), 67–80; English transl., Theory Probab. Appl. **27** (1982), 69–82.
58. ———, *On the absolute continuity of distributions of functionals of random processes*, Teor. Veroyatnost. i Primenen. **27** (1982), 559–566; English transl., Theory Probab. Appl. **27** (1982), 600–607.
59. ———, *Absolute continuity of supremum type functionals of Gaussian processes*, Zap. Nauchn. Sem. LOMI **119** (1982), 154–166; English transl. in J. Soviet Math. **27** (1984), no. 5.
60. ———, *Stratification method for processes with independent increments*, Zap. Nauchn. Sem. LOMI **130** (1983), 109–121; English transl. in J. Soviet Math. **27** (1984), no. 6.
61. ———, *An application of the stratification method to the study of functionals of processes with independent increments*, Teor. Veroyatnost. i Primenen. **29** (1984), 723–734; English transl., Theory Probab. Appl. **29** (1984), 753–764.
62. ———, *Density of the maximum of a Gaussian process*, Teor. Veroyatnost. i Primenen. **29** (1984), 814–815; English transl. in Theory Probab. Appl. **29** (1984).
63. ———, *The probability density of the maximum of a Gaussian process*, Fourth Internat. Vilnius Conf. Probab. Theory Math. Statist., Abstracts of Communications, vol. 2, Inst. Mat. i Kibernet. Akad. Nauk Litovsk. SSR, Vilnius, 1985, pp. 113–114. (Russian)
64. ———, *On the distribution of the maximum of a Gaussian process*, Teor. Veroyatnost. i Primenen. **31** (1986), 134–142; English transl., Theory Probab. Appl. **31** (1986), 125–133.
65. ———, *On the probability density of the norm of a stable vector*, Zap. Nauchn. Sem. LOMI **158** (1987), 105–114; English transl. in J. Soviet Math. **43** (1988), no. 6.
66. ———, *Oscillations and a lower bound of the distribution of the maximum of a Gaussian field*, Zap. Nauchn. Sem. LOMI **177** (1989), 78–82; English transl. in J. Soviet Math. **61** (1992), no. 1.
67. G. V. Martynova, *Limit theorems for functionals of random processes*, Thesis, Leningrad State University, Leningrad, 1987. (Russian)
68. ———, *Strong convergence of distributions of functionals of a sequence of processes linearly generated by independent random variables*, Zap. Nauchn. Sem. LOMI **158** (1987), 122–126; English transl. in J. Soviet Math. **43** (1988), no. 6.

69. ———, *On the strong convergence of distributions of multidimensional functionals*, Rings and Modules. Limit Theorems in Probability Theory, vol. 2, Leningrad State Univ., Leningrad, 1988, pp. 163–171. (Russian)
70. G. M. Molchan and Yu. N. Golesov, *Gaussian stationary processes with asymptotic power law spectrum*, Dokl. Akad. Nauk SSSR **184** (1969), 546–549; English transl. in Soviet Math. Dokl. **10** (1969).
71. S. V. Nagaev, *On probabilities of large deviations for a Gaussian distribution in a Banach space*, Izv. Akad. Nauk UzSSR Ser. Fiz.-Mat. Nauk **1981**, no. 5, 18–21. (Russian)
72. N. V. Norin, *Stochastic integrals and differentiable measures*, Teor. Veroyatnost. i Primenen. **32** (1987), 114–124; English transl., Theory Probab. Appl. **32** (1987), 107–117.
73. S. V. Nagaev and A. N. Startsev, *Asymptotic properties of distributions of an infinite quadratic form of Gaussian random variables*, Limit Theorems for Random Process and Statistical Conclusions, "Fan", Tashkent, 1981, pp. 144–160. (Russian)
74. I. P. Natanson, *Theory of functions of a real variable*, 3rd ed., "Nauka", Moscow, 1974; English transl., Vols. 1 (1st ed., Chaps. I–IX), 2 (2nd ed., Chaps. X–XVI, XVIII), Ungar, New York, 1955, 1961.
75. D. Pap, *On the distribution of the norm of a Gaussian vector in a Banach space*, Litovsk. Mat. Sb. **24** (1984), no. 1, 140–145; English transl. in Lithuanian Math. J. **24** (1984).
76. ———, *Boundedness of the norm density of stable random vectors in a Hilbert space*, Teor. Veroyatnost. i Mat. Statist. **36** (1987), 102–105; English transl. in Theor. Probability Math. Statist. **36** (1988).
77. D. Pap and V. I. Paulauskas, *A remark on the distribution of the norm of a Gaussian vector in the Banach space* $\mathbf{c}_0$, Litovsk. Mat. Sb. **24** (1985), no. 4, 112–117; English transl. in Lithuanian Math. J. **24** (1985).
78. K. R. Parthasarathy, *An introduction to probability and measure theory*, Macmillan of India, Delhi, 1980; Springer-Verlag, New York, 1977.
79. V. I. Paulauskas, *On the closeness of the distributions of two sums of independent random variables with values in a Hilbert space*, Litovsk. Mat. Sb. **16** (1975), no. 3, 177–200; English transl. in Lithuanian Math. J. **16** (1975).
80. ———, *On the rate of convergence in the central limit theorem in certain Banach spaces*, Teor. Veroyatnost. i Primenen. **21** (1976), 775–791; English transl., Theory Probab. Appl. **21** (1976), 754–770.
81. ———, *On the rate of convergence in the Banach space* $\mathbf{c}_0$, Dokl. Akad. Nauk SSSR **254** (1980), 286–288; English transl. in Soviet Math. Dokl. **22** (1980).
82. ———, *On the probability density of the norm of a Gaussian vector in Banach spaces*, Dokl. Akad. Nauk SSSR **266** (1982), 1301–1304; English transl. in Soviet Math. Dokl. **26** (1982).
83. ———, *On the density of the norm of a Gaussian vector in a Banach space*, Abstracts Fourth Soviet–Japan Sympos. Probab. Theory Math. Statist., Tbilisi, 1982, pp. 145–146. (Russian)
84. V. I. Paulauskas and A. Yu. Rachkauskas, *Approximation theory in the central limit theorem. Exact results in Banach spaces*, "Mokslas", Vilnius, 1987; English transl., Kluwer, Dordrecht, 1989.
85. V. I. Piterbarg, *Gaussian random processes*, Itogi Nauki i Tekhniki: Teor. Veroyatnost. Mat. Statist. Teoret. Kibernet., vol. 19, VINITI, Moscow, 1982, pp. 155–198; English transl. in J. Soviet Math. **25** (1983), no. 3. (Russian)
86. ———, *Asymptotic methods in the theory of Gaussian processes and fields*, Moscow State Univ., Moscow, 1988; English transl., Amer. Math. Soc., Providence, RI, 1996.
87. A. P. Robertson and W. J. Robertson, *Topological vector spaces*, Cambridge Univ. Press, Cambridge, 1964.
88. Yu. A. Rozanov, *Infinite-dimensional Gaussian distributions*, Trudy Mat. Inst. Steklov. **108** (1968); English transl., Proc. Steklov Inst. Math. **108** (1971).
89. V. I. Rotar′ and T. L. Shervashidze, *Some estimates for distributions of quadratic forms*, Teor. Veroyatnost. i Primenen. **30** (1985), 549–554; English transl., Theory Probab. Appl. **30** (1985), 585–590.
90. R. Rudzkis, *On the probability of a large excursion of a nonstationary Gaussian process. I*, Litovsk. Mat. Sb. **25** (1985), no. 1, 143–154; English transl. in Lithuanian Math. J. **25** (1985).

91. ______, *On the probability density of a large excursion of a Gaussian process.* I, II, Litovsk. Mat. Sb. **26** (1986), 512–519; **27** (1987), 731–746; English transl. in Lithuanian Math. J. **26** (1986); **27** (1987).
92. A. I. Sakhanenko, *A condition for the existence of a bounded density of the distributions of integral type functionals*, Second Internat. Vilnius Conf. Probab. Theory Math. Statist., Abstracts of Communications, vol. 2, Inst. Mat. i Kibernet. Akad. Nauk Litovsk. SSR, Vilnius, 1977, pp. 149–150. (Russian)
93. ______, *A bound for the probability density of integral type functionals*, Trudy Mat. Inst. Sibirsk. Otdel. Akad. Nauk SSSR **1** (1982), 180–186. (Russian)
94. A. V. Skorokhod, *Stochastic processes with independent increments*, "Nauka", Moscow, 1964; English transls., (a) Clearing House Federal Sci. Tech. Information, Springfield, VA, 1966; (b), *Theory of random processes*, British lending Library Sci. Tech., Boston Spa, Yorkshire, England, 1971; 2nd rev. ed., "Nauka", Moscow, 1986; English transl., Kluwer, Dordrecht, 1991.
95. ______, *On admissible translations of measures in a Hilbert space*, Teor. Veroyatnost. i Primenen. **15** (1970), 577–598; English transl., Theory Probab. Appl. **15** (1970), 557–578.
96. ______, *A note on Gaussian measures in a Banach space*, Teor. Veroyatnost. i Primenen. **15** (1970), 519; English transl., Theory Probab. Appl. **15** (1970), 508–509.
97. ______, *Integration in Hilbert space*, "Nauka", Moscow, 1975; English transl., Springer-Verlag, New York, 1974.
98. Yu. K. Belyaev (editor), *Random processes. Sample functions and intersections*, Collection of translated papers, "Mir", Moscow, 1978. For the contents, see MR **81j**:60049.
99. N. V. Smorodina, *Absolute continuity of distributions of functionals of diffusion processes*, Uspekhi Mat. Nauk **37** (1982), no. 6, 185–192; English transl. in Russian Math. Surveys **37** (1982).
100. ______, *A local invariance principle for processes linearly generated by independent random variables*, Teor. Veroyatnost. i Primenen. **29** (1984), 680–691; English transl., Theory Probab. Appl. **29** (1984), 707–718.
101. ______, *Local limit theorems for functionals of random fields*, Zap. Nauchn. Sem. LOMI **136** (1984), 168–182; English transl. in J. Soviet Math. **33** (1986), no. 1.
102. ______, *Conditions for smoothness of probability densities of functionals of random processes*, Fourth Internat. Vilnius Conf. Probab. Theory Math. Statist., Abstracts of Communications, vol. 3, Inst. Mat. i Kibernet. Akad. Nauk Litovsk. SSR, Vilnius, 1985, pp. 143–145. (Russian)
103. ______, *Conditions for smoothness of the probability density of the* $\mathbf{L}_p$*-norm of a Gaussian vector*, Zap. Nauchn. Sem. LOMI **158** (1987), 161–166; English transl. in J. Soviet Math. **43** (1988), no. 6.
104. ______, *Differential calculus on measurable spaces and conditions for smoothness of probability densities of random variables*, Dokl. Akad. Nauk SSSR **292** (1987), 1053–1057; English transl. in Soviet Math. Dokl. **35** (1987).
105. ______, *Differential calculus in configuration space and stable measures.* I, Teor. Veroyatnost. i Primenen. **33** (1988), 522–534; English transl., Theory Probab. Appl. **33** (1988), 488–500.
106. N. V. Smorodina and M. A. Lifshits, *On the distribution of the norm of a stable vector*, Teor. Veroyatnost. i Primenen. **34** (1989), 304–313; English transl. in Theory Probab. Appl. **34** (1989).
107. V. N. Sudakov and B. S. Tsirel′son, *Extremal properties of half-spaces for spherically invariant measures*, Zap. Nauchn. Sem. LOMI **41** (1974), 14–24; English transl. in J. Soviet Math. **9** (1978), no. 1.
108. A. V. Uglanov, *Surface integrals in a Banach space*, Mat. Sb. **110** (1979), 189–217; English transl. in Math. USSR Sb. **38** (1981).
109. ______, *On the smoothness of distributions of functionals of random processes*, Teor. Veroyatnost. i Primenen. **33** (1988), 535–544; English transl., Theory Probab. Appl. **33** (1988), 500–509.
110. V. V. Ul′yanov, *Bounds of the density of the norm of a Gaussian random variable*, Numerical Methods in Mathematical Physics, "Nauka", Moscow, 1986, pp. 107–109. (Russian)
111. H. Federer, *Geometric measure theory*, Springer-Verlag, New York, 1978.
112. M. I. Freidlin, *The action functional for a class of stochastic processes*, Teor. Veroyatnost. i Primenen. **17** (1972), 536–541; English transl., Theory Probab. Appl. **17** (1972), 511–516.

113. B. S. Tsirel′son, *The density of the distribution of the maximum of a Gaussian process*, Teor. Veroyatnost. i Primenen. **20** (1975), 864–873; English transl., Theory Probab. Appl. **20** (1975), 847–856.
114. ———, *A natural modification of a random process and its applications to random functional series and Gaussian measures*, Zap. Nauchn. Sem. LOMI **55** (1976), 35–63; English transl. in J. Soviet Math. **16** (1981), no. 2.
115. ———, *Addendum to the paper on a natural modification*, Zap. Nauchn. Sem. LOMI **72** (1977), 202–211; English transl. in J. Soviet Math. **23** (1983), no. 4.
116. ———, *Geometrical approach to maximal likelihood estimation for infinitely-dimensional Gaussian location*. I, Teor. Veroyatnost. i Primenen. **27** (1982), 388–395; English transl., Theory Probab. Appl. **27** (1982), 411–418.
117. V. I. Chebotarev, *An estimate for the rate of convergence in the central limit theorem in* $\mathbf{l}_p$, Sibirsk. Mat. Zh. **20** (1979), 1099–1116; English transl. in Siberian Math. J. **20** (1979).
118. T. V. Anderson, *The integral of a symmetric unimodal function over a symmetric convex set and some probability inequalities*, Proc. Amer. Math. Soc. **6** (1955), 170–176.
119. N. Aronszajn, *Theory of reproducing kernels*, Trans. Amer. Math. Soc. **68** (1950), 337–404.
120. A. Badrikian and S. Shevet, *Mesures cylindriques, espaces de Wiener et fonctions aléatoires gaussiennes*, Lecture Notes in Math., vol. 379, Springer-Verlag, Berlin, 1974.
121. S. Banach, *Sur les lignes rectifiables et les surfaces dont l'aire est finie*, Fund. Math. **7** (1925), 225–237.
122. D. Bell, *The Malliavin calculus*, Longman Sci. Tech., Harlow, 1987.
123. K. Bichteler, J. B. Graveraux, and J. Jacod, *Malliavin calculus for processes with jumps*, Gordon and Breach, London, 1987.
124. J.-M. Bismut, *Martingales, the Malliavin calculus and hypoellipticity under general Hörmander's conditions*, Z. Wahrsch. Verw. Gebiete **56** (1981), 469–505.
125. P. Billingsley and F. Topsøe, *Uniformity in weak convergence*, Z. Wahrsch. Verw. Gebiete **7** (1967), 1–16.
126. S. Bochner, *Harmonic analysis and the theory of probability*, Univ. of California Press, Berkeley, 1955.
127. C. Borell, *Convex measures on locally convex spaces*, Ark. Mat. **12** (1974), 239–252.
128. ———, *The Brunn–Minkowski inequality in Gauss space*, Invent. Math. **30** (1975), 207–216.
129. ———, *Gaussian Radon measures on locally convex spaces*, Math. Scand. **38** (1976), 265–284.
130. T. Byczkowski and K. Samotij, *Absolute continuity of stable seminorms*, Ann. Probab. **14** (1986), 299–312.
131. R. H. Cameron and W. T. Martin, *Transformations of Wiener integrals under translation*, Ann. of Math. **45** (1944), 386–396.
132. R. M. Dudley, *Sample functions of the Gaussian process*, Ann. Probab. **1** (1973), 66–103.
133. R. M. Dudley, J. Feldman, and L. Le Cam, *On seminorms and probabilities, and abstract Wiener spaces*, Ann. of Math. **93** (1971), 390–408.
134. A. Ehrhard, *Sur la densité du maximum d'une fonction aléatoire gaussienne*, Lecture Notes in Math., vol. 920, Springer-Verlag, Berlin, 1982, pp. 581–601.
135. ———, *Symétrisation dans l'espace de Gauss*, Math. Scand. **53** (1983), 281–301.
136. X. Fernique, *Intégrabilité des vecteurs gaussiens*, C. R. Acad. Sci. Paris Ser. A **270** (1970), 1698–1699.
137. D. Geman and J. Horowitz, *Occupation times for smooth stationary processes*, Ann. Probab. **1** (1973), 131–137.
138. L. Gross, *Abstract Wiener spaces*, Proc. 5th Berkeley Symp. Math. Statist. Probab., vol. 2, Univ. of California Press, Berkeley, 1965, pp. 31–42.
139. N. Ikeda and S. Watanabe, *Stochastic differential equations and diffusion processes*, Kodansha, Tokyo, and North-Holland, Amsterdam, 1981.
140. K. Itô and M. Nisio, *On the oscillation function of Gaussian processes*, Math. Scand. **22** (1968), 209–223.
141. A. N. Kolmogoroff, *La transformation de Laplace dans les espaces linéaires*, C. R. Acad. Sci. Paris **200** (1935), 1717–1718.
142. J. Kuelbs, *Abstract Wiener spaces and applications to analysis*, Pacific J. Math. **31** (1969), 433–450.

143. J. Kuelbs and T. Kurtz, *Berry–Esseen estimates in Hilbert space and application to the law of the iterated logarithm*, Ann. Probab. **2** (1974), 387–407.
144. H. L. Landau and L. A. Shepp, *On the supremum of a Gaussian process*, Sankhyā Ser. A **32** (1970), 369–378.
145. W. Linde, *Infinitely divisible and stable measures on Banach spaces*, Teubner, Leipzig, 1983.
146. P. Malliavin, *C^k-hypoellipticity with degeneracy*. I, Stochastic Analysis, Academic Press, New York, 1978, pp. 199–214.
147. M. B. Marcus and L. A. Shepp, *Sample behavior of Gaussian processes*, Proc. 6th Berkeley Sympos. Math. Statist. Probab., vol. 2, Univ. of California Press, Berkeley, 1972, pp. 423–442.
148. K. R. Parthasarathy and T. Steerneman, *A tool in establishing total variation convergence*, Proc. Amer. Math. Soc. **95** (1985), 626–630.
149. V. Paulauskas, *On the density of the norm of Gaussian vector in Banach spaces*, Lecture Notes in Math., vol. 990, Springer-Verlag, Berlin, 1983, pp. 179–197.
150. L. Pitt, *Local sample paths properties of Gaussian fields*, Ann. Probab. **7** (1979), 477–493.
151. W. S. Rhee, *On the distribution of the norm for a Gaussian measure*, Ann. Inst. H. Poincaré Probab. Statist. **20** (1984), 277–286.
152. W. S. Rhee and M. Talagrand, *Bad rates of convergence for the central limit theorem in Hilbert space*, Ann. Probab. **12** (1984), 843–850.
153. ———, *Uniform convexity and the distribution of the norm for a Gaussian measure*, Probab. Theory Related Fields **71** (1986), 59–68.
154. M. Ryznar, *Density of stable seminorms*, Bull. Polish Acad. Sci. Math. **33** (1985), 431–440.
155. H. Sato, *Souslin support and Fourier expansion of a Gaussian Radon measure*, Lecture Notes in Math., vol. 860, Springer-Verlag, Berlin, 1981, pp. 299–313.
156. H. Sato and Y. Okazaki, *Separabilities of a Gaussian Radon measure*, Ann. Inst. H. Poincaré Probab. Statist. **11** (1975), 287–298.
157. S. Sawyer, *Rates of convergence for some functionals in probability*, Ann. Math. Statist. **43** (1972), 273–284.
158. I. Shigekawa, *Absolute continuity of probability laws of Wiener functionals*, Proc. Japan Acad. Ser. A Math. Sci. **54** (1978), 230–234.
159. ———, *Derivatives of Wiener functionals and absolute continuity of induced measures*, J. Math. Kyoto Univ. **20** (1980), 263–289.
160. D. Stroock, *The Malliavin calculus and its applications*, Lecture Notes in Math., vol. 851, Springer-Verlag, Berlin, 1981, pp. 394–432.
161. M. Talagrand, *Mesures gaussiennes sur un espace localement convexe*, Z. Wahrsch. Verw. Gebiete **64** (1983), 181–210.
162. ———, *Sur l'intégrabilité des vecteurs gaussiens*, Z. Wahrsch. Verw. Gebiete **68** (1984), 1–8.
163. S. Watanabe, *Lectures on stochastic differential equations*, Springer-Verlag, Berlin, 1984.
164. M. Weber, *Sur la densité de la distribution du maximum d'un processus gaussien*, J. Math. Kyoto Univ. **25** (1985), 515–523.
165. N. Wiener, *Differential space*, J. Math. and Phys. Mass. Inst. Tech. **2** (1923), 131–174.
166. J. Yeh, *Wiener measure in a space of functions of two variables*, Trans. Amer. Math. Soc. **95** (1960), 433–450.
167. ———, *Cameron–Martin translation theorem in Wiener space of functions of two variables*, Trans. Amer. Math. Soc. **107** (1963), 409–420.
168. N. D. Ylvisaker, *The expected number of zeros of a stationary Gaussian process*, Ann. Math. Statist. **36** (1965), 1043–1046.
169. T. Zak and M. Lewandowsky, *On the density of the distribution of p-stable seminorms*, $0 < p < 1$, Proc. Amer. Math. Soc. **100** (1987), 345–351.
170. Yu. A. Davydov, *On the exact continuity of images of measures*, Zap. Nauchn. Sem. LOMI **142** (1985), 48–54; English transl. in J. Soviet Math. **36** (1987), no. 4.
171. V. Yu. Bentkus, *Asymptotic expansions in a local limit theorem in a Hilbert space*, Litovsk. Mat. Sb. **25** (1985), no. 1, 9–22; English transl. in Lithuanian Math. J. **25** (1985).
171. R. E. Zitikis, *Asymptotic expansions in a local limit theorem for the Cramér–Mises–Smirnov statistics*, Thesis, Vilnius, 1988. (Russian)

Index

Selected Titles in This Series

(Continued from the front of this publication)

134 **V. V. Prasolov,** Problems and theorems in linear algebra, 1994
133 **P. I. Naumkin and I. A. Shishmarev,** Nonlinear nonlocal equations in the theory of waves, 1994
132 **Hajime Urakawa,** Calculus of variations and harmonic maps, 1993
131 **V. V. Sharko,** Functions on manifolds: Algebraic and topological aspects, 1993
130 **V. V. Vershinin,** Cobordisms and spectral sequences, 1993
129 **Mitsuo Morimoto,** An introduction to Sato's hyperfunctions, 1993
128 **V. P. Orevkov,** Complexity of proofs and their transformations in axiomatic theories, 1993
127 **F. L. Zak,** Tangents and secants of algebraic varieties, 1993
126 **M. L. Agranovskiĭ,** Invariant function spaces on homogeneous manifolds of Lie groups and applications, 1993
125 **Masayoshi Nagata,** Theory of commutative fields, 1993
124 **Masahisa Adachi,** Embeddings and immersions, 1993
123 **M. A. Akivis and B. A. Rosenfeld,** Élie Cartan (1869–1951), 1993
122 **Zhang Guan-Hou,** Theory of entire and meromorphic functions: deficient and asymptotic values and singular directions, 1993
121 **I. B. Fesenko and S. V. Vostokov,** Local fields and their extensions: A constructive approach, 1993
120 **Takeyuki Hida and Masuyuki Hitsuda,** Gaussian processes, 1993
119 **M. V. Karasev and V. P. Maslov,** Nonlinear Poisson brackets. Geometry and quantization, 1993
118 **Kenkichi Iwasawa,** Algebraic functions, 1993
117 **Boris Zilber,** Uncountably categorical theories, 1993
116 **G. M. Fel′dman,** Arithmetic of probability distributions, and characterization problems on abelian groups, 1993
115 **Nikolai V. Ivanov,** Subgroups of Teichmüller modular groups, 1992
114 **Seizô Itô,** Diffusion equations, 1992
113 **Michail Zhitomirskiĭ,** Typical singularities of differential 1-forms and Pfaffian equations, 1992
112 **S. A. Lomov,** Introduction to the general theory of singular perturbations, 1992
111 **Simon Gindikin,** Tube domains and the Cauchy problem, 1992
110 **B. V. Shabat,** Introduction to complex analysis Part II. Functions of several variables, 1992
109 **Isao Miyadera,** Nonlinear semigroups, 1992
108 **Takeo Yokonuma,** Tensor spaces and exterior algebra, 1992
107 **B. M. Makarov, M. G. Goluzina, A. A. Lodkin, and A. N. Podkorytov,** Selected problems in real analysis, 1992
106 **G.-C. Wen,** Conformal mappings and boundary value problems, 1992
105 **D. R. Yafaev,** Mathematical scattering theory: General theory, 1992
104 **R. L. Dobrushin, R. Kotecký, and S. Shlosman,** Wulff construction: A global shape from local interaction, 1992
103 **A. K. Tsikh,** Multidimensional residues and their applications, 1992
102 **A. M. Il′in,** Matching of asymptotic expansions of solutions of boundary value problems, 1992
101 **Zhang Zhi-fen, Ding Tong-ren, Huang Wen-zao, and Dong Zhen-xi,** Qualitative theory of differential equations, 1992
100 **V. L. Popov,** Groups, generators, syzygies, and orbits in invariant theory, 1992
99 **Norio Shimakura,** Partial differential operators of elliptic type, 1992
98 **V. A. Vassiliev,** Complements of discriminants of smooth maps: Topology and applications, 1992 (revised edition, 1994)

(See the AMS catalog for earlier titles)